Federal Rivers

Federal Rivers

Managing Water in Multi-Layered Political Systems

Edited by

Dustin Garrick

McMaster University, Canada

George R.M. Anderson

New York University, USA

Daniel Connell

Australian National University

Jamie Pittock

Australian National University

Edward Elgar

Cheltenham, UK • Northampton, MA, USA

IWA Publishing

Co-published by IWA Publishing, Alliance House, 12 Caxton Street, London SW1H 0QS, UK Tel. +44 (0) 20 7654 5500, Fax +44 (0) 20 7654 5555
publications@iwap.co.uk
www.iwapublishing.com
ISBN 1843393158
ISBN13 9781843393153

Published by
Edward Elgar Publishing Limited
The Lypiatts
15 Lansdown Road
Cheltenham
Glos GL50 2JA
UK

Edward Elgar Publishing, Inc.
William Pratt House
9 Dewey Court
Northampton
Massachusetts 01060
USA

A catalogue record for this book
is available from the British Library

Library of Congress Control Number: 2013942237

This book is available electronically in the ElgarOnline.com
Social and Political Science Subject Collection, E-ISBN 978 1 78195 505 5

MIX
Paper from responsible sources
FSC® C013056
FSC www.fsc.org

ISBN 978 1 78195 504 8

Typeset by Servis Filmsetting Ltd, Stockport, Cheshire
Printed and bound in Great Britain by T.J. International Ltd, Padstow

Contents

Figures

Tables

Table of conversions

	Litres	Cubic meters	Megalitres	Acre feet
1 litre	1	1×10^{-3}	1×10^{-6}	8.1×10^{-7}
1 cubic metre	1 000	1	1×10^{-3}	8.1×10^{-4}
1 megalitre*	1 000 000	1 000	1	8.1×10^{-1}
1 acre foot	1 233 482	1 233	1.233	1

*1 Gigalitre is 1000 megalitres

Cubic metres are the international standard. *Acre feet* tend to be used in the US. *Megalitres* and *gigalitres* are used in Australia.

Contributors

Shahid Ahmad is a Water Management Expert and Chief Scientist II with the Pakistan Agricultural Research Office based in Islamabad.

José Albiac, University of Zaragoza, Spain
Albiac is a researcher at the Agrifood Research and Technology Center (CITA-DGA) in Zaragoza, Spain. He earned his Bachelor's degree in Economics at the University of Zaragoza, his Master's and his doctoral degree in Agricultural Economics at the University of Illinois. His research specializes in environmental and natural resource economics, environmental and agricultural policies, water management, water scarcity and droughts, irrigation, groundwater and ecosystems protection, nonpoint pollution and climate change.

George Anderson, New York University, US and Canada
Anderson was a career civil servant in the Canadian government, including duties as Deputy Minister of Intergovernmental Affairs and of Natural Resources. He is a past chief executive officer of the Forum of Federations. An Expert with the Mediation Support Unit of the United Nations in 2012–13, he was a resident fellow at the Center on Constitutional Transitions, New York University in 2013–14 and a past Fellow at Harvard's Center for International Affairs. Anderson was educated at Queen's and Oxford and École Nationale d'Administration in Paris. He is author of *Federalism: An Introduction, Fiscal Federalism: A Comparative Introduction*, and editor of *Oil and Gas in Federal Systems* and *Internal Markets and Multilevel Governance.*

Khalid Aziz, Pakistan
Aziz holds a Master's degree in Political Science from Peshawar University and has studied both at Oxford and Cambridge universities. From the latter he obtained an M.Phil in Development Studies in 1981. He joined the Civil Service of Pakistan in 1969 and remained a political agent and a district commissioner. He headed the economic development of NWFP and tribal areas for six years and later became Chief Secretary of NWFP in 1996. One of his outstanding contributions was as negotiator for his province in the distribution of Indus River waters in 1990.

Eugenio Barrios, WWF-Mexico, Mexico
Barrios is a graduate in chemical engineering at the University of Mexico, and earned his Master of Science in Environmental Engineering from the University of Illinois Urbana-Champaign. He has worked for the National Water Commission and the Institute of Engineering at the National University of Mexico, and as an international consultant for the World Bank, the World Meteorological Organization, the Mekong River Commission, and the Government of Panama. In the last 20 years, Barrios has shaped a solid experience in water resource protection projects. Since 2004, he has been Director of the Water Program in WWF-Mexico, and currently he is coordinating the implementation of the national environmental water reserves program in Mexico, and is a member of the Scientific and Technical Review Panel of the Ramsar Convention on Wetlands.

Benedito Braga, University of Sao Paulo, Brazil
Braga holds a Ph.D from Stanford University in water resources. He is President of the World Water Council and Professor of Civil and Environmental Engineering at the University of Sao Paulo. He has served on the Board of Directors of the Brazilian National Water Agency, ANA, from 2001-2009. He was President of the International Water Resources Association (1998-2000) and Vice-President of the World Water Council (2006-2012). Braga chaired the International Committee of the 6th World Water Forum held in Marseille in March 2012. His scientific interests are in the field of systems analysis modeling of hydrologic processes, integrated water resources management and water policy development. He is the author of more than 200 articles and 25 books and chapters of books on the subject of water management at international level. Braga is recipient of the 2002 Crystal Drop Award, given by the International Water Resources Association (IWRA). In 2009 he received the honorary membership from the American Water Resources Association and in 2011 he was awarded the Honorary Diplomate from the American Society of Civil Engineers due to his long and distinguished career in the field of water resources.

Elena Calvo, University of Zaragoza, Spain
Calvo is Professor in the Economics and Business School of the University of Zaragoza. She holds a Ph.D and a B.Sc. in Mathematics from the University of Zaragoza. Her research interests are optimization, numerical methods, and application of mathematical modeling to environmental problems. She works in research areas related to environmental and natural resource economics and policies, water management, nonpoint pollution, forest economics and climate change.

Ana Carolina Coelho, Brazilian National Water Agency, Brazil
Coelho holds a Ph.D in Civil and Environmental Engineering from Colorado State University in Fort Collins (2010), earned as a Fulbright Scholar; a Master's Degree in Environmental and Water Resources Engineering (2010) and a Bachelor's Degree in Architecture and Urban Planning (2002), both from Universidade Federal do Paraná. Coelho is currently a water resources expert at the Brazilian National Water Agency and author of several articles on the subject of water management, in national and international levels. Her fields of expertise include water resources planning and management, institutional analysis, urban and regional planning, decision support systems and GIS.

Daniel Connell, Australian National University, Australia
Connell researches the governance of trans-boundary rivers in the federal systems of Australia, South Africa, United States, Mexico, European Union (Spain), India, China and Brazil. Themes include the impact of the historical legacy, water reform, environmental justice, public participation, cultural change, institutional design, distribution of costs and benefits across borders, water markets and risk created by the interaction of different levels of government. Connell also contributes to the ongoing debate about the future of the Murray Darling Basin – the subject of his book *Water Politics in the Murray-Darling Basin* published in 2007. This has included presentations to government departments and agencies, the Treasury, Environment Water and Heritage, the National Water Commission, Land and Water Australia, the Murray-Darling Basin Commission, the Murray Darling Basin Authority, the rural Labor Caucus, and writing op-ed articles for newspapers and interviews for radio and television. In addition he works with Professor Quentin Grafton, who holds the ANU–UNESCO chair in Water Economics and Transboundary Governance, focusing on the implementation of the Millennium Development Goals in sub-Saharan Africa.

Osman F. da Silva, Brazilian National Water Agency, Brazil
Silva is an oceanographer and holds a Ph.D in Civil Engineering from Campinas State University (Unicamp) and a postdoctoral degree from the São Paulo University, Brazil. He was also a visiting professor at Ohio State University and worked at the Brazilian National Electric Energy Agency. Silva is currently Deputy Manager and water resources expert at the Brazilian National Water Agency (ANA). He is also the author of several articles on water resources management published in national and international journals.

Lucia De Stefano, Botin Foundation and Universidad Complutense de Madrid, Spain
De Stefano is a Research Fellow at the Water Observatory of the Botin Foundation and Associate Professor at Complutense University of Madrid. Her previous post was as a research assistant at Oregon State University, working on global studies of water conflicts and resilience to climate-change-induced water variability, and regional water governance benchmarking. De Stefano has worked as water policy officer at an international environmental NGO and as a water management specialist in the private sector. A hydrogeologist by training, she holds a PhD. on evaluation of water policies. Her main fields of interest are policy evaluation, water planning, groundwater management and the assessment of good governance attributes.

Derick du Toit, Association for Water and Rural Development, South Africa
Du Toit is an ecologist by training but specializes in environmental learning and professional development. He has worked for a number of years at the Desert Research Foundation of Namibia on translating scientific research into learning programmes. He has subsequently worked on education reform in Namibia, Botswana and South Africa. He currently is project manager with the Association for Water and Rural Development (AWARD) in South Africa.

Encarna Esteban, University Defense Centre, Zaragoza, Spain
Esteban is Assistant Professor in the Department of Economic Analysis, University of Zaragoza. She holds a Ph.D in Economics from the University of Zaragoza, a Master's of Science in Agricultural and Resource Economics from the University of Connecticut, and a B.Sc. in Economics specializing in economic analysis, from the University of Zaragoza. Her main area of research covers the study of water bodies, analyzing both quality and quantity aspects. Her work includes environmental economics, natural resources, mathematical optimization, simulation and modeling applied to water problems.

Rodrigo Flecha, Brazilian National Water Agency, Brazil
Flecha is a civil engineer and specialist in environmental and sanitary engineering from Universidade Federal de Minas Gerais. He holds a post-graduate degree in environmental techniques and science from Ecole Nationale des Ponts et Chaussées, École Nationale du Génie Rural, des Eaux et des Forêts and Université de Paris I. He was the Water Resources Management Chief at the Brazilian National Water Agency from 2003 to 2013 and is currently the Water Resources Regulation Chief at the same

agency. He is the author of several articles on water resources management published in national and international journals.

Nelson Freitas, Brazilian National Water Agency, Brazil
Freitas holds a Master's degree in Water Resources Engineering, Planning and Management from Instituto de Pesquisas Hidraulicas – Universidade Federal do Rio Grande do Sul (1996) and a Bachelor's degree in Civil Engineering from Universidade Federal de Santa Maria (1987). He has 25 years of expertise in water resources planning and management, including water supply and treatment, in both private and public sectors, as a team leader and project manager in global organizations. He was an OAS consultant for the Brazilian National Water Plan between 2005 and 2006; the Water Planning Manager at the Brazilian National Water Agency between 2006 and 2010; and, is currently the Water Management Manager at the same agency.

Dustin Garrick, McMaster University, Canada
Garrick is the Philomathia Assistant Professor of Water Policy at McMaster University. His research examines water allocation policy responses to water scarcity and climate risk in large transboundary rivers of federal countries. This research advances theory and evidence on collective action and institutional change. Most recently, he has investigated drought management and water allocation reform in semi-arid rivers of Australia and western North America, focusing at the intersection of water markets and river basin planning. He holds a Ph.D from the University of Arizona, an MPA from Columbia University and a BA from the University of Texas. Prior to joining McMaster University in January 2014, he was a research fellow at the University of Oxford (2011–13) where he remains a visiting scientist in the School of Geography and the Environment. He was a Fulbright Scholar in Australia where he also retains a research associate affiliation of the Australian National University Centre for Water, Economics and Environmental Policy.

Andrea K. Gerlak, International Studies Association and University of Arizona, US
Gerlak is Director of Academic Development with the International Studies Association, and Policy Associate with the Udall Center for Studies in Public Policy at the University of Arizona. Gerlak's research expertise is in institutional analysis and water governance. Her research addresses conditions supporting collective action, the interface between science and policy, institutional change and adaptability, and human rights and development. She has served on the faculty of Guilford College and Columbia University. A few of her recent publications have

appeared in journals such as *Global Environmental Politics; International Environmental Agreements;* and *Water International.*

Tanya Heikkila, University of Colorado Denver, US
Heikkila is Associate Professor at the School of Public Affairs at the University of Colorado Denver. Heikkila's research focuses on analyzing institutions for coordinating and collaborating across natural resource boundaries and for resolving environmental conflicts. This includes studies of interstate water conflicts and cooperation in the US, conflicts and cooperation related to hydraulic fracturing, the organization and performance of large-scale ecosystem restoration programs, the coordination of groundwater and surface water in the western US, and water governance institutions in northeast Brazil and the Mekong Basin. She holds an MPA and Ph.D from the University of Arizona.

Rakesh Hooja
Hooja had been Director of the Indian Institute of Public Administration in New Delhi from 2010 until his untimely death in September 2012. He was educated at the Universities of Rajasthan and East Anglia. He occupied many positions during 36 years in the Indian Administrative Service, including Chief Secretary to the State Government of Rajastan and Joint Secretary, Department of Jammu and Kashmir Affairs, in the Government of India. Hooja authored or edited 31 books, wrote 430 articles and edited various journals. His interests included public administration, local government, and water policy.

Ted Horbulyk, University of Calgary, Canada
Horbulyk is Associate Professor of Economics at the University of Calgary, where his teaching and research interests include environmental and resource economics, cost-benefit analysis and public policy. Horbulyk holds an undergraduate degree in agricultural economics and a doctorate in economics, and he has worked professionally on water resource issues in a number of developing countries. He is a former editor of the *Canadian Journal of Agricultural Economics*, and he previously served as a theme leader in the Research Program on Economics, Policy and Risk funded by the Alberta Ingenuity Centre for Water Research.

Ke Jian, Wuhan University, China
Ke is Associate Professor of Environmental Law at Research Institute of Environmental Law at Wuhan University, and he also works as General Secretary, Committee of Academics, Chinese Association of Environmental and Natural Resources Law. He has rich academic experience as an environmental law scholar in China and overseas. He has studied and researched environmental law at the National University of

Singapore, Sydney University, the University of Regensburg, Harvard University, the French Academy of Sciences and the University of Tennessee. His teaching and research work covers a wide range of environmental law. He has published dozens of articles and books in Chinese and English.

Mujib Khan, Pakistan
Khan is a development expert with over 20 years' experience in public sector management in areas marked with complex social and military conflict, both as a civil servant as well as a consultant, covering both the design and implementation of reforms. He has worked on public policy, governance, and institutional development in the arenas of public administration, public financial management, criminal justice, and anticorruption in South Asia and countries in the former Soviet Union. His recent work has focused on support to the design and implementation of centre-of-government reforms, covering institutional redesign, public financial management, and democratic accountability.

Harlan Koff, University of Luxembourg, Luxembourg
Koff is Professor of Social Sciences at the University of Luxembourg where he coordinates the research axis on Comparative Governance: Europe in the World (European Governance Program). He is founding President of the Consortium for Comparative Research on Regional Integration and Social Cohesion (RISC) as well as coordinator of RISC's Working Group on Development, Equity and Policy Coherence. His research focuses on comparative border politics, comparative regional integration, development and human security. With Carmen Maganda, he is co-editor of the journal *Regions & Cohesion* and he edits the book series on Regional Integration and Social Cohesion for Peter Lang Publishing.

M. Dinesh Kumar, Institute for Resource and Policy Analysis, India
Kumar earned his B-Tech in Civil Engineering in 1988, M.E. in Civil (Water Resources Management) in 1991 and Ph.D in Water Management in 2006. He has 21 years of professional experience in the field of water resources. He is currently Executive Director of Institute for Resource Analysis and Policy in Hyderabad. He has nearly 140 publications to his credit, including three books, two edited volumes, several book chapters, many articles in peer-reviewed journals and research monographs. His latest book is *Managing Water in River Basins: Hydrology, Economics and Institutions*, by Oxford University Press, New Delhi. He has published extensively in international journals, such as *Water Policy, Energy Policy, Water International, Journal of Hydrology, Water Resources Management and Resources, Energy and Development Journal*. Currently, he is Associate Editor of *Water Policy*.

Fang Lan, Shaanxi Normal University, China
Fang is Professor at Shaanxi Normal University. She also serves as Deputy Director of the Northwest Institute of Historical Environment and Socio-Economic Development of the university, and Vice Chairperson of the Shaanxi Association for Yellow River Culture and Economic Development. She has worked on water resource management for more than a decade. Her research field particularly focuses on watersheds conflicts and agricultural water management in northwest China. She studied at the University of Giessen, Germany, where she received her Ph.D in 2003. She also worked at the University of California, Riverside, as a Fulbright scholar.

Elena López-Gunn, ICATALIST and Universidad Complutense, Spain
López-Gunn is founder and manager of ICATALIST, a knowledge broker company currently engaged on projects with FAO and with Repsol, a Spanish energy company. She was previously a Senior Research Fellow at the London School of Economics, working on rights based approach to water in Latin America. Her research and publications record has concentrated on areas related to collective action on common pool resources like groundwater, climate change and adaptation, and studies which adopt a public policy angle on water management issues related to institutional reform.

Carmen Maganda, University of Luxembourg, Luxembourg
Maganda is Research Associate in the Political Science Institute at the University of Luxembourg, Executive Director of the Consortium for Comparative Research on Regional Integration and Social Cohesion (RISC), and co-editor of the journal *Regions & Cohesion* by Berghahn Journals. Formerly she was Coordinator of the Bi-national Border Water Project for the Center for US-Mexican Studies at the University of California San Diego, and Hermès-FMSH Postdoctoral Fellow at Clersé, Université de Lille 1. Maganda has a Ph.D in Social Anthropology from CIESAS-Mexico with a dissertation entitled 'Availability of Water: a Constructed Risk'. Her present research focuses on comparative regional perspectives of water politics/security in border regions, Integrated Water Resources Management in cross-border river basins in Europe and the Americas, and its relationship with development, social participation and the human rights framework.

Mike Muller, Commissioner, National Planning Commission of South Africa, University of Witwatersrand, South Africa
Muller is a Commissioner in South Africa's first National Planning Commission. He was Director-General of the South Africa Department of Water Affairs and Forestry from 1997–2005, responsible for policy

and legislative reform and a range of transformative national water programmes. He has also managed infrastructure and policy programmes at the Development Bank of Southern Africa and water and sanitation programmes for the Mozambique Government. A registered professional engineer, Muller is Visiting Adjunct Professor at Witwatersrand University Graduate School of Public and Development Management and advises a range of organisations on water, infrastructure and development.

Jamie Pittock, Australian National University, Australia
Pittock is Senior Lecturer in the Fenner School of Environment and Society, and Director of International Programs for the United Nations Educational, Scientific and Cultural Organization Chair in Water Economics and Transboundary Water Governance. His current work includes developing research programs that link Australian and southern African expertise to improve management of river basins, green water and agriculture. He is also Program Leader for the Australia and United States – Climate, Energy and Water project of the US Studies Centre and ANU Water Initiative. Prior to joining ANU, Pittock worked for WWF International as director of their global freshwater program on conservation of wetlands, water use in agriculture, and river basin management.

Sharon Pollard, Association for Water and Rural Development, South Africa
Pollard holds a Ph.D in freshwater ecology from the University of Cape Town. Her interests are broad but lie at the interface between society and freshwater sustainability. Important conceptual frameworks for her work include complexity theory and systems thinking and how this is given expression in water governance through integrated approaches such as Integrated Water Resources Management. Equally her interests in praxis have led to a wide range of research that attempts to understand the role of theory (and science) in influencing and learning from the 'real world'. This has meant also a growing interest in social learning. She now co-ordinates various projects that work across multiple scales from the local, such as in the case of wetlands and livelihoods, through to the development of catchment management strategies and international water sharing agreements. These areas of interest have involved the development of systems views to catchments (socio-ecological systems), national guidelines for Catchment Management Strategies and understanding constraints to Integrated Water Resources Management.

J. Owen Saunders, University of Calgary, Canada
Saunders served as Executive Director of the Canadian Institute of Resources Law at the University of Calgary for over two decades until

July, 2012. He is currently Senior Fellow with the Institute and Adjunct Professor in the Faculty of Law at the University of Calgary. He is the author of numerous books and articles on both domestic and international aspects of natural resources law and policy, and is a frequent advisor to governments, international organizations and non-governmental organizations.

Edella Schlager, University of Arizona, US
Schlager is Professor in the School of Government and Public Policy at the University of Arizona. Her research programme focuses on common pool resources and comparative analyses of institutions for governing water. She is co-author of two books, *Common Waters Diverging Streams*, a study of conjunctive water management practices in the western US, and *Embracing Watershed Politics*, an examination of polycentric governance of watersheds. She is also lead editor of *Navigating Climate Change Policy: The Opportunities of Federalism*, an exploration of federal climate change policies and processes.

Andre Silveira, University of Cambridge, United Kingdom
Silveira is a Ph.D student in the Department of Geography, University of Cambridge. He holds a diploma in Political Science and an MA in European Integration Studies. His Ph.D project is concerned with the evolution of institutional structures and processes guiding water quality governance in large cross-jurisdictional river systems, with China's Pearl River Basin and Europe's Rhine River Basin as case studies. He is assistant coordinator of the EU–China River Basin Governance Research Network and has been involved in the policy dialogue component of the EU–China River Basin Management Programme as a consultant.

Preface

The territorial division of power in federations, plus the interconnected politics at the national and regional levels, present a classic governance test for waters shared across multiple political jurisdictions. This is increasingly important as democratic transitions have introduced or invigorated federalism across diverse contexts affecting more than 300 major river basins, including approximately 50 per cent of the world's international rivers.

In April 2012, the University of Oxford hosted an international workshop on *water security and federal rivers*. The workshop convened 35 delegates from 12 countries to exchange lessons learned from water reforms in federal countries. Highlights included the development of a set of case studies anchored in the insights about river basin management and federalism from multiple perspectives across public policy, economics, history and complexity science.

A keynote presentation by Dr Jerry Delli Priscoli of the US Army Corps of Engineers chronicled the history and models of river basin management in the US – the world's oldest federation – to demonstrate the elusive quest for integrated strategies and the significant potential for solutions matched to local and regional circumstances. Case studies across the Americas, Asia, Africa, Europe and Australia highlighted the extent and diversity of federal rivers, identifying alternative pathways to share risks and manage conflicts. The workshop culminated with a preliminary synthesis and comparative reflections by George Anderson, President Emeritus of the Forum of Federations, who noted the need to learn from both success and failure: while some federations have succeeded in putting in place river basin authorities and achieving integrated management, the story more generally is one of failure. He observed that river basin organizations seem to function most effectively when water management is salient enough to merit the organizational effort but not so politically fraught that the conflict must be handled directly by politicians.

The core contribution of this edited volume is to deliver a complementary set of case studies to learn lessons from the water management experiences in nine federal countries and China. The case studies were chosen to develop both the breadth and depth of comparison. Given the spread of federalism, federal countries were chosen from a range of political and environmental settings where water issues have become catalysts for both cooperation and conflict across jurisdictions and between the state and federal levels. In many instances, two case studies are included from each country to provide (1) a broad overview of the federal system of water management coupled with (2) a more detailed case study of a specific river basin or water management 'flashpoint' that illustrates the contested state–federal relations in water management.

The goal is to deliver a reference volume that will motivate further comparative research and exchange among policymakers and practitioners. In doing so, authors from these countries developed case studies to provide a rounded picture of water policy and management in each federation, with a clear presentation of:

- the legal, fiscal and institutional factors that set the framework within which water policy has been developed and managed;
- the key imperatives of rational, sustainable water policy given the supply and demand characteristics of the country and the extent to which these have been developed and recognized within the country; and,
- the political dynamics, which have either aided or impeded the development of good policy or processes (including the role of crises, leadership, recourse to special powers, etc.) and the extent of consensus or conflict around water policy.

While the peculiarities of each system are such that the weight and attention given to any particular factor or issue will vary considerably, the case studies work from a common template to draw out some key lessons.

The editors wish to acknowledge and thank all workshop contributors, as well as Australian National University for sponsoring the workshop with additional support from the Forum of Federations. The workshop was organized by this volume's editors: Dr Dustin Garrick (McMaster University and University of Oxford), Drs Daniel Connell and Jamie Pittock (Australian National University) and George Anderson (New York University).

PART I

INTRODUCTION

1. Federal rivers: a critical overview of water governance challenges in federal systems

Dustin Garrick, George Anderson, Daniel Connell and Jamie Pittock

Sustainable management of rivers, lakes and aquifers is crucial to the well-being of people and the environment (MEA 2005). Farms, cultures, industries, cities and nations have been established along the banks of major river systems. Demand for freshwater and other riverine commodities has increased with population growth and economic development, while climate change and extreme events disrupt hydrological processes and water supply. As a consequence, the World Economic Forum (2013) has identified water supply shocks among its top societal risks for the past three years running. Effective governance of fresh water is therefore a foundation of sustainable and equitable societies.

In this context, the Organisation for Economic Co-operation and Development (OECD) (2012) has identified the global water crisis as a crisis of governance and policy fragmentation. Increasingly, water management challenges involve complex interdependencies between sectors, upstream and downstream jurisdictions and stakeholders at the local, state, national and international levels. These interdependencies pose coordination challenges across political borders – both within and between countries. Federal countries distribute authority between national and state jurisdictions, which complicates water management tradeoffs within river basins shared by multiple territories. This book examines the experience of nine different federal political systems and China in addressing challenges of river basin and water management.

Federal river systems (see Figure 1.1) are major basins within or shared by one of the world's 28 federal countries (Anderson 2008, 2010; Garrick et al. 2013). Shared river basins are a major test of federal systems of governance. Effective management is not the mandate of one level of governance: all levels of governance have key roles, though these will vary from federation to federation. It can be expected that federalism will produce

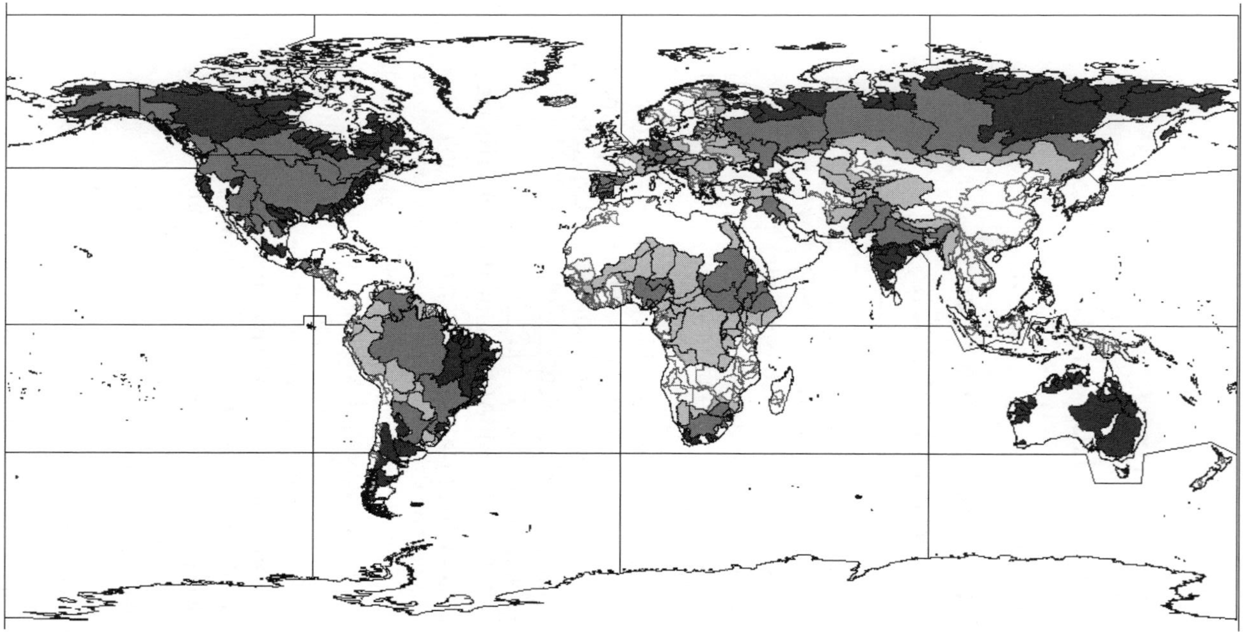

Source: Garrick et al., 2013. Used with permission.

Figure 1.1 Federal rivers. Black refers to domestic rivers falling within a single federal country (domestic federal rivers). Dark grey refers to the federal portion (basin-country unit) of a river shared by two or more countries that include at least one federal country (international federal rivers). Light grey refers to non-federal (unitary) basin-country units of international federal rivers (e.g. Egypt portion of the Nile). White basins (grey borders) refer to domestic rivers in unitary countries or international rivers without federal basin-country units. This demonstrates the global reach of federal rivers across more than 300 of the world's major river basins. (Adapted from Garrick et al. 2013, with permission)

different patterns of decision-making and implementation than unitary regimes and that this will affect outcomes. This is a major issue shaping the chapters in this book.

While there is an extensive literature on the management of river basins across international borders, prior research about river basin management within federal countries has largely focused on individual basins and countries. This volume addresses a number of issues that are key to understanding interjurisdictional water allocation and management in federal countries. What are the salient water management challenges in federal rivers? How does federal design affect the capacity and effectiveness of integrated approaches to water management across multiple jurisdictions? Answering these questions requires an understanding of how federations allocate responsibilities for rivers and how they balance the rights of independent jurisdictions with the need for integrated management. These issues have become more pressing due to the spread of federalism across a diverse range of geographic settings, as well as the growing prevalence of sub-national water conflicts across jurisdictional boundaries.

Equitable risk sharing and conflict resolution are difficult in federal rivers even when constitutional provisions coordinate power sharing and mechanisms are available to manage interjurisdictional disputes over water. The architects of these different federal systems in some cases have elaborated fairly clear allocations of responsibility for river basin management, in other cases not; even when the allocation of responsibilities is clear, the political will and circumstances may make cooperation difficult, with limited coordination of information, weak joint institutions, perverse incentives and insufficient or inappropriate infrastructure across multiple jurisdictions. Fortunately, there are also happier stories and this volume tries to understand the drivers in both cases.

Two overarching themes cut across the country chapters: (1) the increasing need for integrated water resources management and water security in federal rivers, and (2) the capacity of federal water management arrangements to respond to these needs.

1.1 INTEGRATED WATER RESOURCES MANAGEMENT AND WATER SECURITY

Integrated water resources management has received extensive attention as a form of water governance (see Schlager and Blomquist 2008 for a critical review). Following the 1992 International Conference on Water and the Environment in Dublin, global water experts have advocated integrated water resources management (IWRM) – a package of reforms designed to address connections between land and water and encourage participation by multiple stakeholders in water planning (see Conca 2006).[1] River basins have been upheld as the natural unit of water planning and management due to upstream–downstream interdependencies and the connection between land use and water within catchment and basin boundaries. Drought, flooding and increased competition have also prompted calls for river basin management to reduce risks and make tradeoffs at a whole-of-basin scale. However, IWRM has proven elusive in practice. Progress has been limited and contested, leaving the promise of river basin approaches often unfulfilled (Schlager and Blomquist 2008, Connell 2010, 2011).

The divergence of political and biophysical boundaries – across countries and within federations – has been a prime challenge for IWRM. This mismatch makes it more difficult to manage multiple, often conflicting, goals across upstream and downstream jurisdictions within transboundary rivers. Therefore, water policy prescriptions for integrated water resources management are inherently political and have required collective action at multiple scales from water users and state governments to national governments and international river basin commissions and treaty organizations.

Climate change and competition for variable water supplies superimpose the challenge of water security on top of issues of institutional fragmentation in federal rivers. Assumptions of stationarity have been rejected, which means that water management decisions cannot be based on the notion that the future will resemble the past (Milly et al. 2008); the timing, quality and volumes of water flows may change as well as the severity of extreme events like floods and droughts. In this context, water security is increasingly seen as a fundamental goal of integrated water resources management to ensure the availability of water for human development and ecological health along with the capacity to manage and adapt to multiple risks and climate hazards (Grey and Sadoff 2007).

Closed river basins – where downstream commitments are not met due to upstream development – illustrate upstream–downstream interdependencies and the need for a coordinated response in the context of increasing hydroclimatic risks (Falkenmark and Molden 2008). Australia is a striking example that illustrates the connected challenges of integrated water resource management and water security in a federal country facing climate variability and change. Despite the recognized need for integrated river basin management, Australia has experienced entrenched barriers to basin-wide integration within a federal system where powers are divided. The 'millennium drought' (1997–2010) in the Murray–Darling Basin of Australia provided a preview of our changing world (Pittock and Connell 2010). During this prolonged drought, competition for scarce and variable water supplies intensified between production and ecological sustainability.

Australia's drought was then followed with the wet extremes and floods of 2010–13 that replenished wetlands and aquifers but had mixed impacts, including lost lives and economic damage in Queensland, flooded crops and black water events for Murray cod (*Maccullochella peelii*). Water reforms in the Murray–Darling have responded with episodic basin-wide efforts to share risks and benefits that have been marked by high transaction costs and unsatisfactory compromises due to tension between state governments and between the states and federal authorities (Connell 2010, 2011).

The specific issues differ across river systems – for example, semi-arid, subtropical, monsoonal and temperate river systems – and levels of economic development. River basin strategies must therefore be adapted to local and regional contexts. Although river management responses to multiple tradeoffs will invariably involve different pathways across diverse contexts, there is a large scope for mutual learning. Even where there is a recognized need for multi-jurisdictional river basin management, there are many questions about the design of federal arrangements for water management, including: opportunities and limits for river basin governance institutions, the role of subsidiarity (itself a subjective concept) in assigning responsibilities between orders of government, roles of non-government versus government institutions, the merits of polycentric versus more unitary governance structures, and incentives and penalties for implementation of common river management measures. These are the second theme of the forthcoming chapters.

1.2 FEDERAL ARRANGEMENTS FOR WATER MANAGEMENT

Water management in federations is severely complicated by the political partition of the country and the division of powers between the federal and constituent unit governments (Anderson 2008, Elazar 1987). As noted above, the logical management area for water planning and allocation – a major river basin – typically has very different boundaries from those of the constituent units of the federation. Thus federations can have great difficulty adopting cooperative and functional approaches to water policy, especially in situations of severe conflicts caused by competition among multiple values, exacerbated by hydroclimatic variability and associated risks tied to drought, flooding and systemic tradeoffs.

This mismatch between political boundaries and river basins in federal countries poses special challenges for (1) systemic cost and benefit sharing and (2) conflict management between the states of the federation. The literature on international waters has identified the importance of international treaties and river basin organizations to promote cooperation and manage conflicts in shared waters (Wolf 2007, Schmeier 2010). Federal systems often confront similar challenges with conflict management and risk sharing to those arising in international waters despite having an institutional structure and legal regime to share powers. Federal rivers also differ from much studied international rivers in another way: while they face some similar opportunities and constraints on risk sharing and conflict resolution, they operate in a constitutional context in which there is *both* divided sovereignty, as between nations, and united sovereignty, as in unitary regimes. The precise mix and dynamics of these internal arrangements varies greatly across federations.

Federal constitutions allocate varying degrees of authority to their constituent states to manage challenges that span multiple jurisdictions (Bednar 2008). In dealing with river basins that cross state boundaries, they must deal with divided geographic responsibilities (state-to-state) as well as functional responsibilities (federal versus state governments). Each federation is different in its constitutional arrangements, political dynamics and objective challenges in river basin management. What they have in common is the need to operate a complex system of polycentric governance arrangements where there may be strong imperatives to address river management issues that involve several governments.

The global extent of federal rivers invites comparative analysis about federal arrangements to manage shared water resources in many different contexts. A comparative perspective can yield insights about the global picture of federal rivers, the water management challenges they face and the diverse water policy and river basin management approaches developed in response. Authors of the country-level overview chapters develop a concise analysis of the federal system of water management in terms of the type and salience of water challenges, on the one hand, and the characteristics of the federal arrangements for water management, on the other, including the constitutional provisions related water management and the roles of different levels of government and key actors (courts, legislature and so on). The challenges of integrated water resource management and water security are presented in this federal context to identify the primary water issues, institutional mechanisms for interjurisdictional water sharing and conflict resolution, and fiscal arrangements.

1.3 OVERVIEW OF THE BOOK

This volume develops a comparative perspective and set of case studies to assess federal systems of water management. Authors applied multiple analytical frameworks to develop insights at the river and country level. Theoretical approaches and analytical frameworks include: federalism, fiscal federalism, polycentric governance and the politics of scale. These approaches provide a rounded, and complementary, perspective on water management in federal systems and are introduced in more depth by chapter authors.

The introduction has drawn on the other chapters to outline the cross-cutting questions and themes structuring the case studies of water security, river basin management and federalism. Chapter 2 (Pittock) presents a global overview of water security challenges posed by climate change, focusing on institutions needed to coordinate water supply for people and healthy riverine ecosystems in multi-jurisdictional rivers – both nationally and internationally. This overview provides global context and also lessons from Australia's experience in the Murray–Darling Basin, which is on the leading edge of experience with climate-related water security challenges in an historic context of failed policies and ineffective coordination at the whole-of-basin scale.

The rest of the book comprises 17 case studies across ten countries: Australia, Brazil, Canada, China, India, Mexico, Pakistan, South Africa, Spain and the US. The case studies offer an overview of water policy and management in each federation, with a clear presentation of the framework within which water policy has been developed and managed, including the legal authorities, fiscal arrangements and institutional structure for water management. The case studies are organized geographically to offer a global tour across the Americas (Chapters 3–8), Spain (9–10), South Africa (11–12), South Asia (13–15), China (16–18) and Australia (19). Chapter 20 examines human security challenges tied to water management in cross-border regions of multiple federal countries. The concluding chapter draws on the country-specific cases to reflect on lessons and emerging trends in the politics and policy of water management within federal political systems (Chapter 21).

In some countries, two case studies are offered with a case study providing a country-level overview and a second case study analysing river basin management in an important basin within the country. The river basin level case studies include the interstate rivers of the western US, the Saskatchewan River in Canada, three river basins in Mexico, rivers of the Lowveld region of eastern South Africa, the Narmada River in India, the Pakistan portion of the Indus River, the Murray–Darling, and three chapters on rivers in China – a non-federal country with some emerging federal-like characteristics influencing water management. These case studies consider the water security and integrated water resources management challenges for the river basin, along with a series of focal questions:

- Why is the chosen federal river important?
- What are perceived to be the main risks and threats to water security for the major stakeholders?
- What are the ongoing controversies and unresolved issues in the basin? How have they evolved over time?
- What are the legal and policy frameworks and institutional structure, and how are outcomes determined in practice?
- How do the federal and state governments interact within this structure, and how have these interactions evolved?
- What institutional responses are used to promote change and resolve conflicts?

In this context, each river basin case study considers the factors that enable or constrain effective river basin management in federal systems.

1.3.1 Americas

The chapters from the Americas start in the US – the oldest federation. The water management challenges vary from water pollution in the eastern US to water scarcity and climatic variability in the irrigation-dominated region west of the 100th meridian. These contrasting water challenges have historically prevented a unified, national framework for river basin management, despite federal legislation to manage water quality (for example, the Clean Water Act). Gerlak (Chapter 3) traces the evolution of state–federal relations in water management across multiple periods, highlighting the need for a historical perspective to understand the allocation of powers, functions and capacities shaping contemporary river basin management efforts. She shows that contemporary arrangements are marked by 'pragmatic federalism' with collaborative partnerships and coordination across and within state and federal levels in response to integrated water management problems and processes, such as watershed planning.

Chapter 4 (led by Schlager) grounds the American experience in the interstate rivers of the western US to illustrate the role of transboundary compacts in building adaptive capacity to manage historic and projected climate variability and change. A review of multiple interstate rivers identifies ways in which states and the federal government build adaptive capacity through proportional allocation rules and interstate commissions to administer compact agreements governing rivers shared by multiple states.

Canada is richly endowed with water, so that the major issues of water management have typically had more to do with water quality and managing ecosystems than with water scarcity. The federal regime is one of the most devolved in the world and water resources are largely the responsibility of the provinces; the federal government has some significant legislative powers touching water, as well as its spending power, but has tended to use them with restraint. Water politics has tended to have relatively low visibility and to be essentially regional in character. Canada's experience is therefore marked by a more decentralized federal system of water management. Saunders (Chapter 5) demonstrates how the federal government's reticence to exercise its water management powers and other levers of influence has constrained basin management regimes.

The Canadian interstate (inter-provincial) basins with the most acute management challenges are the Mackenzie and Saskatchewan river basins east of the Rockies. The Mackenzie–Athabasca has major issues around water quality and withdrawals because of the oil sands; given their national importance, the federal government has been drawn into the debate and environmental studies. There was agreement in 1969 among the three Prairie Provinces and federal government on water apportionment for the Saskatchewan basin, but this has not provided the mechanism to address some emerging apportionment challenges. The Saskatchewan River of western Canada therefore illustrates the consequences of this fragmentation in Chapter 6 (Horbulyk), where Canada's model of cooperative federalism has failed to build adequate resilience to manage water scarcity (particularly across provinces), droughts, and nutrient loading in Lake Winnipeg. The three provinces operate quite independently of one another in their water management planning.

Mexico is comparatively more centralized in its federal design and has undergone a reform process to decentralize powers and functions in water management over the past two decades. The country spans a range of challenges from the industrializing, semi-arid north, which has a long history of irrigation development, to urbanization in central Mexico and rural development of the subtropical south. A common challenge across these multiple regions is the lack of provision of water for the environment. Barrios (Chapter 7) outlines Mexico's federal system of water management under the 1992 National Water Law, which encouraged decentralization despite the establishment of a powerful national water agency (Comision National del Agua). This chapter emphasizes the challenges of establishing a coherent national policy framework and strengthening river basin organizations despite continuing tensions among the local users, states and federal government. Against this backdrop, new models of water management have been attempted to secure water for the environment through pilot efforts led by World Wide Fund for Nature (WWF) in the Rio Conchos (a subbasin of the Rio Grande/Bravo), the

Copalita–Zimatan–Huatulco river basins and the San Pedro Mezquital River – each of which have included a mix of national and local-indigenous communities to define and manage ecological reserves.

Like Mexico, Brazil's vast territory presents diverse challenges. An innovative federal system of water management has emerged in Brazil in response to disparate water security risks and a constitutional division of powers between rivers designated as 'federal' (which cross state boundaries) and state rivers (which do not). Chapter 8 (led by Coelho) chronicles the reform process leading to the establishment of the National Water Resources Management System in Brazil, which favours economic tools and decentralized participatory management within river basin committees at multiple levels. A recent national 'pact' is attempting to harmonize management across states by using financial incentives to build capacity and empower particular regions of Brazil.

1.3.2 Spain

Spain confronts water stress and intensifying competition for variable water resources in a semi-arid environment. It has a complex multi-level system of water management involving river basin authorities, autonomous communities (states), the national government as well as the supranational water management provisions of the European Union. Chapter 9 (Albiac) documents the deep historical roots of contemporary federalism in Spain dating from the medieval kingdoms of northern and eastern Spain. Constitutionally, responsibility for river basins is vested at the national level for basins that cross the boundaries of multiple autonomous communities. Rivers that fall entirely within one autonomous community are managed by that community. At the national level, the contemporary system of water management is based in a 1985 law. River basin authorities are arms of the national government (for river basins that traverse internal boundaries) but provide the venue for users (particularly irrigation) to make decisions and plans. The European Union exerts increasing influence, principally through the 2000 Water Framework Directive establishing a goal of good ecological status and a corresponding process of river basin planning and management to incorporate EU standards into national and sub-national law.

López-Gunn and De Stefano (Chapter 10) examine contemporary water security challenges in Spain to illustrate changing state–federal–EU relationships under the competing pressures of decentralization and Europeanization of water management. This analysis shows how the definition of water security changes with scale from a more local focus on water storage and irrigation for food production balanced against the

European Union focus on ecology and water quality. Four case studies on water security and decentralization in federal rivers of the Tajo-Segura, Ebro, Jucar and Guadalquivir demonstrate a three-level game across state, federal and European influences on river basin management planning under the Water Framework Directive.

1.3.3 South Africa

South Africa does not call itself 'federal', but the 1996 Constitution has the character of a relatively centralized federation. The Water Act of 1998 reflects the values of the new constitution and the imperative of the transition to democracy after the apartheid era; an important part of its water reform was the need to redress historic inequities and the creation of a domestic and ecological water reserve. Muller (Chapter 11) reviews this water reform and the conscious effort to retain powers and functions at the national level and avoid 'balkanization' at the river basin level. Technical working groups on water supply and sanitation advised a two-level system with a strong national competency to back local and regional delivery. Implementation has been more constrained, however, particularly in the establishment of catchment management authorities to undertake planning, and for the national government to step in if the authorities fail to do so. The South African model avoided the fragmentation of powers and functions to allow infrastructure development and inter-basin tradeoffs.

Chapter 12 (led by Pollard) examines devolution and cooperative governance within the South African context. The establishment of catchment management authorities has proven more difficult to achieve than expected, and exposed challenges of building capacity for effective water governance at more local scales. The experiences of the eastern Lowveld region rivers have illustrated the need to combine formal allocation of powers and functions with informal institutions, legal pluralism and social learning at and within the catchment level. This local experience in a centralized federal system of water management highlights the need for appropriate incentives and self-organization to achieve the objectives of the 1998 Water Act.

1.3.4 South Asia

Asia's experience can be divided between the South Asia federations of Pakistan and India and the Eastern Asia experience in China, which although not a federation, has relevant multi-layered dynamics that show the spread of some federal-like practices even within this authoritarian society.

With its vast population, India's federation (Chapters 13 and 14) shows tremendous diversity – in the size of its federal units (the largest state being over 200 million people, the smallest union territory less than 100,000), in its languages and cultures, and in the development of its regions. Hydrographically, it has both deserts and the rain soaked north-east, but the defining character of the country is the monsoon, which is vital for its overwhelmingly rural population. Increasingly, its politics reflect this diversity, with powerful regionally based parties and strong state governments. Constitutionally, the states have the main responsibility for water and the Union government has never been able to impose a coherent national water policy or even to assert itself in relation to inter-state waters.

A case study of the Narmada River (Chapter 14) illustrates water allocation conflicts among four riparian states. The water dispute tribunal and associated river basin control authority have struggled to manage competing demands and have largely neglected groundwater and land use change. These challenges highlight the need for data sharing between states and an increasing coordination role for the national ministry of water resources to improve understanding of basin hydrology and incorporate data on groundwater use and land use change into updated interstate water sharing agreements.

Pakistan (Chapter 15) has only four provinces, but three of them and over 90 per cent of the population lie within one watershed, the Indus basin. This basin has been subject to massive engineering works, both to deal with the separation of river flows between India and Pakistan and to develop the world's largest contiguous irrigation network managed under a federal arrangement. The Constitution gives the greater role on water management to the federal government, but in practice major decisions on water are so regionally sensitive that the provinces are always deeply involved. After an extended period of inter-provincial tensions around shares, irregular water supplies and power outages, there was an historic Water Apportionment Accord in 1991, agreed by the four provinces with the strong support of the federal government. The implementation of the accord has been hobbled by some optimistic assumptions, different interpretations and weak technical support. The recent constitutional changes, which include strengthened roles for the high level forum of heads of government as well as for the Senate, offer some possibilities for more political dialogue, but a series of major innovations will be needed if the country is to address the challenges it faces.

1.3.5 China

This volume also includes three chapters on China. China is not a democracy, nor is it federal. The constitution plays a minor role in relation to internal governance. However, it is a vast country that necessarily has different regional interests. Therefore, some have suggested that China is quasi-federal and exhibits 'behavioural federalism'. The factors that are producing this result are described in Chapter 16 (Silveira). Using the Pearl River as an example he explains how the national government decision in the 1980s to promote economic development by devolving considerable powers to key provinces selected as policy pilots resulted in an effective handover of responsibility for many of the activities that determine water quality, probably the greatest single water problem in southern China. Water pollution has been a notable by-product of economic development in China. But the policies needed to improve water quality unavoidably affect millions of small and large businesses and clash with provincial policies – and central directives – promoting economic development. The national government has the undoubted legislative power to intervene but the capacity to do so effectively may be in question where powerful stakeholder groups with strong polluting interests have developed in the various provinces.

The complexity of managing in this system characterized by the internal politics of a Communist Party that is increasingly permeable to certain external pressures resulting from economic and social development is investigated further in Chapter 17 (Ke Jian), which examines the case of Tai Lake in eastern China. Here the national government had to intervene in 2007 after the city of Wuxi with one million inhabitants had its drinking water cut off for over a week because of severe pollution. (This event drew attention to a persistent problem which had been intensifying for many years.) The Tai Basin, surrounded by two provinces and Shanghai Municipality and home to just under 40 million people, is one of the most industrialized regions in the country. Despite drastic action by the central government, which closed thousands of polluting businesses and industries at the height of the crisis, the development of a water governance regime that can control pollution without stifling provincial dynamism continues to be an unresolved challenge in the Tai Basin and beyond.

In addition to the challenges resulting from cross-border water pollution, water scarcity also causes conflicts between provincial governments in China. In Chapter 18, Lan Fang examines conflicts between Shaanxi and Hubei provinces which are both being required by the national government to contribute to the middle route of the vast south–north project which will divert water from three points along the Yangtze River to the north of China. At issue is the need to share the costs, direct and indirect, and the compensation being provided by the regions in the north that will benefit. It is an intense struggle between donating and receiving provinces and municipalities with the national government arbitrating the conflict. Provincial governments have the same formal institutional status as national departments so to a significant degree it is a struggle between equals within a bureaucratic and party system rather than a disagreement between junior and seniors within an authoritarian hierarchy.

1.3.6 Australia

Chapter 19 (Connell) provides a report on the ongoing struggle about water policy in Australia's Murray–Darling Basin (MDB). As of December 2012 the national government is implementing a comprehensive basin plan backed by over $12 billion in new investment in infrastructure and water entitlements for the environment. The history of river management in the MDB provides an interesting study in the effective limits of financial power in a federal system. Financial power rests overwhelmingly with the federal government but it has been repeatedly frustrated over the decades by the determination of state governments to use their greater knowledge and administrative capacity in relation to water issues to promote their own often conflicting goals. This chapter places the current debate within its historical context and discusses the new set of institutional arrangements that are emerging in a rather different form than was originally intended by policy makers in the national government but which could prove quite robust and moderately effective.

1.3.7 Comparative Perspective and Conclusions

The final section includes a comparative chapter (Maganda) on human security in cross-border regions of North America, Europe and South America. These cross-border environments involve challenges of human insecurity in meeting basic needs including drinking water and sanitation to support development. The geographic distance of these cross-border regions from the hubs of central governments poses special challenges for federal water management systems. Human security challenges in cross-border regions demonstrate the need for national governments to build capacity in local communities, while respecting their autonomy, similar to the insights emerging from South Africa.

The closing chapter (Anderson) reflects on how the constitutional and political nature of federal systems come together with highly variable water management challenges (even within different regions of the same federation) to affect the policy and institutional response to water management, including in various river basins.

PRELIMINARY REFLECTIONS

The studies in this volume suggest that federal systems have a major advantage in addressing river basin management compared to international rivers: the existence of a national level of government injects a major new actor into conflicts amongst states over the equitable distribution of costs and benefits. However, the legal and financial levers that the national governments have can vary a great deal, given different constitutional arrangements, and the political motivation of national governments to intervene will also vary depending on the national importance of the river basin management issues as well as the political costs and benefits for the government party or parties. Thus in at least some of the cases discussed in this book national governments have proved reluctant to play that role; for example in Canada, the issues of river basin management have not been seen to be so important as to require a major intervention by the federal government, whereas in Spain, in contrast, the issues have been so politically difficult that the national government has played a cautious role in recent years. Even in cases where national governments have been more willing to assert national priorities, as in South Africa, they have still proven difficult to impose in practice. This is even the case in China where the national government is struggling to manage cross-border water quality issues. Probably the most determined example of strong national government intervention discussed in the book is that of Australia where the latest round of reform has resulted in a new institutional rearrangement which now places the national government at the centre of water policy development in the Murray–Darling Basin. It appears that a federal system makes it easier for regional or sectional interests to block change (reform) than do non-federal systems, but it must be noted that federalism is often deeply embedded in the nature of the society – even with its difficulties, it is not an 'optional' choice.

For the researchers contributing to this book, water security and integrated water resources management provided a common reference point. One way to develop a list of threats to water security is to examine the many ways in which people use or relate to water. These could be biophysical, social, cultural, economic, environmental or political. If a particular relationship is important to people, then it probably contains within it dependencies and vulnerabilities that can become significant sources of insecurity under particular circumstances. According to the OECD the primary source of threats to water security comes from inadequate governance systems and the lack of integration, the processes that are used to decide who will get water in what condition, when, where and how: in other words the relationships between different groups of people that determine their access to water or vulnerability to the dangers it creates (such as having to live on the flood plain because they are poor). The goal of improved water security and integrated water resources management implies betterment for society as a whole but the 'fair' distribution of costs, benefits and compensation is crucial if that is to be a reality. A major question for students of federalism, as well as studies of water management, is how federal systems affect the outcomes of struggles about such questions.

NOTE

1. The Global Water Partnership refers to IWRM as a process which promotes the coordinated development and management of water, land and related resources, in order to maximize the resultant economic and social welfare in an equitable manner without comprisiming the sustainability of vital eco-systems.

REFERENCES

Anderson, G. 2008. *Federalism: An Introduction*. Toronto: Oxford University Press.

Anderson, G. 2010. *Fiscal Federalism: An Introduction*. Toronto: Oxford University Press.

Bednar, J. 2008. *The Robust Federation: Principles of Design*. New York: Cambridge University Press.

Conca, K. 2006. *Governing Water: Contentious Transnational Politics and Global Institution Building*. Cambridge, MA: MIT Press.

Connell, D. 2011. The Role of the Commonwealth Environmental Water Holder. In Q. Grafton and D. Connell (eds), *Basin Futures: Water Reform in the Murray–Darling Basin*. Canberra: ANU E-press.

Connell, D. 2010. Managing Climate Variability in Australia, South Africa, United States and Spain, CWEEP Research Paper 10–12.

Connell, D. 2007. *Water Politics in the Murray–Darling Basin*. Sydney: Federation Press.

Elazar, D. 1987. *Exploring Federalism*. Tuscaloosa, AL: University of Alabama Press.

Falkenmark, M. and Molden, D. 2008. Wake up to realities of river basin closure. *International Journal of Water Resources Development* [ISI], 24(2): 201–215.

Garrick D, De Stefano L, Fung F, Pittock J, Schlager E, New M, Connell D. 2013 Managing hydroclimatic risks in federal rivers: a diagnostic assessment. Phil Trans R Soc A 371: 20120415.http://dx.doi.org/10.1098/rsta.2012.0415

Gerlak, A.K. 2006. Federalism and U.S. Water Policy: Lessons for the Twenty-First Century. *Publius: The Journal of Federalism*, 36(2): 231–257.

Grey, D. and Sadoff, C.W. 2007. Sink or swim? Water security for growth and development. *Water Policy*, 9: 545–571.

MEA (Millennium Ecosystem Assessment) 2005. *Ecosystems and Human Well-*

being: Wetlands and Water Synthesis. Washington DC: World Resources Institute.
Milly, P.C.D., J. Betancourt, M. Falkenmark, R.M. Hirsch, Z.W. Kundzewicz, D Lettenmaier and R.J. Stouffer. 2008. Stationarity is Dead: Whither Water Management? *Science*, 1 February 2008: 319(5863): 573–574.
Pittock, J. and Connell, D. 2010. Australia Demonstrates the Planet's Future: Water and Climate in the Murray–Darling Basin. *International Journal of Water Resources Development*, 26(4): 561–578.
Schlager, E. and Heikkila, T. 2011. Left High and Dry? Climate Change, Common-Pool Resource Theory, and the Adaptability of Western Water Compacts. *Public Administration Review*, 71(3): 461–470.
Schlager, E. and Blomquist, W. 2008. *Embracing Watershed Politics*. Boulder, CO: University Press of Colorado.
Schmeier, S. 2010. Governing International Watercourses – Perspectives from Different Disciplines. *Hertie School of Governance – Working Papers, No. 53*, August 2010.
Wolf, A.T. 2007. Shared Waters: Conflict and Cooperation. *Annual Review of Environment and Resources*, 32: 3.1–3.29.

2. Climate adaptation in river management in a post-stationary world

Jamie Pittock

2.1 RIVERS AND CLIMATE CHANGE

Hectare for hectare (acre for acre) the world's freshwater ecosystems have a greater diversity of flora and fauna compared to any other biome (Dudgeon et al., 2006; MEA, 2005). The world's rivers are also the centre of human life: the centre of most major cities and the valleys where agriculture has taken root in support of ever larger populations (Richter et al., 2010; Vorosmarty et al., 2010). Groundwater systems are also of significance for biodiversity and water supply for people but are not considered further in this chapter on rivers.

Because it is central to humanity, the freshwater realm is also more threatened than any other part of our environment. In 2005 the Millennium Ecosystem Assessment (MEA, 2005) concluded that freshwaters have more threatened species per unit area. Two ecosystem services were declared over-exploited: consumption of water and of wild-capture fish from freshwater and marine systems. Further, despite promises to establish ecologically representative reserve systems by national governments under the Convention on Biological Diversity and Ramsar Convention on Wetlands, less extensive protected areas have been established for freshwater and marine ecosystems (MEA, 2005; Pittock et al., 2008).

The assessment goes on to say: 'The most important direct drivers of change in ecosystems are habitat change (land use change and physical modification of rivers or water withdrawal from rivers), overexploitation, invasive alien species, pollution, and climate change. These direct drivers are synergistic' (MEA, 2005: 27).

Climate change is anticipated to be expressed strongly through changes to hydrology (Milly et al., 2005). Importantly, most freshwater management systems were funded on an assumption of stationarity, namely that water flows remain consistent in the long term despite short term variation (Milly et al., 2008). The Intergovernmental Panel on Climate Change outlines how climate change will change precipitation patterns and increase evapotranspiration and thus change river flows (Bates et al., 2008). Many areas will get wetter but these are often places that are already humid or where few people live. Some areas will get drier and these are often mid-latitude river basins that are already water scarce and have large population centres. Diminished glacier and snow melt will reduce the base flow of rivers and increase seasonal variation in flows. In addition to changing average flows, climate change is expected to increase the frequency of extreme floods and droughts. Water quality will diminish due to factors like increased erosion and influx of pollutants due to extreme events and higher temperatures, eutrophication and more extensive cyanobacteria blooms (Pittock, A. 2009).

A number of studies point to international waters that are at particular risk from climate change impacts. A global biophysical assessment by Palmer et al. (2008) projected river discharge under different climate and water withdrawal scenarios to illustrate potential changes in discharge and water stress. They argue that rivers impacted by dams or extensive development will require more management interventions compared to free-flowing rivers. The areas 'likely' to require action are home to one billion people whereas approximately 365 million people live in basins 'almost certain' to require action. The transboundary basins 'almost certain' to require a concerted response include Lake Chad, the Nile, Indus, Ganges–Brahmaputra, Tigris–Euphrates, Colorado and Columbia (Palmer et al., 2008). There are similar assessments at the continental scale, for example, Goulden et al. (2009) summarize climate change impacts on surface water resources in Africa, identifying specific regions with likely substantial decreases and increases in runoff and changes in the frequency and magnitude of floods and droughts.

While these direct impacts of climate change are increasingly well appreciated, less well understood are the likely extensive, negative effects of climate change response policies on freshwater ecosystems (Pittock, 2010, 2011). Many carbon sequestration measures, including carbon capture and storage and forestry plantations consume large amounts of water. In one case study in the Macquarie River in Australia, afforesting 10 per cent of the headwaters is predicted to reduce river flows by 17 per cent (Herron et al., 2002). Low-carbon energy sources often have large water footprints, including technologies like hydropower, solar thermal and 'hot rock' geothermal power stations (Carter, 2011). Further, climate change adaptation measures adopted to date often involve additional consumption, storage and interbasin transfer of water that has perverse impacts on freshwater ecosystems (Barnett and O'Neill, 2010; Hussey and Pittock, 2012).

Until recently many freshwater management experts have regarded climate impacts as secondary to the degradation of freshwater ecosystems from other drivers of change. The focus has been on ameliorating threats driven primarily by growing populations as well as increasing wealth and the concomitant increased demand for water-intensive products like meat. It is now clear that these threats are synergistic and complex, and are sometimes referred to as 'wicked problems' (MEA, 2005; Turnpenny et al., 2009; UNEP, 2007). Adequately responding to these challenges requires higher standards of governance across sectors and across scales than our societies have achieved to date (WWDR, 2012). Much of this volume focuses on institutions at the national scale. To complement this work, this chapter now focuses on the added challenges of better governance of transboundary rivers under climate change, namely the rivers that cross political boundaries, especially of nations and provinces (states).

2.2 THE ADDED CHALLENGE OF MULTI-JURISDICTIONAL RIVERS

The establishment of political boundaries without respect to river basins creates additional challenges and potentially some opportunities for sustainable and equitable governance. Garrick et al. (2013) identify rivers within the territories of federal nations, regional institutions like the European Union with structure that resemble federal governance, as well as others shared between two or more nations. These different circumstances suggest that different degrees of federal governance influence the management of multi-jurisdictional rivers. The institutional implications of subsidiarity and federalism for rivers at the international and then the national scales are considered here in turn.

There are 276 'international' rivers shared between two or more countries (as well as around 300 aquifers). Such international river basins contain 60 per cent of the planet's surface water flow, 50 per cent of the Earth's land surface and 40 per cent of the world's population (Loures et al., 2008). Consequently the sustainable and equitable governance of these shared resources is critical to conservation of biodiversity and generation of services for people. Yet there is strong evidence of rapid degradation with a growing number of shared rivers on every inhabited continent now described as 'closed', that is, all the available water is effectively diverted for human consumption (Falkenmark, 2008).

Many institutions exist for the management of transboundary waters at different scales. Since 1820 nearly 400 treaties have been adopted for cooperative management of shared waters, however only 40 per cent of shared rivers are managed under treaties (Wolf et al., 2005). Further, 80 percent of these agreements involve only two nations meaning that many international rivers do not have treaties that include all basin states, for example, Burma and China have not acceded to the Mekong River Agreement (Loures et al., 2008). A large portion of these agreements are restricted to delimiting borders, dividing water volumes and specific water infrastructure agreements rather than being more comprehensive instruments for managing river basins and their water resources. These treaties were adopted at a time when hydrological stationarity was assumed (Tarlock, 1999–2000). This means that those treaties either must be flexible enough to govern shifting hydrological conditions, or if they are overly prescriptive, they will require amendments. Unfortunately many of the international river treaties do not include the types of procedural mechanisms that enable ready amendment, such as: data sharing, river basin commissions, conflict resolution procedures or review clauses.

There are challenges in managing rivers that cross provincial (state) borders within nations similar to those of international rivers. Garrick et al. (2013) define federal rivers as basins within or shared by one of the world's 28 federal countries. These encompass over 40 per cent of the world's population. They point to additional nations with quasi-federal systems and regions with federal-like river management agreements. Issues of quantity, quality and timing of water supplies, the underlying reasons for water-related controversies, are equally or more prevalent within nations (Wolf et al., 2005). For example, there is a long history of conflict – sometimes violent conflict – over management of rivers between Indian states (Joy et al., 2007). Nations like the United States have agreements between states on management of transboundary waters that were predicated on stationarity, have over-allocated waters and are legally and politically difficult to revise to reflect changing conditions and knowledge (Reisner, 1986).

How then can our societies best respond to these climate-induced changes to water resources that are synergistic with other threats?

2.3 ADAPTATION RESPONSE OPTIONS

2.3.1 Institutional Responses

Dovers and Hezri (2010) develop two key arguments on good practice for climate change adaptation that can inform institutions for managing transboundary river basins. First, they argue that climate change adaptation to different impacts in different places on different sectors has to be embedded in existing institutions at all scales, and should be decentralized (Dovers and Hezri, 2010). Thus, the question for water agreements is whether they enable subsidiarity, that is, facilitate management at the smallest relevant scale within an overarching framework. Second, Dovers and Hezri (2010) also assess the levels of climate change impact and resulting institutional resilience strategies ranging from:

- similar to existing variability, and capable of being managed within existing capacities; to
- significantly exacerbated climate variability, a difficult adjustment but capable of being addressed with further application of existing capacities; and
- climate change and variability beyond current experience and memory, representing a step change that threatens ecosystems and society.

With respect to institutional resilience strategies, the question is whether current treaties on rivers amount to resistance and maintenance of the status quo, or enable change at the margins, or openness and adaptability (Dovers and Hezri, 2010). Each of these strategies has strengths and weaknesses. Treaties that are overly prescriptive (such as in specifying volumes of water) are likely to fail with climate-induced changes. At the very least, useful agreements for transboundary waters must be capable of managing significantly exacerbated climate variability, such as large and more frequent floods and droughts.

In either polycentric or more centralized approaches of climate change adaptation the roles of river basin organizations (whether on international or national scales) are critical. These organizations operate on scales where direct climate change adaptation measures can be tailored to local circumstances. Effective river basin organizations have the attributes needed for adaptive management, including: a sound knowledge base, multi-stakeholder participation, iterative planning and implementation, and source/s of revenue (Pittock, J. 2009). An important question is how can such effective river basin organization be fostered internationally and within federal countries? In the first case study below, it is argued that the entry into force of the Convention on the Law of the Non-Navigational Uses of International Watercourses (New York, 1997) is a key measure for enabling faster development of adaptive river basin management institutions.

2.3.2 Uncertainty, Risk and Maladaptation

Once effective basin scale institutions are in place they then need to consider the types of climate change adaptation interventions they should take in the context of uncertainty as to the scale of the impacts and in managing risks. Different climate change models produce very different forecasts for water resources in many parts of the world (Arnell, 2004). Even well-resourced hydro-climatic modelling may be unable to reduce uncertainty. For example, in the case of the Murray–Darling Basin, CSIRO (2008) modelling of average annual surface water availability in 2030 suggested that the outcome could range between 7 per cent more water to 37 per cent less, which would be magnified down river resulting in a 20 per cent higher outflow through to a 69 per cent decline. The extreme dry scenario in this case is all too plausible when during the 2002–2010 drought inflows fell by more than 60 per cent and the River Murray stopped flowing to the sea (Pittock and Connell, 2010). How should governments respond to this uncertainty? An appropriate risk management approach would be to make plans to manage a less likely but more catastrophic case. However, in this case, the Murray–Darling Basin Authority appears to have misinterpreted the 'median' scenario of a 10–12 per cent loss in average annual surface water availability as 'most likely' and 'balanced' for its planning purposes (MDBA, 2010; Pittock and Finlayson, 2011).

Climate change adaptation should aim to avoid overly-narrow and maladaptation. Overly-narrow adaptation is when an intervention only works with a narrow range of conditions that are likely to be exceeded. One example is that barrages constructed in the 1940s to prevent ingress of sea water from the River Murray's mouth into lakes Alexandrina and Albert. This worked until the 2002–2010 drought when low inflows from the River Murray saw most of the lakes evaporate resulting in hypersalinity and acid drainage (Gross et al., 2012). Further, the barrages are unlikely to resist the high end of the sea level rise forecast by the Intergovernmental Panel on Climate Change for the end of the century (Gross et al., 2012). Maladaptation has been defined as: action taken ostensibly to avoid or reduce vulnerability to climate change that impacts adversely on, or increases the vulnerability of other systems, sectors or social groups (Barnett and O'Neill, 2010). Five distinct types of maladaptation have been recognized, namely actions that, relative to alternatives: increase greenhouse gas emissions, disproportionately impact on the most vulnerable, have high opportunity costs, reduce incentives to adapt, and create path dependencies (Barnett and O'Neill, 2010). The environmental works and measures discussed below are one such example. What then should governments do to sustain river systems through climate change adaptation?

2.3.3 Water Management Responses

Water availability may increase or decrease in different parts of the world with climate change. For conservation of riverine systems, responding to greater water scarcity is the challenge addressed here. A number of types of responses have been tried and can be assessed:

1. *Iterative planning*. Cyclical planning processes are essential for adaptive management institutions, especially for rivers, to enable incorporation of new knowledge and lessons from previous implementation activities (Connor and Dovers, 2004; Pittock, J. 2009). Such planning may occur at different scales and across multiple jurisdictions, requiring coordination of federal, provincial/state and local authorities to deliver integrated approaches.
2. *Cap and trade markets for water*. Water markets have great potential to disconnect water consumption from economic growth (Grafton et al., 2011). During the 2002–2010 drought the ability of farmers to sell water from low value activities and buy water for high value production greatly reduced the socio-economic impact of the drought (Grafton, 2011). However there are many pitfalls in framing new water markets that may have perverse environmental impacts, for instance, by excluding some major water users (NWC, 2011; Young, 2010). Ideally such markets are established at the basin scale, across jurisdictions.
3. *Reallocation of water for the environment*. Reallocation of water in over-exploited systems from consumptive users to environmental flows is one of the most important mechanisms available for conservation of freshwater ecosystems (TNC, 2007). In many rivers enhanced environmental flows may substantially counter the impacts of moderate climate change (Aldous et al., 2011; Pittock et al., 2010). Water may be reallocated on an ongoing basis or seasonally (Pittock, 2013). For best effect this needs to occur at a basin scale. Environmental flow proponents have struggled to identify mechanisms whereby governments will implement such measures systemically and quickly (Poff et al., 2010). Environmental flow rules are at particular risk of institutional failure which demands consideration of measures to ensure their implementation and of alternate ways of spreading risk (Pittock and Finlayson, 2011).
4. *Use of water infrastructure*. Water infrastructure has been proposed to be used in climate change adaptation in two ways. First, it has

been argued that new infrastructure is required to store and distribute water, primarily to supply people (Kabat and van Schaik, 2003; Pittock, 2011). Such infrastructure has also been proposed to enable less water to be used to conserve freshwater biodiversity, so-called 'environmental water demand management' or 'environmental works and measures' (Pittock et al., 2012; Pittock and Lankford, 2010). Use of infrastructure in these ways has high risks of overly-narrow or maladaptation (Barnett and O'Neill, 2010; Nelson, 2010). These risks may be diminished by non-structural alternatives, reuse of existing infrastructure and more incremental development (Hallegatte, 2009). Second, reoperation of water infrastructure may benefit freshwater ecosystems by restoring connectivity (as with fish ladders) and manipulating water temperatures (Pittock and Hartmann, 2011). Any infrastructure development is likely to have implications at a basin scale.

5. *Ecosystem conservation mechanisms.* A range of 'soft path' adaptations involve maintaining or restoring the resilience of river ecosystems (Gleick, 2002), and may include restoration of riparian forests (Davies, 2010), restoration of floodplains (Opperman et al., 2009; Pittock, J. 2009) and removal or reoperation of water infrastructure (Pittock and Hartmann, 2011; Watts et al., 2011). While local benefits can be obtained, for best effect this needs to occur at a basin scale.

The health of river ecosystems is highly dependent on sound management of connectivity from their headwaters to the sea. To be most effective these measures need sound institutions to apply them at a basin scale. A common thread is the need for inter-jurisdictional arrangements and comprehensive frameworks to work synergistically across fragmented institutional landscapes within federal countries. Two cases illustrate this challenge: the UN Watercourses Convention at the international scale and the Murray–Darling Basin in Australia at a more local scale.

2.4 CASE STUDY: UN WATERCOURSES CONVENTION

Given that there are so many international rivers globally that do not have cooperative agreements between basin states, or where only some of the basin states are party to an agreement, there is a strong case for overarching global rules that facilitate sound basin scale agreements. After 37 years of negotiation, the Convention on the Law of the Non-Navigational Uses of International Watercourses (the 'UN Watercourses Convention', or

'Convention') was adopted in 1997 in a UN General Assembly vote: 106 in favour, 57 absent or abstained, three against (Loures et al., 2008). It codifies in international law common measures for cooperative management of shared waters by providing for equitable use of shared rivers balanced by an obligation of member states to do no significant harm (Rieu-Clarke and Loures, 2009). Ratification has been slow. However as at December 2012 there were 27 nations that had acceded which would mean that only eight more ratifications are required: entry into force is now imminent.

The Convention was also negotiated under the paradigm of stationary hydrology, so how then can it aid climate change adaptation? Pittock et al. (2013) have assessed the Convention against ten principles identified for adaptive management of shared rivers and conclude that it performs well on seven and poorly on three. The Convention has effective provisions for subsidiarity, equitable use, no harm, dispute settlement, information and communication. These procedures will be critical to enable governments of different jurisdictions to adjust their institutions in response to climate change impacts. It establishes high level rules for all the rivers without agreements and encourages revision of existing agreements to the minimum and high standards it establishes. However the Convention is largely silent on funding, conjunctive management with groundwater and revision of its provisions. Further institutional reforms are needed to enhance the Convention's governance, including agreements on a secretariat, conferences of the parties and financing. The French Government has proposed to host a first conference of parties suggesting that member states may consider such reforms (WWF, 2012).

Governments like Australia's have argued that the Convention did not need to enter into force to be effective as its principles are already enshrined as customary international law. However until it does enter into force as 'hard' international law, the institutions for governance of these shared rivers are held to the 1997 agreement. Entry into force will enable international law on shared river to evolve once more to embrace new issues like climate change adaptation (Pittock et al., 2013; Rieu-Clarke and Loures, 2009).

There is strong evidence that international treaties on river management that have higher standards are used by governments at larger scales as reasons for instituting sustainability reforms that may not be popular at more local levels. This is aided by the constitutional power afforded to many national governments to implement international agreements. Barrios et al. (2009) give the example in Mexico of a reduction of excessive agricultural water allocations to restore the health of Rio Conchos tributary of the Rio Bravo/Grande predicated on meeting treaty obligations to deliver water to the United States. Ebert et al. (2009) describe the pressure on the nations along the lower Danube River to restore floodplains in order to meet their obligations under the European Union's Water Framework Directive. Thus international agreements can provide a policy reform window for the adoption more sustainable management of multi-jurisdictional rivers when this may not be popular in constituent units of federations.

2.5 CASE STUDY: MURRAY–DARLING BASIN, AUSTRALIA

The Murray–Darling Basin covers a seventh of Australia and its river systems have great environmental and socio-economic values. Governance of the river system has been contested from before the formation of the Australian federation (Connell, 2007). The basin falls within the jurisdictions of one territory and four state (provincial) governments as well as the federal government (Pittock and Connell, 2010). Further, under agreements between these governments various planning and implementation roles have been delegated to the five provincial-scale governments and 19 subsidiary catchment management agencies established under state laws: an arrangement often described as an exemplar of subsidiarity in a federation (Garrick et al., 2012). In practice this nested governance has not realised effective adaptive management as federal and state governments have built catchment management agencies up then torn them down in frequent changes to the agencies' mandates, structures and funding (Pittock et al., in review; Robins and Kanowski, 2011).

The basin is a key place to consider better institutions for climate change adaptation of federal rivers as it is: over-exploited, particularly vulnerable to climatic and hydrologic variability and change, and has been subject to many innovative attempts at policy reform (Pittock and Connell, 2010). The water resources of the basin have been heavily exploited for irrigated agriculture such that average outflows from the River Murray have been reduced by 61 per cent (CSIRO, 2008). Some of the consequences include: desiccation of wetlands, including loss of such biodiversity as floodplain forests and fish; high salinity levels; eutrophication and toxic cyanobacteria blooms; and acid drainage (Pittock et al., 2010). The ecological health of the river system depends on high flow variability, fluctuating between dry periods and floods, which are greatly diminished by agricultural water use (CSIRO, 2008; Overton, 2010).

Already subject to extremes of droughts and floods, the basin's surface waters are especially vulnerable to reductions in precipitation and increases in evapotranspiration (CSIRO, 2008). A number of measures proposed for adaptation of agriculture to drought have been identified as maladaptation to climate change. One example concerns parallel government programmes to subsidize irrigators in the basin, in one case to stay on their farms until conditions return to 'normal' versus another to purchase their water entitlements and cease irrigation to counter over-allocation and climate change impacts (Pittock and Connell, 2010). Another is subsidies for farmers to construct on-farm water storage and reticulation systems that then diminish inflows into rivers (Pittock, 2011; van Dijk, 2006).

The severe impacts of the 2002–2010 drought were catalysts for the Federal Government to legislate to assert greater direct control of water allocations in the Basin through the establishment of a Murray–Darling Basin Authority and its development of a basin plan, due to be adopted in the coming year (Connell and Grafton, 2011). The federal government's constitutional mandate to do this derives substantially from implementation of environmental treaties, especially the Ramsar Convention on Wetlands (Pittock et al., 2010). The authority is required to identify sustainable levels of water diversions that consider emerging risks like climate change impacts in order to sustain key ecological assets (MDBA, 2010). The process is complex, has some benefits for more sustainable management of the basin's environment but there is much to criticize (Connell and Grafton, 2011; Pittock and Finlayson, 2011). Here the key mechanisms that the federal government is proposing or implementing for adjusting governance of water to climate change impacts are briefly reviewed.

Four options for adaptation to climatic variability and change based on manipulating water volumes are being considered by the state and federal governments (MDBA, 2010, 2012; Pittock, in review; Schofield, 2011).

1. *Reallocation of water*. Reallocation of a greater share of the water entitlements from consumptive uses (primarily for agriculture) to sustaining the riverine ecosystems may counter impacts of moderate climate change in the basin (Pittock et al., 2010). The federal government began modest purchases of consumptive water entitlements for the environment in 2008. Scientific assessments conclude that allocation of 4,000–7,600 GL/yr (on average) are needed to sustain freshwater ecosystems in the future (MDBA, 2010; Pittock and Finlayson, 2011). The authority first suggested a reallocation of up to 4,000 GL/yr (37 per cent of diverted waters) but following opposition from some states and irrigators the federal government adopted a plan in 2012

that will reallocate up to 3,200 GL/yr (29 per cent) by 2024 (Minister for Sustainability, 2012). Other provisions mean that the volume of water actually reallocated may be much less (WGCS, 2012). In the original proposal 3 per cent of the reallocation was assigned to managing climate change impacts between 2011 and 2020, a manifestly inadequate amount to manage the risks of severe dry conditions of a decline in water availability by as much as 37 per cent by 2030 – and more in drought – as detailed above (Pittock and Finlayson, 2011). In the adopted plan there is no additional reallocation to manage climate impacts (MDBA, 2012) and available climate impact modelling was not applied to assess the effectiveness of the proposed interventions (Young et al., 2011).

2. *Iterative planning*. Iterative planning could enable periodic reassessment of conditions and new knowledge, including the impacts of climate change and efficacy of adaptation measures, to make appropriate adjustments to the basin plan. An initial five year review is proposed followed by reviews on at least a ten yearly cycle, so this is a positive measure for managing climate change impacts (Pittock, 2013 review). However the stated intention of the next five year review is to consider where to further reduce the proposed reallocation of water to the environment.
3. *Annual water sharing rules*. Ensuring that the environment receives an adequate share of the water available in any one season on a year by year basis may reduce the impacts of dry spells. Although the basin's water entitlement system reduces annual take to all consumptive entitlement holders in equal proportion, the rules discriminate against the environment (CSIRO, 2008; WGCS, 2010). The authority does propose that in future the environment and consumptive water entitlement holders would share losses from low flows equally. Yet it is not clear how the authority will enforce this policy and there are questions as to whether the environment requires more than an equal share (Pittock, 2013; Pittock et al., 2010).
4. *Environmental works and measures*. A series of micro-engineering interventions is being implemented, termed 'environmental works and measures' (also known internationally as 'environmental water demand management' (Pittock et al., 2012; Pittock and Lankford, 2010). These pumps, canals, weirs, and levees are designed to spread smaller volumes of environmental water over larger areas of wetlands to sustain biodiversity (MDBA, 2011). However the risks to flora, fauna and water quality from reduced ecological connectivity of the riverine ecosystem are little appreciated. After spending A$235 million since 2004 at three major wetlands, only a third (36,000 ha) of

> these wetlands could be inundated using the works, representing just 0.6 per cent of the wetlands in the basin (Pittock et al., 2012). While they were originally intended as a last resort measure to maintain refugia in extreme dry spells, politically these works are now being used to justify reduced allocation of water to the environment (Pittock et al., 2012). In this context environmental works and measures are overly narrow and maladaptation.

Alternative measures have been proposed to spread risks and increase the resilience of freshwater ecosystems in the basin to climate change impacts. These involve different but complementary measures apply to rivers on a spectrum of conditions from regulated to free-flowing (Pittock and Finlayson, 2011). Regulated rivers primarily require adequate environmental flows but these need to be enhanced by reoperation of water infrastructure, for instance, to restore fish passage and eliminate thermal pollution from dams. Protection of remaining free-flowing rivers from further development provides different but complementary opportunities for climate change adaptation. All rivers can benefit from other interventions such as re-establishing riparian forests.

The Murray–Darling Basin offers a number of lessons for other countries in federal systems seeking to manage their rivers under conditions of climate change adaptation. Seizing policy reform windows expeditiously and working with policy entrepreneurs is essential to changing management paradigms, but even with a change old ways will persist and need to be managed (as with the Australian drought programmes) (Huitema and Meijerink, 2010). In any of these programmes a reversion to the hydraulic bureaucracy and engineering works is likely to persist even where more effective measures are available (Molle et al., 2010; Pittock et al., 2012). Instituting rigorous, stable and adequately funded subsidiarity measures, involving catchment management agencies, may enhance sustainable and adaptive management. Federal governments can most effectively achieve change with provincial governments by funding transitions but there is a risk of funding being siphoned off into funding political compromises rather than core reforms (WGCS, 2010). All climate change adaptation measures have costs and benefits that need to be rigorously assessed to avoid overly-narrow and mal-adaptation (Pittock et al., 2012). Finally, there may be trade-offs between implementation of desired changes quickly versus establishing institutions for iterative development of sustainable and adaptive measures.

2.6 CONCLUSIONS

Climate change will substantially impact on hydrology and undermine the many river management institutions built on an assumption of stationarity. This risk is acute at the international scale with the 276 rivers shared between two or more countries. As part of a nested governance system global rules for managing such changes are desirable. The UN Watercourses Convention has flaws but may substantially contribute to climate change adaptation by: providing minimum rules for managing shared rivers; creating policy reform windows to revise existing agreements at the bilateral to multi-national scales; and encouraging the creation of more river basin organisations.

River basin management organisations will then face many challenges in managing climate change impacts. All climate change adaptation options have costs, risks and benefits that need to be rigorously considered before implementation. Many popular interventions based on engineering measures or drought subsidies, for example, represent overly-narrow and maladaptation. In federal systems these risks may be minimized by: federal funding for transitions to new management paradigms; subsidiarity; and iterative institutional reviews.

2.7 REFERENCES

Aldous, A., Fitzsimons, J., Richter, B., and Bach, L. (2011). Droughts, floods and freshwater ecosystems: evaluating climate change impacts and developing adaptation strategies. *Marine and Freshwater Research*, *62*(3), 223–231. doi: http://dx.doi.org/10.1071/MF09285.

Arnell, N.W. (2004). Climate change and global water resources: SRES emissions and socio-economic scenarios. *Global Environmental Change*, *14*, 31–52.

Barnett, J. and O'Neill, S. (2010). Maladaptation. *Global Environmental Change*, *20*(2), 211–213.

Barrios, J.E., Rodriguez-Pineda, J.A., and Benignos, M.d.l.M. (2009). Integrated river basin management in the Conchos River basin, Mexico: a case study of freshwater climate change adaptation. *Climate and Development*, *1*(3), 249–260.

Bates, B.C., Kundzewicz, Z.W., Wu, S., and Palutikof, J.P. (eds) (2008). *Climate Change and Water. Technical Paper of the Intergovernmental Panel on Climate Change*. Geneva: IPCC Secretariat.

Carter, N.T. (2011). Energy's water demand: trends, vulnerabilities, and management. Washington DC: Congressional Research Service.

Connell, D. (2007). *Water politics in the Murray–Darling Basin*. Leichardt: The Federation Press.

Connell, D. and Grafton, R.Q. (2011). Water reform in the Murray–Darling Basin. *Water Resour. Res.*, *47*, W00G03. doi: 10.1029/2010wr009820.
Connor, R. and Dovers, S. (2004). *Institutional change for sustainable development*. Cheltenham, UK and Northhampton, MA, USA: Edward Elgar Publishing.
CSIRO (2008). Water availability in the Murray–Darling Basin. A report from CSIRO to the Australian Government. Canberra: CSIRO.
Davies, P.M. (2010). Climate Change Implications for River Restoration in Global Biodiversity Hotspots. *Restoration Ecology*, *18*(3), 261–268. doi: 10.1111/j.1526-100X.2009.00648.x.
Dovers, S.R. and Hezri, A.A. (2010). Institutions and policy processes: the means to the ends of adaptation. *Wiley Interdisciplinary Reviews: Climate Change*, *1*(2), 212–231.
Dudgeon, D., Arthington, A.H., Gessner, M.O., Kawabata, Z.-I., Knowler, D.J., Lévêque, C., and Sullivan, C.A. (2006). Freshwater biodiversity: importance, threats, status and conservation challenges. *Biological Reviews*, *81*(2), 163–182. doi: 10.1017/s1464793105006950.
Ebert, S., Hulea, O., and Strobel, D. (2009). Floodplain restoration along the Lower Danube: a climate change adaptation case study. *Climate and Development*, *1*(3), 212–219.
Falkenmark, M. and Molden, D. (2008). Wake up to the realities of river basin closure. *Water Resources Development*, *24*(2), 201–215.
Garrick, D., Bark, R., Connor, J., and Banerjee, O. (2012). Environmental water governance in federal rivers: opportunities and limits for subsidiarity in Australia's Murray–Darling River. *Water Policy*, *14*, 915–936.
Garrick, D., De Stefano, L., Fung, F., Pittock, J., Schlager, E., Connell, D., New, M., and Connell, D. (2013). Managing hydroclimatic risks in federal rivers: a diagnostic assessment. *Philosophical Transactions of the Royal Society A: Mathematical, Physical and Engineering Sciences*, *371*(2002). doi: 10.1098/rsta.2012.0415
Gleick, P.H. (2002). Water management: soft water paths. *Nature*, *418*(6896), 373.
Goulden, M., Conway, D., and Persechino, A. (2009). Adaptation to climate change in international river basins in Africa: a review [Adaptation au changement climatique dans les bassins fluviaux internationaux en Afrique: une revue] [doi: 10.1623/hysj.54.5.805]. *Hydrological Sciences Journal*, *54*(5), 805–828. doi: 10.1623/hysj.54.5.805.
Grafton, R.Q. (2011). Economic costs and benefits of the proposed basin plan. In D. Connell and R.Q. Grafton (eds), *Basin futures: water reform in the Murray–Darling basin* (pp. 254–262). Canberra: ANU E Press.
Grafton, R.Q., Libecap, G., McGlennon, S., Landry, C., and O'Brien, B. (2011). An integrated assessment of water markets: a cross-country comparison. *Review of Environmental Economics and Policy*. doi: 10.1093/reep/rer002.
Gross, C., Pittock, J., Finlayson, C.M., and Geddes, M.C. (2012). Climate change adaptation in the Coorong, Murray Mouth and Lakes Alexandrina and Albert. *Final report to the National Climate Change Adaptation Research Facility. NCCARF Publication 06/12*. Brisbane: National Climate Change Adaptation Research Facility.
Hallegatte, S. (2009). Strategies to adapt to an uncertain climate change. *Global Environmental Change*, *19*(2), 240–247.
Herron, N., Davis, R., and Jones, R. (2002). The effects of large-scale afforesta-

tion and climate change on water allocation in the Macquarie River catchment, NSW, Australia. *Journal of Environmental Management*, *65*(4), 369–381.
Huitema, D. and Meijerink, S. (2010). Realizing water transitions. the role of policy entrepreneurs in water policy change. *Ecology and Society*, *15*(2), 26 [online].
Hussey, K. and Pittock, J. (2012). The energy-water nexus: Managing the links between energy and water for a sustainable future. *Ecology and Society*, *17*(1), 31 [online]. doi: http://dx.doi.org/10.5751/ES-04641-170131.
Joy, K.J., Gujja, B., Paranjape, S., Goud, V., and Vispute, S. (eds) (2007). *Water conflicts in India: a million revolts in the making*. New Delhi: Routledge.
Kabat, P. and van Schaik, H. (2003). *Climate changes the water rules: how water managers can cope with today's climate variability and tomorrow's climate change*. Delft: Dialogue on Water & Climate.
Loures, F., Rieu-Clarke, A., and Vercambre, M. (2008). Everything you need to know about the UN Watercourses Convention. Gland, Switzerland: WWF International.
MDBA (2010). *Guide to the proposed Basin Plan: overview*. Canberra: Murray–Darling Basin Authority.
MDBA (2011). *The Living Murray Story. One of Australia's largest river restoration projects*. Canberra: Murray–Darling Basin Authority.
MDBA (2012). *Climate change and the Basin Plan*. Canberra: Murray–Darling Basin Authority.
MEA (Millennium Ecosystem Assessment) (2005). *Ecosystems and human well-being: synthesis*. Washington, DC: Island Press.
Milly, P.C.D., Betancourt, J., Falkenmark, M., Hirsch, R.M., Kundzewicz, Z.W., Lettenmaier, D.P., and Stouffer, R.J. (2008). Stationarity is dead: whither water management? *Science*, *319*, 573–574.
Milly, P.C.D., Dunne, K.A., and Vecchia, A.V. (2005). Global pattern of trends in streamflow and water availability in a changing climate [10.1038/nature04312]. *Nature*, *438*(7066), 347–350.
Minister for Sustainability, E., Water, Population and Communities (2012). Water Act 2007. Basin Plan. Canberra: Commonwealth of Australia.
Molle, F., Mollinga, P.P., and Wester, P. (2010). Hydraulic bureaucracies: flows of water, flows of power. *Water Alternatives*, *2*(3), 328–349.
Nelson, D.R. (2010). Adaptation and resilience: responding to a changing climate. *Wiley Interdisciplinary Reviews: Climate Change*, *2*(1), 113–120. doi: 10.1002/wcc.91.
NWC (2011). The National Water Initiative—securing Australia's water future: 2011 assessment. Canberra: National Water Commission.
Opperman, J.J., Galloway, G.E., Fargione, J., Mount, J.F., Richter, B.D., and Secchi, S. (2009). Sustainable floodplains through large-scale reconnection to rivers. *Science*, *326*(5959), 1487–1488. doi: 10.1126/science.1178256.
Overton, I. (2010). Ecological outcomes of flow regimes in the Murray–Darling Basin. *Distilled* (March 2010) [online: http://www.nwc.gov.au/www/html/2750-ecological-outcomes-of-flow-regimes-in-the-murray-darling-basin.asp?intSiteID=2751].
Palmer, M.A., Reidy Liermann, C.A., Nilsson, C., Florke, M., Alcamo, J., Lake, P.S., and Bond, N. (2008). Climate change and the world's river basins: anticipating management options. *Frontiers in Ecology and the Environment*, *6*(2), 81–89. doi: doi:10.1890/060148.

Palmer, M.A., Reidy Liermann, C.A., Nilsson, C., Flörke, M., Alcamo, J., Lake, P.S., and Bond, N. (2008). Climate change and the world's river basins: anticipating management options. *Front. Ecol. Environ.*, *6*, 81–89. doi:10.1890/060148.

Pittock, A.B. (2009). *Climate change: the science, impacts and solutions, 2nd ed.* Collingwood: CSIRO Publishing.

Pittock, J. (2009). Lessons for climate change adaptation from better management of rivers. *Climate and Development*, *1*(3), 194–211.

Pittock, J. (2010). A pale reflection of political reality: integration of global climate, wetland, and biodiversity agreements. *Climate Law*, *1*(3), 343–373.

Pittock, J. (2011). National climate change policies and sustainable water management: conflicts and synergies. *Ecology and Society*, *16*(2), 25 [online].

Pittock, J. (2013). Lessons from adaptation to sustain freshwater environments in the Murray–Darling Basin, Australia. *Wiley Interdisciplinary Reviews: Climate Change, 4*(6), 429-438. doi: 10.1002/wcc.230

Pittock, J. and Connell, D. (2010). Australia demonstrates the planet's future: water and climate in the Murray–Darling Basin. *International Journal of Water Resources Development*, *26*(4), 561–578. doi: 10.1080/07900627.2010.519522.

Pittock, J. and Finlayson, C.M. (2011). Australia's Murray–Darling Basin: freshwater ecosystem conservation options in an era of climate change. *Marine and Freshwater Research*, *62*, 232–243.

Pittock, J. and Finlayson, C.M. (2011). Freshwater ecosystem conservation in the Basin: principles versus policy. In Q. Grafton and D. Connell (eds), *Basin futures: Water reform in the Murray–Darling Basin* (pp. 39–58). Canberra: ANU E-press.

Pittock, J., Finlayson, C.M., Gardner, A., and McKay, C. (2010). Changing character: the Ramsar Convention on Wetlands and climate change in the Murray–Darling Basin, Australia. *Environmental and Planning Law Journal*, *27*(6), 401–425.

Pittock, J., Finlayson, C.M., and Howitt, J.A. (2012). Beguiling and risky: "Environmental works and measures" for wetlands conservation under a changing climate. *Hydrobiologia*. doi: 10.1007/s10750-012-1292-9.

Pittock, J., Hansen, L.J., and Abell, R. (2008). Running dry: freshwater biodiversity, protected areas and climate change. *Biodiversity*, *9*(3 & 4), 30–38.

Pittock, J. and Hartmann, J. (2011). Taking a second look: climate change, periodic re-licensing and better management of old dams. *Marine and Freshwater Research*, *62*, 312–320.

Pittock, J. and Lankford, B.A. (2010). Environmental water requirements: demand management in an era of water scarcity. *Journal of Integrative Environmental Sciences*, *7*(1), 75–93. doi: 10.1080/19438151003603159.

Pittock, J., Loures, F., and Patterson, M. (2013). Governing international watercourses in an era of climate change. In F. Loures & A. Reiu-Clark (Eds.), *The UN Watercourses Convention in force: Strengthening international law for transboundary water management* (pp. 305-320). Oxford: Routledge.

Pittock, J., Marshall, N., Capon, T., Parsons, M., Robertson, A.I., and Casaril, C. (in review). Institutions for effective riparian adaptation.

Poff, N.L., Richter, B.D., Arthington, A., Bunn, S.E., Naiman, R.J., Kendy, E., and Warner, A. (2010). The ecological limits of hydrologic alteration (ELOHA): a new framework for developing regional environmental flow standards. *Freshwater Biology*, *55*(1), 147–170. doi: doi:10.1111/j.1365-2427.2009.02204.x.

Reisner, M. (1986). *Cadillac desert. The American west and its disappearing water (revised and updated)*. New York: Penguin Books.

Richter, B.D., Postel, S., Revenga, C., Scudder, T., Lehner, B., Churchill, A., and Chow, M. (2010). Lost in development's shadow: the downstream human consequences of dams. *Water Alternatives*, *3*(2), 14–42.

Rieu-Clarke, A. and Loures, F.R. (2009). Still not in Force: Should States Support the 1997 UN Watercourses Convention? *Review of European Community & International Environmental Law*, *18*(2), 185–197. doi: 10.1111/j.1467-9388.2009.00640.x.

Robins, L. and Kanowski, P. (2011). 'Crying for our Country': eight ways in which 'Caring for our Country' has undermined Australia's regional model for natural resource management. *Australasian Journal of Environmental Management*, *18*(2), 88–108. doi: 10.1080/14486563.2011.566158.

Schofield, N. (2011). Climate change and its impacts – current understanding, future directions. In Q. Grafton and D. Connell (eds), *Basin futures: Water reform in the Murray–Darling basin* (pp. 81–100). Canberra: ANU E Press.

Tarlock, A.D. (1999–2000). How well can international water allocation regimes adapt to global climate change. *J. Land Use & Envtl. L.*, *15*(423), 423–450.

TNC (2007, 23 October 2008). World Leaders Sign Brisbane Declaration to Protect the Planet's Freshwater Systems, from http://www.nature.org/initiatives/freshwater/press/press3195.html.

Turnpenny, J., Lorenzoni, I., and Jones, M. (2009). Noisy and definitely not normal: responding to wicked issues in the environment, energy and health. doi: 10.1016/j.envsci.2009.01.004. *Environmental Science & Policy*, *12*(3), 347–358.

UNEP (2007). *Global environmental outlook GEO4 environment for development*. Nairobi: United Nations Environment Program.

van Dijk, A., Evans, R., Hairsine, P., Khan, S., Nathan, R., Paydar, Z., Viney, N., and Zhang, L. (2006). *Risks to the shared water resources of the Murray–Darling Basin*. Canberra: Murray–Darling Basin Commission.

Vorosmarty, C.J., McIntyre, P.B., Gessner, M.O., Dudgeon, D., Prusevich, A., Green, P., and Davies, P.M. (2010). Global threats to human water security and river biodiversity. *Nature*, *467*(7315), 555–561.

Watts, R.J., Richter, B.D., Opperman, J.J., and Bowmer, K.H. (2011). Dam reoperation in an era of climate change. *Marine and Freshwater Research*, *62*(3), 321–327. doi: doi:10.1071/MF10047.

WGCS (2010). Sustainable Diversions in the Murray–Darling Basin. An analysis of the options for achieving a sustainable diversion limit in the Murray–Darling Basin. Sydney: Wentworth Group of Concerned Scientists.

WGCS (2012). Does a 3,200Gl reduction in extractions comnined with the relaxation of eight constraints give a healthy working Murray–Darling Basin River system? Sydney: Wentworth Group of Concerned Scientists.

Wolf, A.T., Kramer, A., Carius, A., and Dabelko, G.D. (2005). Managing water conflict and cooperation. In L. Starke (ed.), *State of the world 2005: A Worldwatch Institute Report on progress toward a sustainable society* (pp. 80–95). New York: W.W. Norton.

WWDR (2012). Managing Water under Uncertainty and Risk, *4th World Water Development Report*. Paris: UNESCO.

WWF (2012). WWF welcomes French support of UN Watercourses Convention. *Media release 12 March 2012*. Retrieved 14 April 2012, from http://m.

mediterranean.panda.org/news/?203850/WWF-welcomes-French-support-of-UN-Watercourses-Convention.

Young, M.D. (2010). Environmental effectiveness and economic efficiency of water use in agriculture: the experience of and lessons from the Australian water reform programme *Sustainable management of water resources in agriculture*. Paris: Organisation for Economic Co-operation and Development.

Young, W.J., Bond, N., Brookes, J., Gawne, B., and Jones, G.J. (2011). Science review of the estimation of an environmentally sustainable level of take for the Murray–Darling Basin. Final report to the Murray–Darling Basin Authority. Canberra: CSIRO.

PART II

AMERICAS

3. Federalism and US water policy

Andrea K. Gerlak

3.1 INTRODUCTION

Water resource management in the United States today is increasingly complex and fragmented as strategies are pursued across multiple decision forums and institutional arrangements (Adler 2000; Stakhiv 2003; Gerlak 2008). Accumulated mandates, authorities, practices, and habits of generations of governmental participation in water resources characterize US water policy. At the federal level alone there are 20 or more federal agencies with some responsibility in water resources (AWRA 2007), working across more than 100 federal water programs for water quality and quantity, water supply, navigation, hydropower, recreation, climate change, natural hazards management, and integrated water resources management (US ACE 2011, p. 20). As part of this complex web of federal involvement in water resource management and use, many congressional committees are involved in legislating, funding, and overseeing the water-related activities of the numerous federal agencies (Cody et al. 2012, p. 1). At last count, more than 100,000 local water-related entities and more than 300 departments in 50 states have water-related functions (Dworsky et al. 1991).

The roles of all levels of government in water resources within the US have evolved over time to account for new challenges and incorporate a broader range of stakeholders (White 2000; Deason et al. 2001; ICWP 2006). The relative capabilities of the actors also have been changing as state, local governments, and non-governmental organizations have improved their resource base and increased their abilities in multiple areas of water management (Bell and Johnson 1991; Rogers 1993; Feldman 2007; Mayor's Water Council 2011). Increasingly, water management can be characterized as highly pragmatic, emphasizing collaborative partnerships, adaptable management strategies, and place-based processes (Gerlak 2006).

We can better understand the historical role that the federal government has played with regard to water policy by observing the eras and patterns of federalism in the US with regard to water resource management

(Lepawsky 1950; Engelbert 1957; King 1959; Hart 1966; Sabatier et al. 2005a). Analogous to a river or stream, the relationship of the federal government to other interested parties has alternatively ebbed, flowed, dried up, and overflowed (Gerlak 2006, pp. 1–2). Navigation of the historical streams of US water policy is key to understanding the allocation of powers and functions that bear on river basin management. A federalism lens shines a spotlight on the flow of money from the federal government to states and local communities. It also reveals how past decisions and behaviors in water policy impact current as well as potential future arrangements in US water policy.

3.2 STREAMS OF US WATER POLICY

The history of the federal government's relationship to the states and local communities and sharing of power in water policy is not static but forever evolving and changing. The eras of federalism can be divided into five streams: from state-based federalism to centralized federalism to cooperative federalism to devolution to pragmatic federalism. This chapter traces federal–state relations and finds intergovernmental relations related to water policy have evolved into a pragmatic federalism that is place-based, collaborative, and experimental.

First Stream: State Ingenuity and Independence (Late 1700s to Early 1900s)

The first stream, which encompasses the birth of the US in 1776 to Theodore Roosevelt's presidency in the early twentieth century, is characterized by state-based federalism. States led the canal-building era of the nineteenth century and flood control on the Ohio and Mississippi rivers (Lepawsky 1950). In 1784 the first interstate commission – the Bi-State Commission – was established when Maryland and Virginia joined together to develop the Potomac River for navigation purposes (ICWP 2006). As water resource issues developed, they were resolved primarily at the state or local level (Holmes 1972).

At the request of the federal government, the 1808 Gallatin Report represented the first national effort at regional strategic planning in water resources management. By the mid-nineteenth century the US Army Corps of Engineers (Corps) expanded their interests in river and canal development, navigation, and flood control. The Corps became involved in large-scale river basin planning in the Mississippi River Basin in 1850 for flood prevention (Clarke and McCool 1996, pp. 20–21). The Mississippi River

Commission, established in 1879, is thought to be the first federal commission with multiple objectives in navigation and flood control. Yet, the federal government continued to defer to states in areas of water allocation. The practice of prior appropriation, or 'first in time, first in right', adopted in Western mining camps became custom and would eventually find its way into the territorial and state codes in the West.

Second Stream: Federal Development and Dominance (1900–1960)

During the second stream of federalism the federal government increased its role in water management in an era characterized by centralized federalism. With President Theodore Roosevelt's support, Congress passed the Reclamation Act of 1902 authorizing federal funds for constructed reservoirs and water distribution facilitation in 16 western states. By the 1920s the federal government also emerged with active roles in hydropower development, navigation, and flood control. Under the Federal Power Act of 1920 the Federal Power Commission was charged with responsibility for licensing non-federal power developments on navigable waters in the public domain and managing the sale of surplus power generated from federal dams. For example, with the Rivers and Harbors Act of 1925, the Corps was authorized to survey all navigable waters and formulate general plans for irrigation, navigation, power production, and flood control.

But federal water resource development reached its zenith in the New Deal era (Rogers 1993, p. 50). Comprehensive river basin development gave birth to multi-purpose planning and development with the Tennessee Valley Authority in 1933 (Lilienthal 1944). Federal agencies like the Bureau of Reclamation and the Corps received commensurate powers to implement national water management and development goals. Political scientist and federalism scholar Daniel Elazar (1965, pp. 18–19) refers to this era as 'crisis-oriented centralism'; an era where the federal government became a dominant party in the American federal system. The shift toward multiple-purpose endeavors intensified jurisdictional problems between federal agencies and federal and state actors.

In the 1930s and 1940s Congress ratified several interstate compacts. For example, Congress ratified a compact between New York and New Jersey in 1936 to establish the Interstate Sanitation Commission. In 1937, Minnesota, North Dakota, and South Dakota entered into a compact for management of the Red River. In 1947, the New England Water Pollution Control Commission was established by Congress with jurisdiction over the interstate waters of New England and New York. Federal responsibility for flood control was centralized with the Corps with cost share requirements established with affected states. Although these regional committees were designed to be forums for improving interagency coordination and planning, their limited authority along with interagency differences made it difficult for the committees to resolve water issues (Featherstone 2001, p. 276).

Concern with deteriorating water quality and drought conditions in the northeast US in the late-1950s prompted the federal government to emphasize river basin commissions. The Delaware River Basin Commission, designed to better coordinate and plan within the basin, marked the first time the federal government partnered with states to form such a commission. Many commissions struggled to coordinate and overcome political disagreements between agencies and levels of government (Feldman 1992, pp. 11–12). Responding to such deficiencies and lack of overall success, in 1961 Congress created the Water Resources Council – an interagency data-gathering and policy body designed to provide presidents with water policy advice.

In the post-war era, the US Department of Justice embraced a staunch federal supremacy position regarding western water rights. Fearing federal encroachment and the 'nationalization' of western waters, western states united together to ensure that state courts – as opposed to federal ones – would be the forum for adjudicating federal water claims. Fearing potential loss of water allocation, or even worse a federal regulatory approach to determine and administrate water rights, states representatives determined to adjudicate water rights on their own (Thorson et al. 2005). As a result, many western states embarked on data gathering campaigns designed to shore up state water rights against outside threats.

Third Stream: Development Doubts and Environmental Concerns (1960s to 1980s)

In the third stream, cooperative federalism, or shared federal–state authority, characterized water policy in the US. Numerous pieces of legislation, like the Wild and Scenic Rivers Act and the Endangered Species Act, attention drew to fish and wildlife concerns, recreation, wetland protection, and quality control. The Clean Water Act of 1972 established a permit system for the discharge of pollutants to be administered by US Environmental Protection Agency (EPA). One observer commented that this act represented the first time that state actions had ever been subject to such complete federal control (Lieber 1975, p. 9). While the federal government demanded massive improvements in municipal sewage treatment from the states, it also allocated vast sums of money to pay for these improvements (Landy 1999, p. 248). Further, the National Environmental

Policy Act created opportunities for greater public involvement in water and natural resource-related decision-making. Increasingly, federal interests in water management began to revolve around environmental protection and recreational uses (McGinley 1983).

Several federally-created commissions in the 1950s recommended strengthening the role of states in water resource development. These include the President's Water Resources Policy Commission (1950), Missouri Basin Survey Commission (1953), Commission on the Executive Branch of Government (1955), the Commission on Intergovernmental Relations (1955), and the Presidential Advisory Committee on Water Resources Policy (1955). Congress also encouraged states to more actively participate in water resource decision-making. The Water Resources Research Act of 1964 charged a college or university in each state, appointed by the state's governor, to conduct research on water resources, initiating the creation of water research centers at universities across the United States. The Water Resources Planning Act of 1965 put the federal government and states on equal standing in the newly established river basin commissions designed to conduct comprehensive, coordinated joint plans and provided financial assistance to states to enhance planning. The Water Resources Council, a cabinet-level interagency planning and coordinating body worked to create a coordinated framework for water management. The National Water Commission, created by Congress in 1968, foresaw a shift of focus to conservation and water quality in their 1973 report and recommended greater local control of water management and protection (NWC 1973).

Despite calls for greater state management of water resources, in 1963 with *Arizona* v. *California* (373 US 546) the US Supreme Court for the first time recognized federal proprietary water rights. This ruling extended the reserved rights doctrine to include other federal reservations, such as wildlife refuges, and maintained that the intention of the US was to reserve sufficient water to meet the needs of all reservations in the future as well as at the time they were created. Most disturbing to western states was the ruling that Congress has the power to dam and store a navigable river, and distribute water to the states and users within each state. With such broad navigational powers, Congress could allocate to intrastate users in disregard of state law.

Fourth Stream: Devolution and Penny-pinching (1980s)

The new federalism of the Reagan era ushered in the fourth stream which is characterized by an emphasis on increasing state responsibilities, such as cost-sharing in water projects, but also a marked a reduction in federal funding. Sympathetic to calls from industry to roll back environmental regulations and return power to states and local communities, Reagan's brand of 'new federalism' in the 1980s has been critically dubbed 'cooptive' federalism, or 'fend-for-yourself' federalism (Shannon 1989, pp. 119–122). Reagan sought to 'de-institutionalize most federally encouraged regional bodies' and weaken regional river basin planning by withdrawing federal funds and participation (Walker 1983, p. 13). He quickly defunded the Water Resources Council and River Basin Commissions in 1982, signalling that states would have to go it on their own.

Demonstrating this shift, amendments to the Clean Water Act in 1987 phased out the construction grants program and replaced it with state water pollution control revolving fund, commonly known as the Clean Water State Revolving Fund, a new funding strategy addressing water quality needs by building on EPA-state partnerships. Congress also established the National Estuary Program in US EPA, a non-regulatory, collaborative, watershed approach that brings together private and public stakeholders from all levels of the federal system to outline action items for protecting coastal water quality (Bearden 2001).

State and federal officials 1983 signed the Chesapeake Bay Agreement in 1983, pledging cooperation to restore and protect the bay. They created a regional institution, the Chesapeake Bay Commission, an advisory body of state legislators, federal agency officials, and citizen representatives. Cooperative agreements between states and the Department of Interior for habitat restoration were reached in the Upper Colorado Basin in the late-1980s to bring together federal and state officials with environmental and resources users to coordinate management along this segment of the river.

State institutional and administrative capacity continued to grow and beginning in the 1980s some 30 states responded to growing concerns about groundwater depletion and quality, creating new programs and adopting new legislation (Blomquist 1991). Power issues continued to plague federal–state relations in the late-1980s and early-1990s. With *California* v. *Federal Energy Regulatory Commission* (877 F. 2d 743 (9th Cir. 1989), aff'd 110 S. Ct. 2024) the US Supreme Court ruled that state law cannot supplement federal flow requirements, striking a blow to many western state water officials. By the late-1980s the federal government continued to withdrew funds sewage treatment plants, water development projects, dam safety programs, and water data collection (Bell and Johnson 1991).

Fifth Stream: Restoration and Collaboration (1990s–Present)

The devolution that began under President Reagan continued through the Clinton White House but with a marked emphasis on restoration and collaboration. The watershed approach, advanced as an integrated approach to controlling pollution sources, was designed to improve coordination within and across levels of government (NRC 1999). US EPA worked to improve federal–state relations with a focus on sustainability at the community level (Kraft and Scheberle 1998). The Clinton Administration's 1998 Clean Water Action Plan sought to better coordinate federal water quality efforts aimed at non-point sources of pollution with an emphasis on 'partnering' with state, tribal and local governments and offering federal grants and technical assistance. States would no longer have to go it on their own. Neither would tribes. Embracing a settlement approach initiated in the late-1970s, the Clinton Administration advanced settlement negotiations with tribes as an alternative to litigation.

Some notable congressional legislation in the early 1990s marked a significant shift in water policy. The Reclamation Projects Authorization and Adjustment Act of 1992 decreased water flow to irrigation in California's Central Valley Project (CVP), resulting in a 20 percent reduction in supply to agriculture and dedicating 800,000 acre-feet of CVP yield for environmental protection. In addition, this legislation changed Glen Canyon Dam's operation from maximizing power generation during peak flows to protecting and improving the natural and cultural resources of Grand Canyon National Park and the Glen Canyon National Recreation Area. Under the Adaptive Management Plan for the dam, simulated floods mimic historic flows by releasing waters downstream through the canyon to improve habitat, protect archaeological sites, and enhance recreation. With amendments to the Safe Drinking Water Act in 1996, Congress gave states more funding to comply with environmental standards and flexibility to exercise authority over drinking water standards and their enforcement (Kincaid 1998, p. 22). It provided much needed relief for small local governments that faced the choice of bankruptcy or the abandonment of their drinking water supply systems (Zimmerman 2005, p. 361). Most states run the National Pollutant Discharge Elimination System of the CWA's permitting program for point source discharges and many states continued to demonstrate significant advances around improved groundwater planning and management (Ashley and Smith 1999).

The Department of the Interior under the Bush Administration continued to advance the adaptive management approach and proceed with tribal negotiations. Bush's 2004 executive order creating a cabinet-level interagency task force composed of federal agencies, states, local government, tribes, and other interests to move toward more formal integration and heightened federal–state collaboration in the Great Lakes. President Obama also furthered cooperative efforts in the Great Lakes (Bentley 2012) as well as the Florida Everglades (Liston 2012). Increasingly, federal agencies are coordinating on climate change policies and strategies through the Interagency Climate Change Adaptation Task Force (ICCATF 2011). A few partnership between the National Oceanic and Atmospheric Administration, the Army Corps and the US Geological Survey aims to better leverage science around water resources management to 'improve water resource forecasts, streamline access to water information, and enable more informed decision making' (USGS 2011). In recent years, both the American Water Resources Association and the Corps initiated policy dialogues (e.g., AWRA 2007, 2008) and listening sessions (2011), respectively, to further a national dialogue on water management in the US

3.3 TODAY'S PRAGMATIC FEDERALISM

Federal–state relations around water policy have evolved from more state-based federalism that characterized the late-eighteenth century and nineteenth century to a federally-dominant centralized federalism and more cooperative federalism in the early- and late-twentieth century, respectively. Today's water policy can be seen as pragmatic federalism characterized by: (1) emphasizes collaborative partnerships, (2) relies on adaptable management strategies, and (3) is problem- and process-oriented.

Collaborative Partnerships

One important feature of pragmatic federalism is the development of collaborative partnerships between federal, state, and local agencies that transcend geo-political boundaries and focus programs and projects on a hydrologic basis. In part, collaborative arrangements can be seen as an alternative to earlier more centralized approaches to manage water resources (Karkkainen 2002). There are countless examples of collaborative efforts occurring on a more local small-scale watershed level, where citizens initiate collaboration (Kenney and Rieke 1997; Sabatier et al. 2005b). Large-scale collaborative efforts have also been on the rise in a variety of river basins across the country, illustrated by the Everglades Restoration Plan, the Chesapeake Bay Program, the Coastal Louisiana initiative, the Gulf of Maine initiative and the Great Lakes Regional Collaboration (Gerlak and Heikkila 2006; Doyle and Drew 2009).

In some cases, collaborative efforts have entailed new institutions or revisions to recognize and reorganize the restoration effort's complex and multi-jurisdictional management requirements. Some research suggests that widespread public awareness about the nature and extent of the problems are closely linked to institutional formation and evolution (Heikkila and Gerlak 2005). So too are federal regulatory laws like the Endangered Species Act and the Clean Water Act important triggers in collaborative efforts.

In absence of these larger collaborative institutions we see the development of cooperative agreements and programs. Cooperative agreements between states and the Department of Interior for habitat restoration were reached in the Big Bend reach of the Platte River in Nebraska in 1997 (Freeman 2007). States have also initiated collaborative efforts for planning and management activities within a watershed, illustrated by Colorado's efforts in the Animas River (Koontz et al. 2004). Given the diversity in institutional design, the role of the federal government varies considerable in collaborative efforts – from enforcer and regulator to project planner, partner or funding source (Doyle and Drew 2008, p. xi).

Adaptable Management Strategies

Adaptive management strategies are another key feature of pragmatic federalism. In 1992 the National Research Council endorsed adaptive management as a valuable tool for ecosystem restoration (NRC 1992, p. 41). With a focus on experimentation and 'learning by doing', this management strategy recognizes uncertainty and promotes flexible decision-making that can be adjusted in the face of uncertainties. Central to the adaptive management model is careful monitoring and an iterative learning process. An interdisciplinary approach, adaptive management seeks action in the face of limitations on scientific knowledge and the complexities of large ecosystems (Holling 1978).

Embraced on a large scale in the management of the Columbia River in the Pacific Northwest (Lee 1993), an adaptive management approach can be found in several ecosystems across the US, including the Florida Everglades, Missouri Dam and reservoir system, Upper Mississippi, Coastal Louisiana, and Glen Canyon Dam on the Colorado River ecosystem (NRC 2004). Water scholar David L. Feldman (2007, p. 53) suggests that adaptive management has been somewhat ironically legitimized through declining federal interest in water supply issues, as demonstrated by a withdrawal of federal funding for river basin commissions, and a decline in support for dam construction.

More recently, adaptive management has evolved from a more practical management approach for ecological systems into an interdisciplinary field of research and action, often known under the broader term of adaptive governance aimed to better develop governance concepts that can handle the complexity and unpredictability of social-ecological systems

(Termeer et al. 2010). Adaptive governance calls for greater attention to the institutional dimensions of watershed governance (Scholz and Stifel 2005; Folke et al. 2005). In the case of Glen Canyon Dam, Congress mandated a process of adaptive management for the Colorado River with the Grand Canyon Protection Act of 1992. Under this legislation, the effects of dam operations on downstream resources would be monitored and assessed. Similarly, in Florida's Everglades the Comprehensive Everglades Restoration Plan embraces an adaptive management strategy. But adaptive management is thought to still be largely untested in practice in large-scale collaborative restoration efforts across the country (Doyle and Drew 2008, p. xiii) with considerable debate over what it means in practice for collaborative restoration efforts (Susskind et al. 2010; US GAO 2003, p. 4).

Problem- and Process-oriented

Finally, pragmatic federalism is characterized by a more problem- and process-oriented focus. Increasingly, water policy is focused on specific problems with new integrated and coordinated processes designed to tackle a targeted problem. Under a watershed context, the underlying planning is conducted in partnership between federal, state, and local entities where cross-sector needs can be evaluated in a more comprehensive and place-based manner rather than on a stove pipe basis (AWRA et al. 2008, p. 3). Processes are designed to be more open and transparent.

Efforts underway in the Gulf of Mexico region demonstrate the changing nature of governmental roles and capabilities and the evolution to more place-based comprehensive planning and restoration efforts that cut across a variety of agency programs, policies, and missions (Day et al. 2007). For example, the Mississippi River/Gulf of Mexico Watershed Nutrient Task Force brings together five federal agencies, 12 states, and the tribes within the Mississippi/Atchafalya River Basin to collaborate on the design of a scientific assessment of hypoxia and a plan for reducing and mitigating it in the Gulf of Mexico.

Notably, tribes are active participants in these place-based and problem-focused collaborative efforts in basins across the country including the Columbia, Rio Grande, Missouri and Everglades systems (McCool 2006–2007). This is not to say that a more problem- and process-oriented approach has eliminated conflicts in water management. In the upper Mississippi River, for example, basin stakeholders continue to disagree over both the nature of the problem and the appropriate set of solutions for water management in the region (Drew 2008). Dozens of lawsuits have been filed over the past decade in the Everglades despite the large-scale collaborative processes in place there. Most recently, lawsuits have been filed in the Chesapeake Bay concerning new proposed EPA water quality rules. The tenor of the dispute is in part a reflection of conflict over the role of the federal government in state and local water management (Wheeler 2012). Further, states continue to disagree with the federal government over the federal–state cost sharing arrangement illustrated by the recent dispute involving the restoration of wetlands destroyed by construction of the Mississippi River-Gulf Outlet (Schleifstein 2012).

3.4 CONCLUSION

A review of the streams of water policy illustrates how federal authority over water expanded greatly in the twentieth century, encompassing navigation, flood control, hydropower, water quality, irrigated agriculture, and public health concerns. Governmental response is often reactionary: devastating floods give way to flood control policies, dramatic species declines result in restoration efforts, and severe degradation in waterways trigger pollution controls. The result has been a patchwork of laws, regulations, and programs that are at times contradictory – but always incomplete and conflicting. Although we witness a relative shift to address emerging water issues at the state and regional level today, federally derived standards, scientific and technical assistance, and funding remain constants in the federal–state relationship (ICWP 2006, p. 25; AWRA et al. 2008, p. 3).

Water policy today can be seen as pragmatic federalism, emphasizing collaborative partnerships, and relying on adaptable management strategies with a focus that is problem- and process-oriented. In part, today's federalism is a response to past policies – and signals an effort to address the fragmentation of the American watershed. There are potential great benefits to pragmatic federalism. It involves a level of planning that has not existed since efforts in the 1960s. It promises greater accessibility to environmental and more local interests, and attempts better integration of water quality and quantity concerns. Of course, it is not without its critics. Some are calling for a new national water policy in the US that is more coherent and integrated, and can help to restructure and streamline federal water policy (Christian-Smith and Gleick 2012). Others point to the need for federal leadership in facilitating and funding initiatives a multi-agency, multi-stakeholder bottom-up process in water management (Loucks 2004). Still others argue for a new form of interagency approach involving the states (Whipple et al. 1999), or a new federal coordinating water policy body to better tackle emerging resources management issues (Schad 1998). The real test of pragmatic federalism may likely be its ability to solve a particular watershed's ecological problems and better coordinate stakeholders and program activities, thereby overcoming the policy fragmentation that has become all too common in our nation's water policy.

3.5 REFERENCES

Adler, Robert W. (2000), 'Watershed and the Integration of US Water Law and Policy: Bridging the Great Divides', *William and Mary Environmental Law and Policy Review* 25, 1–67.

American Water Resources Association (AWRA) (2007), *Setting a Direction for Water Policy: Suggestions and Guidance for Improving the Nation's Water Policies*, Washington, DC: Third National Water Resources Policy Dialogue.

American Water Resources Association (AWRA) (2008), *Fourth National Water Resources Policy Dialogue*, Washington, DC: Fourth National Water Resources Policy Dialogue.

American Water Resources Association (AWRA), Environment and Water Resources Institute of the American Society of Civil Engineers (EWRI/ASCE), and the National Wildlife Federation (NWF) (2008), *Summary Fourth National Water Resources Policy Dialogue*, Washington, DC: AWRA.

Ashley, Jeffrey S. and Zachary A. Smith (1999), *Groundwater Management in the West*, Lincoln, NE: University of Nebraska Press.

Bearden, David M. (2001), *National Estuary Program: A Collaborative Approach to Protecting Coastal Water Quality*, Washington, DC: Congressional Research Service.

Bell, Craig D. and Norman K. Johnson (1991), 'State Water Laws and Federal Water Uses: The History of Conflict, the Prospects for Accommodation', *Environmental Law* 21, 63–64.

Bentley, Chris (2012), 'An Updated Great Lakes Accord', *New York Times*, September 10.

Blomquist, William (1991), 'Exploring State Differences in Groundwater Policy Adoption, 1980–1989', *Publius: The Journal of Federalism* 21 (Spring 1991), 101–115.

Christian-Smith, Juliet and Peter H. Gleick (2012), *A Twenty-first Century US Water Policy*, New York: Oxford University Press.

Clarke, Jeanne Nienaber and Daniel McCool (1996), *Staking out the Terrain: Power Differentials among Natural Resources Management Agencies*, Albany, NY: State University of New York Press.

Cody, Betsey A., Judy Schneider, Mary Tiemann, and Grace Relf (2012), *Selected Federal Water Activities: Agencies, Authorities, and Congressional Committees*, Washington, DC: Congressional Research Service.

Day, John W. Jr., D.F. Boesch, E.J. Clairain, G.P. Kemp, S.B. Laska, W.J. Mitsch, K. Orth, H. Mashriqui, D.J. Reed, L. Shabman, C.A. Simenstad, B.J. Streever, R.R. Twilley, C.C. Watson, J.T. Wells, and D.F. Whigham (2007), 'Restoration of the Mississippi Delta: Lessons from Hurricanes Katrina and Rita', *Science* 315, 1679–1684.

Deason, Jonathan P., Theodore M. Schad, and William Sherk (2001), 'Water Policy in the United States: A Perspective', *Water Policy* 3, 175–192.
Doyle, Mary and Cynthia A. Drew (eds) (2008), *Large-scale Ecosystem Restoration: Five Cases from the United States*, Washington, DC: Island Press.
Drew, Cynthia (2008), 'The Upper Mississippi River and the Army Corps of Engineers' New Role' in Mary Doyle and Cynthia A. Drew (eds), *Large-scale Ecosystem Restoration: Five Cases from the United States*, Washington, DC: Island Press, pp. 229–268.
Dworsky, Leonard, David Allee, and Ronald North (1991), 'Water Resources Planning and Management in the US Federal System: Long Term Assessment and Intergovernmental Issues', *Natural Resources Journal* 31(3), 475–547.
Elazar, Daniel J. (1965), 'The Shaping of Intergovernmental Relations in the Twentieth Century', *The Annals of the American Academy of Political and Social Science* 359 (May).
Engelbert, Ernest A. (1957), 'Federalism and Water Resources Development', *Law and Contemporary Problems* 22, 325–350.
Featherstone, J.P. (2001), 'Existing Interstate Compacts: The Law and the Lessons', *Toledo Journal of Great Lakes' Law, Science and Policy* 4, 271–282.
Feldman, David L. (1992), *Water Resources Management: In Search of an Environmental Ethic*, Baltimore, MD: Johns Hopkins Press.
Feldman, David L. (2007), *Water Policy for Sustainable Development*, Baltimore, MD: Johns Hopkins University Press.
Folke, C., T. Hahn, P. Olsson, and J. Norberg (2005), 'Adaptive Governance of Social-ecological Knowledge', *Annual Review of Environment and Resources* 30, 441–473.
Freeman, David M. (2007), 'Negotiating for Endangered and Threatened Species Habitat in the Platte River Basin' in Mary Doyle and Cynthia A. Drew (eds), *Large-Scale Ecosystem Restoration: Five Cases from the United States*, Washington, DC: Island Press, pp. 59–88.
Gerlak, Andrea K. (2006), 'Federalism and US Water Policy: Lessons for the Twenty-first Century', *Publius: The Journal of Federalism* 36(2), 231–257.
Gerlak, Andrea K. (2008), 'Today's Pragmatic Water Policy: Restoration, Collaboration, and Adaptive Management along US Rivers', *Society & Natural Resources* 21(6), 538–545.
Gerlak, Andrea K. and Tanya Heikkila (2006), 'Comparing Collaborative Mechanisms in Large-scale Ecosystem Governance', *Natural Resources Journal* 46(3), 657–707.
Hart, Gary Warren (1966), 'Creative Federalism: Recent Trends in Regional Water Resources Planning and Development', *U. Colo. L. Rev.* 39.
Holling, C.S. (ed.) (1978), *Adaptive Environmental Assessment and Management*, New York: John Wiley and Sons.
Holmes, B.H. (1972), *A History of Federal Water Resources Programs, 1800–1960*, Washington, DC: US Department of Agriculture, Economic Research Service.
Interagency Climate Change Adaptation Task Force (ICCATF) (2011), *National Action Plan: Priorities for Managing Freshwater Resources in a Changing Climate*, Draft. June 2.
Interstate Council on Water Policy (ICWP) (2006), *Interstate Water Policy Solutions for the New Millennium*, Rockville, MD: Interstate Council on Water Policy.
Karkkainen, Bradley C. (2002), 'Collaborative Ecosystem Governance: Scale, Complexity and Dynamism', *Virginia Environmental Law Journal* 21.

Kenney Douglas and Betsey Rieke (1997), *Resource Management at the Watershed Level: An Assessment of the Changing Federal Role in the Emerging Era of Community-based Watershed Management*, Springfield, VA: National Technical Information Service, Western Water Policy Review Advisory Commission.

Kincaid, John (1998), 'The Devolution Tortoise and The Centralization Hare', *New England Economic Review* (May/June), 13–40.

King, Dominic B. (1959), 'Federal–State Relations in the Control of Water Resources', *University of Detroit Law Journal* 37.

Koontz, Tomas, Toddi A. Steelman, JoAnn Carmin, Katrina Smith Korfmacher, Cassandra Moseley, and Craig W. Thomas (2004), *Collaborative Environmental Management: What Roles for Government?*, Washington, DC: Resources for the Future Press.

Kraft, Michael E. and Denise Scheberle (1998), 'Environmental Federalism at Decade's End: New Approaches and Strategies', *Publius: The Journal of Federalism* 28(1), 131–146.

Landy, Marc K. (1999), 'Local Government and Environmental Policy' in Martha Derthick (ed.), *Dilemma of Scale in America's Federal Democracy*, Cambridge: Cambridge University Press, 1999, pp. 227–260.

Lee, Kai N. (1993), *Compass and Gyroscope: Integrating Science and Politics for the Environment*, Washington, DC: Island Press.

Lepawsky, Albert (1950), 'Water Resources and American Federalism', *American Political Science Review* 50, 631–649.

Lieber, Harvey (1975), *Federalism and Clean Waters: The 1972 Pollution Control Act*, Lexington, MA: DC Heath and Company.

Lilienthal, David E. (1944), *TVA: Democracy on the March*, New York: Harper and Row.

Liston, Barbara (2012), 'Obama administration pledges $80 million more to Everglades restoration', *Reuters*, July 13. At http://www.reuters.com/article/2012/07/13/us-usa-everglades-restoration-idUSBRE86C1AG20120713.

Loucks, Daniel P. (2004), 'Federal Leadership in Managing America's Rivers', *Water Resources Update* 127, 108–111.

Mayor's Water Council (2011), *Newsletter of the Mayors Water Council of The United States Conference of Mayors, Spring 2011*, Washington, DC: The United States Conference of Mayors.

McCool, Daniel (2005), 'The River Commons: A New Era in US Water Policy', *83 Texas Law Review: 1903*.

McCool, Daniel (2006–2007), 'Rivers of the Homeland: River Restoration of Indian Reservations', *16 Cornell Journal Law and Public Policy* 539.

McGinley, Patrick (1983), 'Federalism Lives! Reflections on the Vitality of the Federal System in the Context of Natural Resource Regulation', *Kansas Law Review* 32.

National Research Council (NRC) (1992), *Restoration of Aquatic Ecosystem – Science, Technology and Public Policy*, Washington, DC: National Academy Press.

National Research Council (NRC) (1999), *New Strategies for America's Watersheds*, Washington, DC: National Academy Press.

National Research Council (NRC) (2004), *Adaptive Management for Water Resources Project Planning*, Washington, DC: National Academy Press.

National Water Commission (NWC) (1973), *New Directions in US Water Policy: Summary, Conclusions and Recommendations from the Final Report of Water*

Policies for the Future, Washington, DC: Final Report to the President and to the Congress of the United States by the National Water Commission.

Rogers, Peter (1993), *America's Water: Federal Roles and Responsibilities*, Cambridge, MA: MIT Press.

Sabatier, Paul A., Chris Weible, and Jared Ficker (2005a), 'Eras of Water Management in the United States: Implications for Collaborative Watershed Approaches', in Paul A. Sabatier, Will Focht, Mark Lubell, Zev Trachtenberg, Arnold Vedlitz and Marty Matlock (eds), *Swimming Upstream Collaborative Approaches to Watershed Management*, Cambridge, MA: MIT Press, pp. 23–52.

Sabatier, Paul A., Will Focht, Mark Lubell, Zev Trachtenberg, Arnold Vedlitz, and Marty Matlock (2005b), *Swimming Upstream: Collaborative Approaches to Watershed Management*, Cambridge, MA: MIT Press.

Schad, T. (1998), 'Water Policy: Who Should Do What?', *Water Resources Update* 111, 51–61.

Schleifstein, Mark (2012), 'Corps approves $2.9 billion MR-GO restoration plan, but demands state pay 35 percent of cost', *The Times-Picayune*, October 3.

Scholz, John T. and Bruce Stiftel (eds) (2005), *Adaptive Governance and Water Conflict: New Institutions for Collaborative Planning*, Washington, DC: Resources for the Future.

Shannon, John (1989), 'The Return to Fend-for-Yourself Federalism: The Reagan Mark', in ACIR *Readings in American Federalism: Perspectives on a Decade of Change*, Washington, DC: Advisory Commission on Intergovernmental Relations, pp. 119–122.

Stakhiv, Eugene Z. (2003), 'Disintegrated Water Resources Management in the US: Union of Sisyphus and Pandora', *Journal of Water Resources Planning and Management* 129(3), 151–154.

Susskind, Lawrence, Alejandro E. Camacho, and Todd Schenk (2010), 'Collaborative Planning and Adaptive Management in Glen Canyon: A Cautionary Tale', *Columbia Journal of Environmental Law* 35(1), 1–54.

Termeer, C.J.A.M., A. Dewulf, and M. van Lieshout (2010), 'Disentangling Scale Approaches in Governance Research: Comparing Monocentric, Multilevel, and Adaptive Governance', *Ecology and Society* 15(4), 29. Accessed from: http://www.ecologyandsociety.org/vol15/iss4/art29/.

Thorson, John E., Ramsey Kropf, Dar Crammond, and Andrea Gerlak (2005), 'Dividing Western Waters: A Century of Stream Adjudications, Part I', *University of Denver Water Law Review* 8(2), 355–461.

US Army Corps of Engineers (US ACE) (2010), *Building Strong Collaborative Relationships for a Sustainable Water Resources Future, National Report: Responding to National Water Resources Challenges*, Washington, DC: US Army Corps of Engineers, Civil Works Directorate.

US Army Corps of Engineers (US ACE) (2011), *National Collaborative Water Resources Conference Proceedings*, Washington, DC: US Army Corps of Engineers, Civil Works Directorate.

US General Accounting Office (US GAO) (2003), *South Florida Ecosystem Restoration: Task Force Needs to Improve Science Coordination to Increase the Likelihood of Success*, Washington, DC: General Accounting Office.

US Geological Survey (USGS) (2011), *New Federal Agreement to Support Water Resources Management*, Media Advisory. In partnership with the National Oceanic and Atmospheric Administration, and US Army Corps of Engineers. Accessed from http://www.usgs.gov/newsroom/article.asp?ID=2790&from=rss.

Walker, David B. (1983), 'The Federal Role in the Federal System? A Troublesome Topic', *National Civic Review* LXXII (January), 13.
Wheeler, Timothy B. (2012), 'Court hears legal challenge to Chesapeake Bay "pollution diet": Farmers, builders square off against EPA, environmentalists, local officials', *The Baltimore Sun*, October 4.
Whipple Jr., William, Donald Duflois, Neil Grigg, Edwin Herricks, Howard Holme, Jonathan Jones, Conrad Keyes Jr., Mike Ports, Jerry Rogers, Eric Strecker, Scott Tucker, Ben Urbonas, Bud Viessman, and Don Vonnahme (1999), 'A Proposed Approach to Coordination of Water Resource Development and Environmental Regulations', *Journal of American Water Resources Association* 35(4), 713–717.
White, Gilbert (2000), 'Water Science and Technology', *Environment* 42(1), 30–39.
Zimmerman, Joseph F. (2005), 'The Nature and Political Significance of Preemption', *PS: Political Science and Politics* 38(3), 359–362.

4. Water scarcity, conflict resolution, and adaptive governance in federal transboundary river basins

Edella Schlager and Tanya Heikkila

4.1 FEDERALISM AND RIVER BASIN GOVERNANCE

Federalism is characterized by multiple, independent, jurisdictions that exercise concurrent authority over citizens regarding a diversity of issues.[1] Numerous scholars have noted the advantages of federal systems. Federalism 'significantly broadens the perspectives brought to bear on an issue' (Shapiro 2011: 40), which produces several benefits, from governments acting as 'laboratories', developing and testing different policies; to redundancy, which allows for failure without the entire system collapsing. And, from learning and error correction to the effective representation of many and diverse interests.

Of course, this description of federalism is too panglossian. Governments in a federal system are no more immune to public goods problems and collective action dilemmas than are individuals. The promise and opportunity of federalism is only to be realized if governments are supported and encouraged to cooperate and coordinate to realize shared benefits, and if competition and conflict among governments is kept within productive bounds and not allowed to become destructive. In other words, as Bednar (2009) argues, opportunism among governments in federal systems must be checked.

Bednar (2009: 9) defines opportunism as when a government or governments in a federal system pursue their own interests at the expense of the general welfare. Opportunism may take at least three forms: encroachment – when a higher level government undermines the authority of a lower level government; shirking – when governments fail to follow through with their commitments and responsibilities to the union; and burden shifting – when governments impose externalities on one another (Bednar 2009: 9).

Checking opportunism among independent and autonomous governments first requires its detection through common information and consensus that an opportunistic act occurred. Then compliance must be realized through persuasion, incentives, or sanctions. The erring state must be brought back into the federal community. Thus, a federal system works to the extent that it establishes mechanisms and safeguards to detect and limit opportunism among its members.

As Bednar (2009) points out, however, a 'robust federation' requires more than just compliance, it also must be adaptable. The initial distribution of authority among governments may be ill conceived, or environmental circumstances change and new and unanticipated problems emerge. Bednar (2009: 3) explains, 'The federal system needs a procedure for adapting the distribution of authority even as it upholds the existing rules.' Adapting to changing circumstances requires the ability to monitor, learn, and experiment with different strategies and rules, and adopt those that appear to produce better outcomes. Adaptive capacity in a federal system is a characteristic of governments and the institutional linkages that support cooperation and coordination among governments.

Distinguishing between acts of opportunism and acts of productive adaptation is challenging. Too strictly limiting opportunism, which limits experimentation, undercuts adaptation resulting in a brittle federal system. Being too permissive towards adaptation, undermines stability and encourages opportunism which may lead to system collapse. The benefits of federalism are realized to the extent that the relations among governments are structured so as to dampen opportunism while allowing for innovation and change. This requires finding the appropriate balance between self-rule and shared rule with the balance shifting under changing circumstances (Elazar 1987).

River basins in federal systems are well suited for exploring issues of adaptation and opportunism. River basins are complex adaptive systems (Schlager and Blomquist 2008) encompassing numerous common pool resources tied together at varying temporal and spatial scales. Streams, wetlands, groundwater basins, forests, riparian habitats and their many and diverse inhabitants form complex, interacting relations across ecosystems. The biophysical diversity is often matched by social, economic, and institutional diversity with a host of governments and organizations representing different communities of place and of interests. These overlapping and concurrent governments pursue a variety of goals, from stable water supplies and drought mitigation to habitat protection and river restoration. In other words, a federal governance system is built around a changing and dynamic biophysical system, and as the two systems interact, issues of adaptation and opportunism will constantly emerge. Whether and how they are resolved will determine the sustainability of both.

Following Bednar's (2009) argument, this chapter explores three closely related questions in the context of federal, transboundary river basins: (1) How does the institutional design and structure of polycentric river basin organizations address and dampen opportunism, particularly shirking and burden shifting? More specifically, is opportunism more problematic in river basins in which the governing arrangements are not well matched to the physical setting, or provide for little flexibility in responding to disturbances? (2) As problems and conflicts among governments are addressed over time, are the capacities of polycentric river basin organizations to adapt to changing hydrologic regimes enhanced? That is, as states respond to problems and conflicts whose sources are either in changing biophysical settings or in opportunism issues, do they invest in institutional capacity that allow them to better manage both challenges? (3) How is adaptation, both to existing and anticipated hydrologic variability a function of institutional design? These questions will be examined in the context of a particular form of river basin management common among interstate rivers in the western US, the interstate river compact.

4.2 THE HYDROLOGIC SETTING OF THE WESTERN US

The climate and hydrology of most of the western United States is dry (Andreadis and Lettenmaier 2006). On average, west of the 100th meridian, states receive less than 500 mm of moisture per year (except coastal region of the Pacific Northwest), with areas of the Southwest and Intermountain West receiving less than 250 mm per year. Variability of precipitation and streamflows is quite high, both intra- and inter- annually. Annually, a significant proportion of precipitation in the West falls and accumulates as snow in the high mountain ranges, where it is 'stored' until spring when it melts, feeding mountain springs and rivers that typically experience their highest flows in late spring. Water tends to be more plentiful in winter and spring, when demand for it is lower and scarcer in summer and fall when demand is much higher. Inter-annually, much of the West is prone to long-term droughts lasting up to a decade, such as those of the 1930s, 1950s, 1980s and more recently in the early part of the twenty-first century.

In order to put the limited and highly variable water supplies to productive economic use federal and state governments, as well as local water user associations invested billions of dollars developing extensive water storage and conveyance infrastructure across the West. This infrastructure captures and stores snow melt and spring runoff, making it available to water users, sometimes hundreds of miles away from a river, during the summer months when demand is high. The economies of western states rely heavily on mountain snowpack and surface water reservoirs, both of which are sensitive to climate change (Climate Change Science Program 2008).[2]

What might water users and managers expect to see in the coming decades from the effects of climate change on the hydrology of the West? There is reasonable scientific evidence to suggest that much of the intermountain and southwestern US will experience a decrease in water supplies in the coming 20 to 50 years (Climate Change Science Program 2008). For instance, streamflows have been declining across much of the western US since the 1980s (Climate Change Science Program 2008). In addition, declining mountain snowpack, which has been attributed to increased temperatures, has led to earlier spring snowmelt in many basins – regions where agriculture is often dependent on the timing of these flows for irrigation. Since 1980, the West has also experienced increased variability in streamflows between April and September, which is the prime agricultural season. Moreover, there is some evidence that the western United States has experienced both an increase in the severity and duration of droughts over the past 100 years (Andreadis and Lettenmair 2006). In addition climate change models simulating runoff conditions for the period 2041 to 2060 (relative to the period 1901 to 1970) predict a likely decrease in streamflows ranging from 10 to 25 percent in much of the Southwest and Intermountain West (Climate Change Science Program 2008). Thus, the hydrologic regime unfolding in the West is one that is drier, more variable, and with less storage in the form of snowpack and less reliable storage in the form of surface reservoirs.

4.3 FEDERAL WATER GOVERNANCE AND WESTERN INTERSTATE RIVER COMPACTS

River basin governance in the US federal system is characterized by the national and state governments exercising concurrent and overlapping authorities. In general states have taken the lead in water allocation issues, devising, administering and enforcing property rights systems in all forms of water. This is reflected in most states' constitutions that place ownership of waters with the state as well as granting constitutional protections to property rights in water. The national government has taken the lead on environmental issues, from water quality to endangered species, and in providing for large infrastructure projects. Of course, these authorities overlap and impact each other, leading to conflict, coordination, and cooperation among the states and the national government. For instance, even though the national government builds dams and reservoirs, the water captured is governed by states' water rights systems. Conversely, most states have their own endangered species protection and water quality laws that work in tandem with national laws. Shapiro (2011) describes the US federal system as polyphonic rather than dualistic. Governments work in concert, admittedly sometimes off key, rather than in distinct policy realms, and river basin governance certainly reflects this reality.[3]

Beginning in the late 1800s, water conflicts began erupting among water users along states' boundaries. Economic development in upstream states led to increasing water use and river and stream depletions. Water users in downstream states sought to protect their water rights by filing suits in federal district courts seeking cross-boundary water rights adjudications. Such suits turned into lengthy affairs and hardly settled the conflicts as not all water rights claimants participated in the proceedings. In addition, as the national government began investing in large scale surface water projects, states were encouraged to settle their cross boundary water conflicts so as to allow for the orderly development of river basins.

Water users and state water officials began searching for regional solutions to cross border water conflicts. The US Constitution provides three avenues by which the waters of interstate rivers may be allocated among states. First, Congress and the President, through the legislative process, may allocate waters. Of the three allocation mechanisms this one is the least used, as states most directly affected by an allocation have the least control over the process because representatives and senators of all states and their various interests and constituents participate in devising the allocation. Second, the states in a river basin may petition the US Supreme Court to decree a water allocation. The Supreme Court has issued few decrees and, in general, discourages states from pursuing them. Consequently, the third avenue, the interstate river compact, is the primary institutional arrangement used by states to cooperate, coordinate, and jointly govern shared river basins. The compact clause of the US Constitution allows states to engage in 'treaties' specifying water allocation rules and to establish a governing structure through which state representatives administer compact requirements. Known as compact commissions, they typically utilize unanimity decision rules, call upon the

Table 4.1 Overview of interstate river compacts

Compact	Year Signed	Member States	Collective Choice Body
Arkansas	1949	Colorado, Kansas	Commission
Bear	1955, 1980	Idaho, Utah, Wyoming	Commission
Belle Fourche*	1943	South Dakota, Wyoming	State Water Officials
Big Blue	1971	Kansas, Nebraska	Commission
Costilla Creek	1944, 1963	Colorado, New Mexico	Commission
Klamath (Upper)	1956	California, Oregon	Commission
La-Plata*	1922	Colorado, New Mexico	State Water Officials
Pecos	1949	New Mexico, Texas	Commission
Republican	1943	Colorado, Kansas, Nebraska	Commission
Rio Grande	1938	Colorado, New Mexico, Texas	Commission
Snake (Upper)	1949	Idaho, Utah, Wyoming	State Water Officials
South Platte*	1923	Colorado, Nebraska	None
Upper Niobrara*	1962	Nebraska, Wyoming	State Water Officials
Yellowstone	1950	Montana, Wyoming (North Dakota, non-voting)	Commission

Note: * Denotes where member states do not actively meet or administer the compact.

water expertise in state agencies, and have the authority to devise rules and regulations to realize the water allocation rules. Commissions vary in the number of state representatives allowed, the position of an independent chairman, monitoring and enforcement mechanisms, and conflict resolution mechanisms. Finally, compacts may only be dissolved through the unanimous consent of their members.

From the early 1920s to 1970, the western states engaged in numerous interstate river compacts. This chapter is based on a study of 14 of them. See Table 4.1 for an in depth description of the compacts.[4]

Interstate river compacts are governing arrangements that allow states to realize a variety of shared benefits and public goods, such as stable, predictable, and enforceable water rights as well as large surface water projects that provide water supplies and flood control. How well compacts are matched to their hydrologic environment, how adaptable they are, and how well they limit opportunism is addressed in the next section.

4.4 THE ADAPTIVE CAPACITIES OF INTERSTATE RIVER COMPACTS

Adaptive capacity of governing arrangements may take two forms. First, access, use and allocation rules may be purposefully designed to match different circumstances so that water users and managers may respond at the operational level, without having to revise or create new rules. For instance, farmers in some irrigation systems have devised different water allocation rules for different 'water scenarios'. When water is abundant, farmers may take as much water as they want. When water is scarce, more restrictive allocation rules are followed such as taking turns diverting water (Coward 1979; Ostrom 1990; Tang 1992). In these settings, resource users operationally switch among allocation rules designed to match different resource settings.

Second, adaptability may be realized by collective choice mechanisms that allow resource users, public officials, and others to create new, and revise existing, allocation rules in light of changing circumstances. For instance, the California legislature has enabled citizens and local jurisdictions to create different forms of local and regional water districts. The governing boards of the water districts devise and revise rules partly in response to changing water scarcity (Blomquist 1992). In this section, both forms of adaptability will be explored, beginning with the operational level.

4.4.1 Operational Level Adaptation to the Historic Hydrologic Setting

This subsection addresses the first research question: how does the institutional design and structure of polycentric river basin organizations address and dampen opportunism, particularly shirking and burden shifting? More specifically, is opportunism more problematic in river basins in which the governing arrangements are not well matched to the physical setting, or provide for little flexibility in responding to disturbances?

Compact water allocation rules direct water managers on how and when to store water, when to release water, and how and when to administer water rights. In Table 4.2, the multiple dimensions of compact allocation rules among each of the 14 river basin compacts are displayed. The compact allocation rules reflect a particular hydrologic environment and water infrastructure built to harness that environment. Most compacts have distinct allocation rules for different sources of water along the river. In addition, half of the compacts include different rules for particular

Table 4.2 Summary of water allocation rules for 14 interstate compacts

Compacts	Number of rule defined seasons	Predominant Water Rule Type	Number of types of water governed[a]	Specific Users covered by Compact
Arkansas (Colorado, Kansas) – 1949	3	fixed	3	yes
Bear (Wyoming, Idaho, Utah) – 1958, Amended 1980	3	Proportionate	2	yes
Belle Fourche (Wyoming, South Dakota) – 1943	0	Proportionate	2	yes
Big Blue (Kansas, Nebraska) – 1971	2	Fixed – minimum downstream	2	yes
Costilla Creek (Colorado, New Mexico) – 1944, Amended 1963	2	Proportionate	2	no
Klamath (Oregon, California) – 1956	no	prior appropriation	1	yes
La Plata (Colorado, New Mexico) – 1922	2	proportionate	1	no
Pecos (New Mexico, Texas) – 1949	0	Fixed – minimum downstream	3	no
Republican (Colorado, Kansas, Nebraska) – 1943	0	Fixed – all states	2	yes
Rio Grande (Colorado, New Mexico, Texas) – 1938	0	Proportionate	2	no
Snake (Wyoming, Idaho) – 1949	0	Proportionate	2	yes
South Platte (Colorado, Nebraska) – 1923	2	Fixed – minimum downstream	1	yes
Upper Niobrara (Wyoming, Nebraska) – 1962	2	Basin wide prior appropriation	3	yes
Yellowstone (Wyoming, Montana) – 1950	0	proportionate	2	yes

Note: a The types of water governed by compacts include surface water, groundwater, reservoir water, flood waters, and salvaged water.

seasons, with the summer season – when water demand is highest – being the most restrictive. Finally, some compacts also place spatial or temporal boundaries around who is governed by the water allocation rules. For purposes of this chapter, however, the type of water sharing rule, whether fixed or proportionate will be the lens through which operational level adaptation will be examined.

Fixed water allocation rules

Fixed water allocation rules take three forms, either a fixed amount of water is allocated to each state, or a fixed minimum stream flow must cross state boundaries (typically during a defined irrigation season), or a fixed amount of water must annually cross state boundaries. Fixed water allocation rules place the risk of water shortage on a specific state rather than sharing shortages among states. A state bears the cost of water shortage in order to provide benefits for other states. Consequently, all else being equal, fixed water allocation rules tend to be the least adaptive to diverse hydrologic events, and thus may more often be associated with compact compliance issues.

Of the five compacts that use some form of fixed water allocation rule, three have experienced sufficiently severe compliance issues that the downstream state filed and won a US Supreme Court case, requiring costly compliance actions on the part of the upstream state. The two compacts that have not experienced interstate conflict may in part be attributed to the relatively limited scope of the minimum stream flow rules. In the South Platte and Big Blue river basins, the volume of water covered by the rules is a smallish proportion of the overall volume of water in the rivers, thus providing the upstream states with substantial buffers against variability.

Proportionate water allocation rules

In contrast to fixed allocation rules, proportionate rules share the risk of water variability among all member states. Some compacts, such as the La Plata, establish a fixed proportionate rule. In this case, the upstream state of Colorado must share 50 per cent of the available streamflows during the irrigation season with its downstream neighbor, New Mexico. Other compacts, such as the Rio Grande, use a variable proportionate rule with the sharing rule varying based on the volume of water available. The sharing of water variability among states appears to dampen severe compliance issues. Only one of the compacts using a proportionate water allocation rule has used the US Supreme Court to address a compliance issue. The remaining proportionate rule compacts have not experienced significant compliance issues.

Water allocation rules do exhibit differing degrees of adaptability to variable water conditions. Compacts based on proportionate water allocation rules exhibit greater adaptability – all member states share the risk of reduced water supplies. In contrast, compacts based on fixed water allocation rules exhibit less adaptability – a single state bears the risk of reduced water supplies. In such compacts, the state that bears the brunt of water scarcity must strictly regulate, even reduce, the water allocations for their own citizens in order to provide water to the citizens of other states, something that states are reluctant to do. Consequently, opportunism emerges – states attempt to burden shift – delegating water scarcity to other states in violation of compact rules. How that opportunism is dealt with is the subject of the next subsection.

4.4.2 Collective Choice Level Adaptation to Historic Hydrologic Setting

In this subsection research question two is addressed. As problems and conflicts among governments are addressed over time, are the capacities of polycentric river basin organizations to adapt to changing hydrologic regimes enhanced? That is, as states respond to problems and conflicts whose sources are either in changing biophysical settings or in opportunism issues, do they invest in institutional capacity that allow them to better manage both challenges?

Among the 14 compacts included in this study, 28 compact related conflicts occurred among ten compacts (Schlager and Heikkila 2009). In addition, six compact commissions addressed 32 problem events. Problems are events in which commission members agree that an issue must be addressed and debate and discussion occurs around how to resolve it. In contrast, conflicts involve disagreement over either issues, solutions, or both, and may be addressed in a variety of venues, not just compact commissions. As illustrated in Table 4.3, the overwhelming numbers of problem events and conflicts involve water allocation issues, and many fewer conflicts involve water quality or habitat and species issues. Given that almost all of the compacts were established to equitably allocate water among states, conflicts over water allocation are to be expected.

How conflicts and problems are resolved, and for conflicts, in what venues, begin to reveal the adaptive capacity of these federal arrangements for governing river basins. By definition, problem events are addressed by commissions, but compact related conflicts may be addressed and resolved in a variety of venues – courts, compact commissions, federal and state water agencies, and state legislatures. Likewise, conflict resolution may involve a variety of outcomes. One outcome may be the status quo. A party may challenge the interpretation of a compact rule and a court may affirm the accepted interpretation and thus maintain the status quo. Or, resolving a conflict may involve actors changing their strategies within the context of existing rules. A water agency may change when it releases

Table 4.3 Types of compact related problems and conflicts

Compact	Problems				Conflicts		
	# of Problem-Solving Events	Water Allocation	Habitat/ Quality	Other	Conflict Events Addressed by Commission / # of conflicts	Water Allocation	Habitat/ Quality
Arkansas	6	4	0	2	2/5	5	0
Bear	11	8	1	2	4/7	6	1
Big Blue	1	0	1	0	0/0	0	0
Costilla	5	4	0	1	1/1	0	1
Klamath	0	0	0	0	0/5	0	5
La Plata	0	0	0	0	0/1	1	0
Pecos	–				0/1	1	0
Republican	0	0	0	0	0/1	1	0
Rio Grande	3	2	0	1	1/3	3	0
Snake	0	0	0	0	0/1	1	0
Yellowstone	6	5	0	1	0/3	2	1
Total	*32*	*23*	*2*	*7*	*28*	*20*	*8*

water from an upstream reservoir for transport and storage to a downstream reservoir in order to maintain in-stream flows. Rule changes may also be the outcome of resolving conflicts. Rules regulating groundwater pumping may be adopted for all wells in a compact basin in order for the state to meet its water delivery requirements. Finally, compacts themselves may be revised to resolve conflicts among their members.

Conflict resolution

Of the 28 conflicts, 24 were addressed by courts (n=15) and compact commissions (n=9). Furthermore, court resolved conflicts almost exclusively raised questions of opportunism. Of the 15 court addressed conflicts consider the seven conflicts that resulted in status quo outcomes. In six of those seven conflicts, private parties challenged decisions made either by compact commissions or state or federal water agencies acting in a compact governed basin.[5] The conflicts resulting in status quo outcomes may be thought of as 'accountability' conflicts. The authority of compact decision making bodies and public agencies to make specific decisions was reviewed by an independent third party and found appropriate.

In another seven of the 15 court cases rule changes were the outcome of the conflicts and claims of opportunism are much more pronounced. For instance, in four US Supreme Court-addressed conflicts the downstream states claimed that water users in upstream states used more than the states' water allotment, depriving the downstream states and water users of their lawful share of water from the river. For a decade or more, the downstream states repeatedly raised compliance issues before the compacts' commissions, but the upstream states, through their veto power, prevented commissions from taking action.[6] In three of the cases the court found in favor of the downstream state, in the fourth case, the upstream state voluntarily acted to comply with the compact without the court having to issue a ruling.

This last case provides an excellent example of how in addressing opportunistic behaviors, adaptive capacity to better respond to environmental disturbances may be a beneficial side effect. In the 1950s, the Rio Grande River Basin was gripped by drought, causing the two upstream states (Colorado and New Mexico) to under-deliver their compact water requirements and to accumulate water debits that they would have to make up for at some point in the future. By the early 1960s, the drought had eased, but Colorado continued to under-deliver water supplies to the two downstream states with debits reaching almost 1 million acre feet of water. In 1965, New Mexico and Texas sued Colorado before the US Supreme Court. The states entered into negotiations and devised an agreement in which Colorado committed to meet its compact requirements with no allowance for debits or credits and to strictly administer water rights in the Rio Grande River Basin in the state. In order to meet the requirements of the agreement, Colorado invested in more personnel in order to more carefully administer surface water rights. Furthermore, the Colorado State Water Engineer imposed a well moratorium in the Rio Grande Basin and over time developed regulations on existing wells. Colorado eventually erased all water debits and the suit was dismissed. While primarily a conflict around Colorado shirking its compact duties and harming the downstream states, the resolution of the conflict also led to increased capacity in Colorado to better administer and regulate its waters, thus potentially increasing its adaptive capacity to respond to environmental disturbances.

The conflicts addressed by compact commissions also reveal that opportunism issues predominate. Six of the nine conflicts addressed by commissions resulted in adopting rule changes around how the compacts were to be administered. For instance, after the Rio Grande Compact court case discussed above was settled and Colorado was once again in compliance, the three member states experienced conflict over an appropriate water accounting system. To that point, each state used a different accounting method for tracking debits and credits with states arguing over which method was appropriate. The Rio Grande Compact Commission developed an accounting procedure for tracking states' water debits and credits that all member states agreed to abide by, thus limiting accusations of opportunism; however, with a workable accounting system the commission also invested in adaptive capacity.

Problem solving

Turning from conflicts to problem solving events suggests a greater emphasis on adaptation and a lesser emphasis on opportunism. Conflicts tend to raise issues around compliance whereas problem events tend to raise issues around mismatches between compact rules/administration and the biophysical setting. Many of the problems addressed by commissions centered on issues that if addressed, would allow the compact to be better matched to the hydrologic setting and to better meet the goals of the compact.

Among the 32 problem events addressed by commissions, the most common solution was to change compact operations rules. Rules were changed 18 times. For instance, the Big Blue Compact not only allocates water, but it also attends to water quality issues. In the 1980s, water quality issues emerged in the Kansas portion of the basin and Kansas requested that the commission address the issues. Both states agreed that water quality was a problem and the commission began to explore ways of addressing it. One action that the commission took was to change monitoring rules to establish a water quality monitoring network.

Another piece of evidence suggesting that problem solving tends to focus on revising the administration of the compacts to better fit the biophysical environment is the fact that resolving problems spurred much more investment in the capacity of compacts to govern than did conflicts. In resolving problems, commissions created ten new committees and a third of them were made permanent. In addition to information gathering, which was common in relation to many problems, commissions collaborated with outside agencies and actors to address problems. Furthermore, commissions invested in new forms of monitoring in resolving problems. In contrast, conflicts led to many fewer investments in capacity building. In the cases where commissions resolved conflicts the most common response was to gather information, however, no permanent committees were created or new staff hired.

Collective choice adaptability varies among the compacts and has changed over time as problems and conflicts are addressed. Conflicts appear to revolve around claims of opportunism, and in resolving such claims rules are changed to better ensure compliance with the agreements. Problems appear to involve issues related to the biophysical setting and responses are geared at better coping with hydrologic variability. In so doing, compacts invest in their capacity to govern, thus strengthening their ability to adapt to future challenges.

4.5 CONCLUSION

River basins are social-ecological systems. In a federal setting, sustainably governing river basins involves two significant challenges. First, the governing system must be well matched to the ecological, or biophysical, system that is dynamic and changing. Meaning that the governing system must be adaptable and changing as well. Second, the governing system must be well designed so that the multiple, independent, but overlapping governments that are exercising concurrent authorities are productively engaged in providing shared goods and benefits while avoiding collective action dilemmas. Ensuring a stable but dynamic and adaptable governing system is not an easy accomplishment. As Bednar (2009) points out, distinguishing between acts of opportunism and acts of productive adaptation will be challenging. Too strictly limiting opportunism undercuts adaptation resulting in a brittle federal system. Being too permissive towards adaptation encourages opportunism which may lead to system collapse.

In the case of interstate river compacts in the western US, both brittle systems and more balanced systems are observed. The brittle systems, or compacts, are those that are grounded in water allocation rules that attempt to fix a specific hydrologic system in place. Saddled with attempting to realize a fixed and unchanging hydrologic system, the member states are trapped, unwilling to renegotiate the compacts for fear of losing precious water rights, but unable to make an unworkable system work.

The more balanced systems are those that are grounded in proportionate water allocation rules that are more adaptable to hydrological variability. With workable water allocation rules, the compact commissions have engaged in addressing conflicts and solving problems, the results of which have been to both better match the governing system to the biophysical system while also dampening acts of opportunism.

How well will compacts respond to future challenges, particularly climate change impacts? The climate change trends discussed earlier will create mismatches between compact water allocation rules and their hydrologic settings which are likely to be accompanied by compliance issues. The compacts that are likely to feel the impacts of climate change first are those with fixed water allocation rules, and as discussed earlier, these are the very compacts that are currently experiencing repeated compliance issues because their allocation rules so tightly tie them to a particular hydrologic regime. In addition, compacts with minimal administrative capacities to implement water allocation rules and revise operations in light of experience are also likely to struggle to respond and adapt to climate change impacts.[7] This failure to invest in administrative capacity largely occurred because water supplies exceeded water demands in these basins. Water users were consistently receiving their accustomed shares of water thus providing little pressure on compact commissions to invest in costly administrative systems. The compacts with inadequate administrative capacity face very difficult challenges, responding to climate change impacts while at the same time investing in the capacity to implement existing arrangements. Claims of non-compliance are likely to become more common as water scarcity increases.

How might compacts be revised to better match the hydrologic settings they will be operating within in the future while also tending to opportunism? The most obvious option – reopening the compacts and developing different water allocation rules that better match the water setting – is the least likely to be pursued. States will be attempting to re-allocate less water. How states will gain sufficient support to reduce or eliminate water rights is unclear. Also, as discussed above, the different processes for negotiating and re-negotiating compacts present states with the worrisome possibility of losing control of the process. The hurdles that states face in re-designing compacts are quite high and are unlikely to be attempted.

A more promising alternative, although not without its own challenges, is investing in more transparent monitoring and more sophisticated hydrological modeling of rivers so that states can learn about and anticipate changes in hydrology. This would be a valuable step forward because as hydrologic systems of the western US change and become drier and more variable, downstream states will find it difficult to determine whether compact violations are due to upstream states allowing additional water development, failing to effectively limit intra-state water use, a changing hydrologic regime, or a combination of all three. That is, a downstream state will have an increasingly difficult time distinguishing between climate change impacts and acts of opportunism. Better monitoring and modeling will work to dampen compliance conflicts and will support efforts to adapt the compact to changing climate conditions.

Another alternative that states may pursue is to devise extra-compact arrangements to address climate impacts. In a federal system, states and citizens may respond to problems by creating new agreements and possibly new governments. For instance, new arrangements have been created in two compact river basins – the Platte and the Klamath – both in response to endangered species. In each case, the agreements recognize and leave unchanged the compacts' water allocation rules. Rather, state and local governments, and various private associations and organizations, commit to taking actions to recover endangered species and restore habitat.

Among the 14 river compacts in this study about half face serious challenges from climate change – the three with mismatched water allocation rules and the four with limited or no governing capacity. The remaining seven compacts, while experiencing conflicts and problems, have both operational and collective choice levels of adaptability that can be used and built on to innovate responses to changing hydrologic regimes.

4.6 NOTES

1. Bednar (2009: 18–19) argues that for a governing system to be considered federal it must exhibit three characteristics: (1) territory that is divided into mutually exclusive states which cannot be unilaterally abolished; (2) independence among the states and national government, which is protected constitutionally, and (3) governments directly govern their citizens. If two of the three conditions are met the system is considered quasi-federal.
2. A third form of storage that is prevalent throughout the West, and something that water users have increasingly relied on over the past five decades are groundwater basins (Blomquist et al. 2004).
3. See Gerlak (this volume) for excellent overviews of the US federal system of river basin governance.
4. For this study, we collected and coded data on compact rules, the collective choice arrangements that administer the compacts (typically commissions), and the operational decisions and actions of the compact administration over time. The data were collected by examining the compact agreements, which are formally established under state statutes, reports or meeting minutes from the compact governing bodies and negotiating committees, as well as other historical records available on each compact (for example, legislative histories). We also identified and coded data on both the conflicts and the problem solving events related to each compact (starting with the inception of each compact). Compact "problems" differ from conflicts in that they are defined as an issue that members of the commission agree needs to address and member states find jointly-agreed upon solutions to the problem. Conflicts, on the other hand, are defined as debates or disagreements over the cause of a problem and/or the appropriate solution to the problem. Conflicts and problem solving were identified through compact commission meeting minutes and annual reports, as well as historical newspaper stories and court cases, supplemented by relevant secondary literature (e.g. books and journal articles). In coding conflicts and problems, we also have identified the means by which they are resolved and the outcomes or solutions produced. For in depth descriptions of data collection see Schlager and Heikkila (2009) and Heikkila et al. (2011).
5. In the remaining conflict the lawsuit was dismissed because of the failure to include an indispensable party.
6. To be fair to the upstream states, the conflicts involved genuine differences of interpretation of the compact and each state's compact requirements. For instance, in the Republican River Compact case, Kansas claimed that groundwater use should be counted in each state's water allotment; however, Nebraska argued that the compact did not cover groundwater, only surface water, and by that measure it was within its

compact requirements. Whether the compact governed groundwater was the key issue to be decided by the Supreme Court.

7. These compacts are the Belle Fouche, Upper Snake, Upper Niobrara, and Yellowstone.

4.7 REFERENCES

Andreadis, K.M. and D.P. Lettenmaier. 2006. "Trends in 20th century drought over the continental United States", *Geophysical Research Letters*, 33.

Barnett, Jack. 2006. Bear River Compact Commission, Engineer-Manager. Personal Communication.

Barnett, Tim and David Pierce. 2008. "When Will Lake Mead Go Dry?" *Water Resources Research*, 44.

Bear River Compact Commission. 1958–1970. Annual Meeting Minutes.

Bennett, Lynn and Charles Howe. 1998. "The Interstate River Compact: Incentives for Noncompliance", *Water Resources Research* 34(3): 485–495.

Bednar, Jenna. 2009. *The Robust Federation*. Cambridge: Cambridge University Press.

Blomquist, William. 1992. *Dividing the Waters*. Oakland, CA: ICS Press.

Blomquist, William, Edella Schlager and Tanya Heikkila. 2004. *Common Waters, Diverging Streams: Linking Institutions and Water Management in Arizona, California, and Colorado*. Washington, DC: Resources for the Future Press.

Brikowski, T.H. 2008. "Doomed Reservoirs in Kansas, USA? Climate Change and Groundwater Mining on the Great Plains Lead to Unsustainable Surface Water Storage" *Journal of Hydrology* 354: 90–101.

Climate Change Science Program. 2008. *The Effects of Climate Change on Agriculture, Land Resources, Water Resources, and Biodiversity*. A Report by the US Climate Change Science Program and the Subcommittee on Global Change Research. Washington, DC: US Environmental Protection Agency.

Committee on the Scientific Bases of Colorado River Basin Water Management. 2007. *Colorado River Basin Water Management: Evaluating and Adjusting to Hydroclimatic Variability*. Washington, DC: National Academies Press.

Coward, E. Walter. 1979. "Principles of Social Organization in an Indigenous Irrigation System" *Human Organization* 38(1): 28–36.

Dietz, Thomas, Elinor Ostrom, and Paul C. Stern. 2003. "The Struggle to Govern the Commons" *Science* 302: 1907–1910.

Elazar, Daniel. 1987. *Exploring Federalism*. Tuscaloosa, AL: University of Alabama Press.

Gerlak, Andrea. 2005. "Federalism and US Water Policy: Lessons for the Twenty-first Century" *Publius* 36(2): 231–257.

Gerlak, Andrea and Tanya Heikkila. 2007. "Collaboration and Institutional Endurance in US Water Policy" *PS: Political Science and Politics* (January) 55–60.

Heikkila, Tanya, Edella Schlager, and Mark Davis. 2011. "The Role of Cross-scale Institutional Linkages in Common Pool Resource Management: Assessing Interstate River Compacts" *Policy Studies Journal* 39(1): 121–145.

Jibson, Wallace. 1997. "The History of the Bear River Compact" The Bear River Commission.

Kansas Nebraska Big Blue River Compact Commission. 2004. Thirty-first Annual Report.
Knox, Kenneth. 2003. "The Costilla Creek Compact" *University of Denver Water Law Review* 6: 453–490.
Milly, P.C.D. et al. 2008. "Stationarity is Dead: Whither Water Management?" *Science* 319: 573–574.
Muys, Jerome, George William Shirk, and Marilyn O'Leary. 2007. "Utton Transboundary Resources Center Model Interstate River Compact" *Natural Resources Journal* 47: 18–115.
Nebraska Department of Natural Resources. June 2003. "Summary of Republican River Compact Litigation Settlement" Pamphlet.
Nebraska Department of Natural Resources. 2006. Annual Evaluation of Availability of Hydrologically Connected Groundwater Supplies: Determination of Fully Appropriated. Big Blue River Basins. Lincoln, NE.
Ostrom, Elinor. 1990. *Governing the Commons*. Cambridge: Cambridge University Press.
Ostrom, Elinor. 2005. *Understanding Institutional Diversity*. Princeton, NJ: Princeton University Press.
Republican River Compact Administration. 2002. Accounting Procedures and Reporting Requirements.
Schapiro, Robert. 2011. "The Varieties of Federalism". In Edella Schlager, Kirsten Engel, and Sally Rider (eds) *Navigating Climate Change Policy: The Opportunities of Federalism*. Tucson, AZ: The University of Arizona Press.
Schlager, Edella and William Blomquist. 2008. *Embracing Watershed Politics*. Boulder CO: The University Press of Colorado.
Schlager, Edella and Tanya Heikkila. 2009. "Resolving Water Conflicts: A Comparative Analysis of Interstate River Compacts" *Policy Studies Journal* 37(3): 367–392.
Schlager, Edella and Tanya Heikkila. 2011. "Left High and Dry? Climate Change, Common Pool Resource Theory, and the Adaptability of Western Water Compacts" *Public Administration Review* 71(3): 461–470.
Schlager, Edella, Kirsten Engel, and Sally Rider. 2011. *Navigating Climate Change Policy: The Opportunities of Federalism*. Tucson, AZ: The University of Arizona Press.
Shapiro, Robert. 2011. "The Varieties of Federalisms". In Edella Schlager, Kirsten Engel, and Sally Rider (eds) *Navigating Climate Change Policy: The Opportunities of Federalism*. Tucson, AZ: The University of Arizona Press, pp. 35–47.
Stern, Paul et al. 2002. "Knowledge and Questions After Fifteen Years of Research". In Elinor Ostrom et al. *Drama of the Commons*. Washington, DC: National Academy Press, pp. 445–489.
State of Wyoming. 2001. Bear River Basin Water Plan. Wyoming Water Development Commission. http://waterplan.state.wy.us/plan/bear/bear-plan.html.
State of Wyoming. 2002. Powder/Tongue River Basin Plan Final Report. Wyoming Water Development Commission Basin Planning Program. http://waterplan.state.wy.us/plan/powder/finalrept/finalrept.html.
State of Wyoming. 2003. Snake/Salt River Basin Water Plan. Wyoming Water Development Commission. http://waterplan.state.wy.us/plan/snake/finalrept/finalrept.html.

Tang, S.Y. 1992. *Institutions and Collective Action: Self Governance in Irrigation*. Oakland, CA: ICS Press.
Woodhouse, C.A. and J.T. Overpeck. 1998. "2000 years of drought variability in the central United States" *Bulletin of the American Meteorological Society* 79, 2693–2714.

5. Managing water in a federal state: the Canadian experience

J. Owen Saunders

5.1 INTRODUCTION

5.1.1 General

The governance of water resources in the Canadian federation reflects an approach to natural resources management under which the national government largely defers to the provinces as the primary resource managers. This chapter, first, explores the constitutional and political roots of provincial supremacy in water resources management; second, illustrates through two examples how this supremacy may lead to poor policy outcomes for the management of shared water basins; and finally, offers some suggestions as to how Canada's decentralized approach to federalism can be reconciled with the need to reflect national interests in interjurisdictional water management.

5.1.2 Canada's Freshwater Endowment

Canada is one of the most richly-endowed countries in the world with respect to its freshwater resources. Approximately one-half of 1 per cent of the world's population enjoys 20 per cent of its freshwater resources and 7 per cent of the world's renewable freshwater resources.

Both the nature of water resources and the management challenges they present vary significantly from region to region. Canada's five major drainage basins (there are numerous, sometimes significant, sub-basins) flow into the Pacific Ocean, the western Arctic Ocean, Hudson Bay, the Atlantic Ocean and (of very minor significance) the Mississippi River system. The six eastern provinces rely primarily on the Atlantic basin (including notably the Great Lakes and St Lawrence River), although northern rivers in Quebec and Ontario that flow into Hudson Bay have great significance for hydroelectric development. The population centers of the three Prairie Provinces (Alberta, Saskatchewan and Manitoba) are located primarily in the Hudson Bay Basin (though the Arctic Basin, dominated by the Mackenzie River and its tributaries, is of growing significance because of the rapid resource development in the northern Prairies, notably with the Alberta oilsands). Finally, the West Coast province of British Columbia is overwhelmingly reliant on the Pacific drainage basin, though in the northeast of the province the Arctic Basin holds important hydroelectric capacity. Much of Canada's freshwater is far from the major centers of population, with 60 per cent of its rivers flowing north – away from the 80–90 per cent of its population living within 300 kms of the United States (Nowlan, 2005).

In addition to ten provinces, Canada has three federal territories – Yukon, the Northwest Territories and Nunavut, the latter of which is almost entirely in the Hudson Bay drainage basin and the former two which lie within the Arctic Basin. The three territories are at different stages of constitutional devolution, including with respect to their water management responsibilities. While in many respects they enjoy responsibilities broadly comparable to the provinces, there are important exceptions; moreover, their devolved responsibilities are the result of federal legislation, and are not constitutionally guaranteed. Therefore, the primary emphasis in this chapter is on the provincial and federal roles in water management.[1]

Just as the nature of Canada's water resources varies significantly from region to region, so do its water management challenges. In eastern Canada and British Columbia, water quantity has not typically figured as a major worry, although in the Great Lakes region variations in lake levels have sometimes generated concerns related especially to recreational and navigation uses. Generally, though, concerns in these provinces have focused on water quality issues (drinking water quality is a particular challenge for Aboriginal communities across Canada).

By contrast, in the Hudson Bay Basin shared by the Prairie Provinces, the water management challenges posed have traditionally been related primarily to water quantity, although water quality has emerged as an issue, especially because of intensive livestock activities and oil and gas development.

Similarly, in the Arctic Basin parts of the Mackenzie system face challenges both of water quality and water quantity. These are aggravated by the differing needs of the various jurisdictions in the basin – with British Columbia having a special interest in the use of the basin for hydroelectricity generation, Alberta relying on the basin for the development of other natural resources (especially petroleum and forestry) and the Northwest Territories relying on the Mackenzie for in-stream uses associated with traditional Aboriginal lifestyles.[2]

The one unifying theme that does seem to characterize Canadians' use of their freshwater resources is profligacy. Canadians rank among the highest users of freshwater in the world, ranking second only to the United States in the OECD in per capita use, with consumption 65 per cent above the average in the OECD (Nowlan, 2005).

5.2 THE CONSTITUTIONAL CONTEXT FOR WATER MANAGEMENT IN CANADA

The Canadian Constitution in general divides legislative powers between the federal and provincial levels of government into two discrete lists, with each level of government sovereign in its own sphere, and with any residual powers resting with the federal government. (There are also a few concurrent powers, of which only agriculture has relevance to water management.)

On the face of it the Constitution provides important powers to the federal government through which it may influence water management. Of these, the most obvious relate to navigation and fisheries, but other important levers are found in the federal responsibility for criminal law, trade and commerce, and interprovincial and international works and undertakings (for example, canals). It is an oddity of Canadian constitutional development that the federal government has plenary authority to implement treaties concluded by the British Empire on Canada's behalf (so-called 'Empire treaties', which were concluded up until the 1920s) but no such power to implement treaties that it has subsequently concluded in its own right should they intrude on provincial jurisdiction. Of most relevance for this chapter, the International Boundary Waters Treaty of 1909 is an Empire treaty and as such affords the federal government significant legislative powers for its implementation.

While no one head of legislative power would enable the federal government to legislate comprehensively with respect to water resources management, an expansive reading of the various individual powers might well provide considerable room for the federal government to act. However, the tendency of constitutional jurisprudence, especially in recent decades, has been to give a narrow reading to the specific powers and to restrict them to their most obvious intent. So, for example, federal pollution-related legislation passed under the authority of the fisheries power has been struck down as overreaching where there was no demonstrated link between the water pollution and harm to fish or fish habitat [*Fowler* v. *R.*, [1980] 2 S.C.R. 213]. The Supreme Court required a similar demonstrable link between a water policy measure and a specific head of power with respect to the invocation of the navigation power [*Northwest Falling Contractors* v. *R.*, [1980] 2 S.C.R. 292].

In addition to specific heads of legislative authority, the federal government has other broad powers under the Constitution, the most important being the general power to legislate for the 'peace, order and good government' of Canada. This power has a chequered history of judicial interpretation, which has evolved to embrace a number of branches, with the 'national concern' doctrine being the most relevant for water management. Whatever the theoretical potential of this doctrine, the federal government in recent decades has been hesitant to invoke it explicitly as a basis for federal regulation of natural resources management, and has preferred to rely primarily on specific heads of federal power. Even when the federal government has indicated a willingness to assert a stronger role in water management, it does not always follow through. For example, in the *Canada Water Act* of 1970 [R.S.C. 1985, c. C-11] the federal government claimed the authority to take strong and, where necessary, unilateral water management measures in basins where water quality had become a matter of national concern. In practice, however, those provisions of the *Act* have never been invoked. Similarly, an ambitious federal agenda for action in the *Federal Water Policy* of 1987 (Environment Canada, 1987) has gone largely unimplemented. More typically when the federal government wishes to pursue water management goals, it employs its spending power – usually in the form of cost-shared programs negotiated with the provinces.[3]

The Constitution provides other, even more sweeping powers on the part of the federal government that need not be discussed here. While these powers were frequently used in the first several decades of confederation, today they have fallen into desuetude and their invocation would be regarded as politically explosive, and potentially even as breaching established constitutional convention. This last point – the political impossibility of invoking certain constitutional authorities that formally exist in the written Constitution – reflects a broader and vital dimension of Canadian federalism: in many respects the actual mechanics of Canadian federalism are influenced more by what has been termed the 'political constitution' than by the written Constitution. The core of Canada's constitutional federalism is still the 1867 legislation passed by the British Parliament (and incorporated into the new Constitution in 1982 with all its peculiarities intact). The original drafters emphasized a strong federal government, and in the early days of confederation the federal government used all its powers, so that the regime was sometimes called 'quasi-federal'. However, the relative positions of the federal and provincial levels of government have changed significantly over the years, with the federal government becoming much more respectful of provincial autonomy, both for political reasons and as a result of constitutional jurisprudence. In the result, the Canadian federal system has evolved into one of the most decentralized in the world. One consequence of this evolution is the inability – or at least unwillingness – of successive federal governments to assert a strong role in natural resources management, including water management.

In contrast to the federal role, provincial governments are seized with much broader authority over water management. This authority is both proprietary and legislative in nature. The provinces own the great bulk of public lands within their borders (albeit there are some important federal proprietary interests as the result of lands held for Aboriginal people and through ownership of national parks, defense lands, etc.). The proprietary rights that attach to provincial ownership are buttressed by legislative rights, both specifically with respect to the management of public lands (and resources), and more generally with respect to matters of a local or private nature, including property and civil rights. It is thus at the provincial level that one finds the broad legislative authority to manage water resources in a comprehensive manner. This dominant role of the provinces has been strengthened over the years as a result of their acquisition of expertise and skilled personnel to manage their natural resources. Thus they have the 'boots on the ground' required to implement resource management initiatives. By contrast, the federal government has not developed the same level of expertise, and in many cases depends on cooperation with provincial officials to implement even its own initiatives. Over the past 20 years, especially, the federal government for fiscal reasons has shed much of its policy expertise in a number of areas, including water management.

In sum, there are significant powers available to each level of government with respect to water management. This has generated an extensive literature analysing the potential reach of federal powers in particular (Gibson, 1968; Alhéritière, 1976). While there is little consensus among authors as to the precise limits of federal or provincial authority (Bankes et al., 1986, at 77), in practice this has proved of little significance given the willingness of the federal government to defer to provincial interests.

5.3 THE NATURE OF FEDERAL INVOLVEMENT IN WATER MANAGEMENT

Given Canada's vast size, and the fact that most water issues are primarily of local or regional interest, the Canadian constitutional position may be considered largely appropriate for such a federation. In effect, the default position for water management in Canada is provincial regulation.[4] There are, however, water management issues that are clearly of national or interprovincial concern. A central question, therefore, relates to the when and how of federal intervention, and in particular the criteria that influence the decision to act or not to act. In many cases the constitutional powers available to both levels of government are broad enough to allow the federal government significant discretion as to whether to intervene. While the nature of federal involvement is harder to predict in these gray areas, some observations can be made about the relevant factors that influence the federal decision on whether to act.

First, federal action in water management almost always proceeds from a stance that is reactive rather than proactive: a reluctance to regulate water. This reflects the role of the provinces as primary regulators of their natural resources. Second, the federal government is most likely to intervene when it sees a clear link to a specific head of constitutional power. Finally, and related to the first point, the federal response is often opportunistic: the government will be most ready to act when it feels there is a current in public opinion that it can profit from or must respond to. These observations can be illustrated by reference to two recent examples of significant Canadian water management issues: water exports and the management of the Mackenzie Basin.

5.3.1 Water Exports

The perceived threat of water exports to the United States has exercised Canadian public opinion on a number of occasions since the 1980s.[5] In the period leading up to the 1988 federal election, which was fought primarily on the issue of the proposed Canada–US Free Trade Agreement, the Progressive Conservative government, in an attempt to defuse public concern that the Agreement might facilitate water exports, introduced legislation to, in its view, clarify that this was not the case. The bill died when the election was called and was not re-introduced after the re-election of the government; moreover, the agreement was ratified without any specific amendment to address water exports. A few years later, the issue arose again during the negotiation of the North American Free Trade Agreement (NAFTA). In the 1993 federal election campaign the opposition Liberals attacked parts of the agreement, which had yet to be ratified, and promised to 'fix' them as a condition of ratification. Their most important commitment was to exclude from the NAFTA the bulk export of water (although the Conservative government maintained that this already fell outside the scope of NAFTA). The Liberals won the election, but proceeded to ratify NAFTA as negotiated, subject only to a separate unsigned statement by the parties (outside the treaty) to the effect that water in its natural state was not the subject of the trade undertakings agreed to. The precise legal status of this statement has been the subject of speculation.

Public concern over water exports came to the fore again in 1998 with the issuance of a provincial license by Ontario for tanker exports of water from Lake Superior. Despite the fact that the license was subsequently withdrawn – and that the export plan as conceived was both absurdly naïve and never remotely likely to come to fruition – the fact that such licenses could apparently be readily obtained created a national outcry for action by the federal government. The (Liberal) federal government responded to this public pressure with a number of measures. Of most interest for this chapter was the legislation it introduced to address the issue. This consisted essentially of an amendment to the existing legislation implementing the Canada–US International Boundary Waters Treaty (IBWT). The amendment included a prohibition, with some limited exceptions, on the removal of *boundary* waters out of the basin in which they were located – that is waters falling within the ambit of the Empire treaties clause in the Constitution.[6] Thus the government addressed only a small part of the problem and did not deal with the potential for export of other waters (whether transboundary or purely domestic). Such a partial solution is clearly not consistent with what water managers unconstrained by political considerations would consider an appropriate holistic response. The federal initiative appears even more tepid when one considers that any practical threat of water exports is – for reasons of simple geography – almost certainly to come not from boundary waters but from either transboundary river systems or (especially for tanker exports) coastal freshwater lakes and rivers that do not fall within the ambit of the IBWT.

The federal elections of 2006 and 2008 returned Conservative minority governments. In the Throne Speech of 2007 the new government undertook to bring forward a national water policy, followed by subsequent commitments to enact legislation to prohibit bulk water exports. These commitments were widely viewed as attempts by the federal government to burnish its environmental credentials in the context of other widely criticized changes to environmental legislation. While the federal government has still not produced a national water policy, it has twice introduced legislation with respect to limiting water exports.

In May 2010 a government bill was introduced with the purported aim of limiting bulk water exports, although on closer examination it clearly would have had no practical effect on any conceivable export scheme. The bill was introduced against the backdrop of the April 2010 federal budget, which had included significant amendments to environmental legislation that received a scathing reaction from the environmental community. The bill died with the calling of the 2011 federal election.

That election yielded a Conservative majority government. In December 2011, a government MP, with the clear support of the government, introduced a private member's bill which closely resembled the 2010 government bill, but with changes such that it did indeed command broad support as a serious effort to address the issue (the bill is also noteworthy in that it is clearly not based on a narrow interpretation of the federal government's constitutional authority but rather is rooted implicitly in the general peace, order and good government power).[7] Given its multi-party support it received unanimous approval in Parliament and ultimately received royal assent on June 19, 2013. While this might be seen as a step towards a more proactive role in water management by the federal government, the bill can also be seen partly as a response to a competing private member's initiative from a Liberal MP. Moreover, the initiative must also be judged in light of other contemporaneous government environmental initiatives – notably, limitations on the scope of the federal role in environmental assessments and fisheries regulation – that were strongly criticized by the environmental groups.[8]

In sum, the federal initiatives on water exports have typically not been proactive but have come in response to specific triggering events, and have typically been minimalist and rooted in a clear single power under the Constitution (the Empire treaties clause). The one exception is the most recent bill. However, even this apparent deviation from the 'norm' can be explained by some other significant factors at work: the strong and long-standing political and public consensus in favor of a restriction on water exports; the introduction of a competing bill by a non-government MP which would clearly appeal to this broad sentiment; the relative quiescence of the provinces on the matter (accounted for by a sentiment in provincial governments also largely opposing water exports); and the interest of the federal government in burnishing its environmental credentials in light of other measures it was pursuing.

5.3.2 Mackenzie Basin Management

The Mackenzie River system flows through three provinces and two territories for over 4000 km in northern Canada before emptying into the Arctic Ocean. The major portion of the Mackenzie Basin, however, is located in three jurisdictions, the provinces of British Columbia and Alberta, and the Northwest Territories. The tenth largest river basin in the world (1.8 million km^2), it comprises about 20 per cent of Canada's landmass (Mackenzie River Basin Board, 2003). The basin's 400,000 people include a significant Aboriginal population, especially in the northern areas. There are obvious resource-use conflicts in the basin, which is the site of significant hydroelectric development (especially in British Columbia) and the subject of increased pressures on both water quality and water quantity in the face of the rapid development of Alberta's massive oilsands deposits. At the same time, Aboriginal populations in the north are highly dependent on the basin's waters for their traditional lifestyles; unfortunately, they are largely downstream users of the resource.

While the Mackenzie is the sort of interjurisdictional system with competing resource uses that is most in need of an overarching arrangement that would balance various interests, the actions taken to date towards a comprehensive plan for the management of the Mackenzie Basin are disappointing. The deficiencies in governmental measures reflect the unwillingness of successive federal governments to articulate a strong voice, even where it seems clearly appropriate. Multi-jurisdictional negotiations on the Mackenzie began in the early 1970s but an intergovernmental agreement on the use of the basin [Mackenzie River Basin Transboundary Waters Master Agreement, *in effect* July 24, 1997] did not come into effect until 1997. The agreement commits the governments of Canada and the five basin provinces and territories to such worthwhile principles as maintenance of the integrity of the aquatic ecosystem, intergenerational sustainability, reasonable use (or more precisely the avoidance of unreasonable harm) and duties to notify and consult – in effect the core of what would be considered a 'modern' instrument for sharing the waters of an interjurisdictional basin. Beyond these general principles, however, the agreement offers little comfort to downstream jurisdictions. For example, it lacks a process for binding dispute resolution and the articulation of specific substantive obligations is to be negotiated through bilateral subsidiary agreements between the parties. After over a decade, however, the only such bilateral agreement concluded is that between the Yukon and Northwest territories, which have no shared waters of great significance. Finally, should other bilaterals be successfully concluded (negotiations are still in the early stages as of this writing), they will likely not be legally binding.

The federal government would almost certainly have the constitutional authority – under its power in relation to peace, order and good government – to intervene legislatively with respect to the Mackenzie, but its role has largely been quiescent, and at most facilitative in encouraging negotiations. Given that the upstream jurisdictions have no incentive to act against their own self-interest in the development of the basin's water resources, it is hardly surprising that negotiations have produced such unimpressive results to date. Prospects for significant future commitments by the upstream provinces remain uncertain, especially given that climate change is likely to generate above-average adverse effects on flows to the Mackenzie.

This lack of federal interest in pursuing an overarching vision for the Mackenzie contrasts with the limited initiatives it *has* pursued in the basin. Once again, these are largely rooted in very specific interests – in particular, those falling under the federal jurisdiction over fisheries. Thus, the federal government has negotiated an approach to water use in the Lower Athabasca (a sub-basin of the Mackenzie system) with Alberta and industry to address the implications of oilsands activities on water quantity and fisheries (although even this involvement has been criticized as overly deferential to provincial resource managers). More recently, it has been involved in demanding better monitoring of the water quality impacts of oilsands operations, but tellingly this occurred only in response to the work of independent scientists and the public outcry over the obvious deficiencies in the monitoring regime.

5.4 EMERGING POLICY CHALLENGES AND OPTIONS

Canadian water managers must confront significant new challenges, some of which are conventional in nature but of a quantitatively different magnitude than experienced in the past. An obvious example is the increasingly apparent conflict between energy security and water security. The threat that rapid development of Alberta's oilsands presents for both water quality and quantity in the Mackenzie Basin is fast becoming a national issue, as well as the object of significant debate within the province. More generally, as the West becomes a more dominant engine for economic growth in Canada, there will be increasing pressure on water resources in the arid Prairies. In addition to addressing the conventional pressures on water resources policy that arise from a growing population and economy, Canadian water managers in the coming decades will have to confront the impacts associated with climate change, which include predicted significant negative effects on river flows in the Prairie Provinces. The outlook for eastern Canada is more mixed, but again the forecast is for lower water levels in the Great Lakes region (Environment Canada, 2008). Thus, it is increasingly important that the federal and provincial governments arrive at an accommodation of responsibilities that allows for policy responses that are based on sound principles of water management rather than on the protection of jurisdictional turf.[9]

A federal response to these challenges could draw on an array of options. However, two useful first steps can be suggested – one at a broad conceptual level and the other much more immediately practical and specific. With respect to the former, it is well past time for the federal government to articulate its vision of its role in Canadian water management. The 1987 *Federal Water Policy* (Environment Canada, 1987) promised much and received broad bipartisan support, but has been left almost entirely unimplemented. The current Conservative government at one point undertook to develop a new federal water policy, but so far that commitment has fallen by the wayside so that federal water policy continues to develop in an *ad hoc* and largely reactive fashion. Without an overarching statement of how the federal government sees its responsibilities in water management, one can hardly expect this approach to change.

As to specific initiatives, the federal government should assume the normal, primary role of a national government in most federations in facilitating, directly or indirectly, the resolution of disputes between sub-national entities. For water resource management, this could mean inducing or requiring management structures for interjurisdictionally shared waters. One attraction of such an approach is that a framework for principled negotiations is more likely than intergovernmental litigation to generate outcomes acceptable to both parties. Both options have been exercised in the United States, but neither has been pursued with any great vigor in Canada where there is not only a federal disinclination to impose solutions, but also a tradition of not litigating intergovernmental disputes. This creates little incentive for upstream provinces to seriously address interjurisdictional sharing of water resources. As noted in the context of the Mackenzie Basin, this unwillingness of upstream provinces to constrain their discretion on the use of the resource is likely to be exacerbated as the effects of climate change become increasingly apparent.

How, then, can the federal government act to resolve festering interprovincial (and provincial–territorial) differences while respecting the strong deference in Canadian federalism to the role of the provinces as managers of their resources? The preferable option would both minimize federal intrusion and provide sufficient incentives for the parties (but especially upstream provinces) to reach an agreement. An option that came close to satisfying these dual requirements was proposed in the report of the Inquiry on Federal Water Policy of 1985 (Canada Inquiry on Federal Water Policy, 1985). This proposal was that the federal government would provide a fallback should provincial or territorial governments be unable to reach an agreement on shared water resources. Where 'reasonable efforts' at agreement failed, and upon the receipt of a complaint from one of the parties, the federal government would set up a board (including representation from all affected parties) to make recommendations on the dispute, and its resolution by the federal government would be based on these recommendations. The constitutional authority for this would be the 'national concern' branch of the federal peace, order and good government power. Ideally of course the very existence of such a mechanism would lead to a political resolution of most disputes and largely obviate the need for its invocation.

5.5 CONCLUSION

Canadian federalism is highly decentralized, especially with respect to natural resources management, including water management. This decentralization is not a necessary – and certainly not intended – result of the Constitution adopted in 1867, but reflects a consensus that has developed over time that provinces are best placed to manage their own resources. One need not question the appropriateness of a strong provincial voice in resource management in suggesting that there are also good reasons for a strong and consistent federal role with respect to those aspects of resource management that have national dimensions, notably for the management of waters that cross provincial or international boundaries. Unfortunately, successive federal governments of differing political complexions have been consistent in their reluctance to pursue a principled and coherent approach to interjurisdictional water management.

Given the likelihood of increased stresses on Canadian water resources in coming decades, this reluctant and reactive federal approach may well be tested. This chapter has suggested that there are options available to the federal government to exercise a national voice in a way that is consistent with recognition of provincial primacy in water resources management.

5.6 NOTES

1. Although the focus of this chapter is on surface water, Canada has significant groundwater resources, which are not evenly distributed across the country. Approximately 30 per cent of Canadians (two-thirds of whom are in rural areas) depend upon groundwater to provide their drinking water (Côté, 2006). Despite its importance, Canada's groundwater endowment is in many respects poorly understood, and the knowledge varies significantly as between regions and aquifers.
2. The Yukon and Saskatchewan also share the basin, but have far less significant interests in basin management than the other three jurisdictions.
3. The two most significant ongoing examples of the federal spending power are the Fresh Water Quality Monitoring program and the National Hydrometric Program, both managed by Environment Canada. The department has been the subject of heavy criticism by the Auditor General's Office for its management of both these programs (Auditor General, 2010 Fall).
4. The position is somewhat different for the three federal territories that are currently at different stages in receiving powers from the federal government.

5. An earlier outbreak of public concern over a series of proposed mega-projects for water export in the 1960s had largely evaporated as the obvious impracticality of the projects became apparent.
6. The IBWT distinguishes between boundary waters – that is, waters, such as the Great Lakes, along which the international boundary passes – and waters that cross the boundary. The most significant provisions of the treaty relate to the former. The willingness of the federal government to take initiatives on boundary waters (and its concomitant reluctance to pursue initiatives in other water bodies) reflects its clear constitutional authority to implement Empire treaties.
7. Bill C-383, *An Act to amend the International Boundary Waters Treaty Act and the International River Improvements Act*, 1st Sess., 41st Parl., 2011. The bill still does not address the possible export of water by tankers from waters that are neither boundary nor transboundary – for example, coastal lakes or streams. In this respect, it has been the subject of some criticism that it is not ambitious enough.
8. The federal budget of 2012 contained a number of important resource-management provisions including, *inter alia*, new environmental assessment legislation that reduces the scope of and number of federal assessments, and amendments to the *Fisheries Act* which limit the application of the fish habitat protection provisions (Heelan, 2012). These follow on earlier initiatives affecting other federal powers such as navigation; the net effect of all these changes was to reduce significantly the federal voice in natural resources management – and water management in particular.
9. Although this chapter is critical of the timidity of the Canadian federal government's view of its role in water management, this emphasis on a functional approach also suggests that on occasion the federal government should defer to provincial initiatives where the provinces are best placed to manage an issue, even where the Constitution might suggest a primary federal role. One example of this is the cross-border initiative between Canadian provinces and American states in the Great Lakes Basin that produced the Great Lakes Charter of 1985 – a 'soft law' document that is significant in its own right and that has led to subsequent fruitful water management initiatives in the basin.

5.7 BIBLIOGRAPHY

Alhéritière, D. (1976). *La gestion des eaux en droit constiutionnel canadien* (Québec: Éditeur official du Québec).

Bankes, Nigel D., Constance D. Hunt and J. Owen Saunders (1986). "Energy and Natural Resources: The Canadian Constitutional Framework", in Mark Krasnick, Research Coordinator, *Case Studies in the Division of Powers* (Toronto: University of Toronto Press) 53.

Canada. Office of the Auditor General of Canada (2010 Fall). *Report of the Commissioner of the Environment and Sustainable Development to the House of Commons*, Chapter 2, Monitoring Water Resources.

Canada. Environment Canada (1987). *Federal Water Policy*.

Canada. Environment Canada (2008). *Climate Change Overview, The Science of Climate Change*.

Canada. Inquiry on Federal Water Policy (1985). *Currents of Change, Final Report, Inquiry on Federal Water Policy*. Environment Canada, Ottawa.

Canadian Water Issues Council (2008). *A Model Act for Preserving Canada's Waters*. Munk Centre for International Studies, University of Toronto, Toronto.

Côté, François (2006). Freshwater Management in Canada: IV. Groundwater. Library of Parliament, Parliamentary Information and Research Service, PRB 05-54E (February).

Gibson, D. (1968). "The Constitutional Context of Canadian Water Planning", 7 *Alta. L.Rev.* 71.

Heelan, Brenda (2012). "An Overview of Bill C-38: The Budget Bill that Transformed Canada's Federal Environmental Laws", Environmental Law Centre (Alberta), online at: http://environmentallawcentre.wordpress.com/2012/08/10/an-overview-of-bill-c-38-the-budget-bill-that-transformed-canadas-federal-environmental-laws/.

Nowlan, Linda (2005). "Buried Treasure, Groundwater Permitting and Pricing in Canada", Research Report.

Mackenzie River Basin Board (2003). *Mackenzie River Basin, State of the Aquatic Ecosystem Report 2003*. Mackenzie River Basin Board Secretariat, Fort Smith, NT.

Saunders, J. Owen (1988). *Interjurisdictional Issues in Canadian Water Management*. Canadian Institute of Resources Law, Calgary.

6. Resilience of river basin governance institutions in the Saskatchewan River Basin of Western Canada

Ted Horbulyk

6.1 INTRODUCTION

River basin governance institutions can be assessed according to their current effectiveness and their resilience to adapt to new challenges and opportunities. In the Canadian case examined here, a multi-government network of agencies addresses their ongoing responsibilities with relatively few significant problems. Careful examination of the larger governance structure raises concern about a lack of resilience in the face of impending challenges.

This chapter describes the model of governance that operates within the Saskatchewan River Basin and introduces three water policy challenges. This basin is mainly situated in three Western Canadian provinces: Alberta, Saskatchewan and Manitoba.[1] These provinces share responsibility with the federal government for specific aspects of water management. There are roles and responsibilities for aboriginal populations, regional and municipal governments, watershed management organizations, non-governmental organizations and some role for broader public participation.

The hydroclimatic features of the basin are described in the next section, followed by the current water governance model. Then the potential effectiveness and resilience of governance institutions are assessed in relation to three selected water management challenges. The first is year-to-year water allocation, including the regulation of transboundary flows. The second concerns each jurisdiction's response to common risks and variability, such as multiyear droughts. The third challenge concerns water quality and Lake Winnipeg, where nutrient loading is a serious ecological problem.

6.2 AN OVERVIEW OF THE SASKATCHEWAN RIVER BASIN

The Saskatchewan River Basin occupies more than 400,000 km^2; an area larger than the UK or Germany. The basin drains the eastern slopes of the Rocky Mountains across the prairie region to Lake Winnipeg, after which the waters eventually reach the North Atlantic and Arctic oceans. Elevations range from over 3000 m above sea level at the continental divide to about 215 m at Lake Winnipeg (Partners for the Saskatchewan River Basin, 2009).

Annual precipitation ranges from over 1.5 m in the mountains to about 30 cm in the rain shadow of the Rockies. Large areas of the prairie experience average precipitation of 40–50 cm/year that supports extensive non-irrigated agriculture, yet irrigated agriculture is the largest consumptive use of water in the semi-arid, South Saskatchewan River sub-basin. Runoff from the mountains makes up more than 90 per cent of the flow volumes in the North and South Saskatchewan rivers. Numerous reservoirs and control structures provide water storage and reduce flooding. Away from the mountains, spring rains and snowmelt are the largest sources of runoff to rivers, yet in many areas, annual evaporation exceeds precipitation. About 44 per cent of the basin's entire area does not contribute to surface runoff in an average year (Partners, 2009, pp. 12–13). A large portion of the basin has been water-stressed even without further effects of climate change. Water quantity constraints differentiate this basin from much of the rest of Canada, where concerns about water quality are more common.

The population of the basin is about 3 million (2006 data) with more than 80 per cent living in Alberta. The cities of Calgary and Edmonton (in Alberta) represent over two-thirds of the basin's population, which is more than 85 per cent urban. All of the provinces report growing urban populations.

The most important human use of water is for irrigated agriculture, followed by municipal uses, industrial use, petroleum extraction and livestock water uses. The largest municipal users are near the head of the basin and, following tertiary or quaternary sewage treatment, they return most of their water withdrawals to the rivers near the points of extraction. Consumptive uses are reducing the basin-wide flow by 10–20 per cent relative to the estimated natural flow levels. Groundwater usage accounts for less than 3 per cent of consumptive use, and most of this is for rural uses (Partners, 2009, pp. 20–22). An emerging and controversial use of water is for hydraulic fracturing to enhance the extraction of oil and gas. Important non-consumptive water uses include hydroelectricity generation, the sport and commercial fisheries, recreation and environmental services.

Numerous threats to water security in the basin could affect water users. Land use changes, such as conversion of wetlands and vegetation, arise from extensive agricultural production, forestry, and the continuing expansion of urban and periurban development. There are perceived risks of droughts and floods associated with the current and historical climate regime, and some pressures for additional investment in water control infrastructure. There are problems with deteriorating water quality that are accentuated in periods of low or high surface flow. There are concerns about endocrine-disrupting substances that are not removed by current water treatment processes. Introduced and invasive species of plants, animals, fish and microorganisms could have significant effects on the economy and ecology.

Climate change is expected to have a direct effect on basin hydrology, as well as to hasten the changes associated with many of these other threats. According to Byrne et al. (2010, p. 69) climate change will result in lower snow accumulations in the mountains, and in a shift in the timing of precipitation and streamflow runoff from the late spring and early summer to earlier in the spring. Rainfall is expected to become more intense but less frequent. Summers will be warmer and longer, increasing the demand for irrigation withdrawals.

6.3 THE MODEL OF WATER GOVERNANCE

Canada's provinces are assigned the major roles relating to water use and management under the Canadian Constitution. Importantly, the Constitution also assigns both specific and general powers to the federal government and protects some rights that have been allocated to First Nations and Métis people.[2] The provinces have delegated certain provincial powers and responsibilities to municipal and regional governments and to watershed planning groups and advisory councils.

In a mode of operations referred to as cooperative federalism, the federal and provincial levels of government coordinate many of their actions related to water management through numerous federal–provincial agreements on water management and cost sharing. A number of authors have identified a high degree, nation-wide, of jurisdictional fragmentation with respect to water. They suggest there is high potential value to water governance from greater inter-governmental harmonization (Bakker and Cook, 2011).

The following description of water governance will proceed from the highest level of government (federal) to the lowest (local), then address the federal-provincial and inter-provincial agreements. The Constitutional powers of each level of government and their influence on federal and provincial approaches to water policy are described in a separate chapter (see Chapter 5).

6.3.1 The Federal Government Role

In practice, the federal government has an *independent* role in water governance in the basin mainly through its work with fisheries protection, through its funding of agricultural water supply and drainage, and through meeting the water and environmental management needs for federal lands and First Nations reserves. The federal government has a *collaborative* role in water governance through its participation in federal–provincial agreements.

Historically, the federal government has not had a single water ministry or interministerial agency to coordinate federal actions with respect to water. The Environment Ministry takes the lead role on issues of surface water, whereas the Natural Resources Ministry focuses on groundwater. Three other ministries with important roles are Agriculture and Agri-Food Canada, Fisheries and Oceans Canada and Health Canada.

There are numerous explanations for the observed reluctance of successive federal governments in Canada to play a larger role in water governance, especially when compared with other federal states. Gibbins (2012) argues that, given the geographic scale of the country, problems of water supply and water security often affect only specific locations—unlike the pan-Australian drought of the 2000s—and are viewed as local or regional problems. Historically, water issues have not been of high economic or social concern for Canadians. For the most part, Canadians have enjoyed an abundance of water, albeit with diverse concerns about water quality. Water issues do not have strong international trade implications that invoke a federal response. Bakker and Cook (2011, p. 277) argue that a larger federal role could conflict with provinces' fiscal interests in resource extraction. The expected costs of federal action are too concentrated relative to diffuse potential benefits.

6.3.2 First Nations and Métis Roles

Since the Constitution Act of 1982 contained a new clause that recognizes the original and treaty rights of the aboriginal peoples of Canada, including those pertaining to the use or ownership of lands and waters, aboriginal peoples now claim a role in water management decisions. According to Walkem (2007, p. 306) indigenous peoples' understanding of their rights is much larger than that currently recognized by Canadian law and does not depend upon Canadian legal or political recognition. It is unclear how such competing views might be reconciled or resolved.

Aboriginal peoples' rights to water and other resources have not been well defined in Canadian legislation or jurisprudence. Many issues remain disputed or are the subject of larger mediation processes, such as land claims. The courts have found that the protections described in the Constitution Act give governments 'a duty to consult' aboriginals prior to taking any actions that might affect their rights.

All aboriginal peoples make up about 6 per cent of the total population of Alberta and about 15 per cent of Manitoba and Saskatchewan, with more than 70 per cent living off reserves. About 40 per cent of the total aboriginal population in these provinces is Métis, and less than 1 per cent is Inuit. There are 181 Indian bands in the three provinces and nine Métis settlements; their reserve and settlement lands cover about 0.1 per cent of the provinces' area (Statistics Canada, 2008).

In practice, First Nations and Métis peoples are not playing a significant role in water governance at the basin scale in the Saskatchewan River Basin. Many First Nations manage water use at the community level, similar to local governments, and some groups are important users of water for irrigation.

Each of the provinces has invited First Nations' participation in advisory councils, but their role is limited given its advisory nature. Some First Nations groups are reluctant to participate in informal local processes. Their concern is that some government agency, having encouraged local First Nations' participation in these informal processes, might declare that its government's more formal obligations to consult or to accommodate have thereby been satisfied. These government agencies might then take a freer hand in subsequent policy decisions that could adversely affect First Nations.

6.3.3 The Provincial Government Roles

The Constitution Act assigns to the provinces both the ownership of the resource and the authority to legislate concerning water. Even though the provinces of Alberta, Saskatchewan and Manitoba did not exist when these provincial powers were first set out in 1867, each province was later granted all the same rights relating to water (and other) resources as the other provinces. Since the federal government had, in the intervening years, granted specific rights to others to use the water in this basin and had signed other related treaties and agreements, the terms of this resource transfer bound the provinces to uphold those obligations.

Alberta and Saskatchewan were created as provinces in 1905 and

Manitoba's boundaries were extended substantially in 1912. Even so, the federal government retained full control of natural resources in these provinces until 1930. After 1930, the federal government played an important role in water management, notably through the Prairie Farm Rehabilitation Administration that was established in 1935 to finance projects to conserve water and alleviate drought (Percy, 2012, p. 7).

The provinces are now the owners of the groundwater and surface water resources (except for those originating on or passing through federal and First Nations lands). The provinces regulate (and sometimes operate) water control infrastructure (including irrigation and drainage), fisheries, flood protection, the protection of source waters and drinking water supplies, and water quality. They establish the terms of water use through water rights, permits, enforcement and pricing.

Water permits and licensing in these provinces borrow heavily, with some significant legal differences, from similar systems of appropriative rights in the western United States. These provinces have granted some limited riparian water rights to those lands adjoining a waterway. In addition, provinces have granted general rights to use specified volumes of water for domestic uses and for livestock watering, independent of seniority or priority. However, most of the water use is allowed through a system of seniority-based water allocations that specify which volumes of surface water or groundwater can be used for which purposes on a given landholding. These licenses were granted from as early as 1894. In periods of water scarcity or shortage, each more senior licensee is allowed to take up to the full license entitlement (without rationing or reduction) so that any imposed restriction is passed to those whose historical dates of license are more recent. In times of shortage there is no explicit provision for proportional sharing, or for prioritizing water allocations according to their current values or to the types of water use.

In the South Saskatchewan River sub-basin of Alberta, provincial legislation has empowered 13 irrigation districts. Many of these institutions hold large and senior water licenses. All of these districts exercise their authority to provide water to irrigators or others. This water is stored and conveyed through works and infrastructure that is managed by each district, which may collect water fees from individual irrigators. Although their client base historically was irrigators, there has been an increase in some rural residential populations and rural industries, such as water-intensive food processing. As a result, some of these districts are now faced with roles and challenges similar to those of municipal water purveyors.

The Alberta government's lead agency on water resource management is the Ministry of Environment and Sustainable Resource Development. This ministry is currently coordinating a province-wide land-use planning program that follows an earlier program of water and watershed management planning.

As part of Alberta's vision for water management (Alberta Environment, 2003), the government has created an Alberta Water Council with representatives from government and from aboriginal, industry and environmental groups. This council oversees a province-wide watershed planning process that is undertaken by a series of 11 watershed planning and advisory (WPA) councils. These councils are supported in some of their work by government staff, but are arm's-length, community-based groups. These WPA councils include members from local governments, major water users (irrigators, industry), environmentalists and the public at large. Among other obligations, these councils prepare 'state of the basin' reports and integrated watershed management plans, which they implement and monitor. This work involves diverse watershed stewardship groups who operate at the level of local watersheds, as well as involving the public directly.

In Saskatchewan, the government's lead agency on water issues is the Saskatchewan Watershed Authority. A framework for watershed management established in 2002 sets out a standardized process for developing and implementing watershed management plans with public participation (Saskatchewan Watershed Authority, 2002). There is a province-wide Saskatchewan Watershed Authority Advisory Committee and, at the sub-basin level, watershed advisory (WA) committees have memberships similar to the WPA councils in Alberta. These Saskatchewan WA committees work with technical staff from government and elsewhere to develop action plans that these WA committees will then implement with support from the Saskatchewan Watershed Authority. In some cases, the plans developed by three or four WA committees working at sub-watershed levels have been amalgamated into a single, larger plan for implementation purposes.

In Manitoba, the Department of Conservation and Water Stewardship plays the lead role in water resources planning and management. Manitoba has a Manitoba Water Council that oversees the process of integrated watershed management planning (Manitoba, 2003). At the watershed level, water planning authorities with diverse membership prepare 'state of the watershed' reports and source water assessments leading to finalized watershed plans. Manitoba has had a system of conservation district authorities that predates these water planning authorities, and the planning process integrates the expertise and resources of the established conservation districts along with broader public input.

For the entire Saskatchewan River Basin spanning all three provinces, there is no apparent way that these provincial planning processes aggregate or integrate priorities, either within each province or across the provinces. In Alberta and Saskatchewan, the South and North Saskatchewan rivers' sub-basins are addressed by separate plans. In Manitoba, much of the Saskatchewan River Basin above Lake Winnipeg falls into one planning area, but no plan provides a comprehensive basin-scale perspective.

6.3.4 The Local Government Role

In Canada, cities have no independent jurisdiction over water issues and are able to operate only within those limits prescribed for them by each province. Municipal and regional governments in this basin attend to such concerns as water supply, sanitation, management of urban runoff, and programs related to water conservation and awareness.

Some cities are large water rights holders and high-volume water users. Other cities with growing populations are struggling to secure rights to water volumes that will be sufficient to meet future needs without having to cap the cities' commercial development and populations. Each city's choices with respect to managing the riparian zone, limiting the spatial extent of growth, and treating wastewater effluent to a high standard can have a significant and cumulative effect on downstream communities.

6.4 FEDERAL–PROVINCIAL AND INTER-PROVINCIAL AGREEMENTS

Saunders (1988, p. 37) describes Canada's situation as one in which the federal and provincial governments have interwoven Constitutional responsibilities for water resources, with evident ambiguity about the allocation of some governmental powers. Historically, the federal government has been unwilling to legislate unilaterally or to submit matters to the courts for adjudication. The preferred approach has been one of constructing federal-provincial agreements that set out the respective sharing of responsibilities or funding. In this basin, the most important such agreement is the 1969 'Canada–Alberta–Saskatchewan–Manitoba Master Agreement on Apportionment.'[3]

This master agreement establishes an equitable basis for sharing water volumes on all of the eastward flowing rivers that cross inter-provincial boundaries among these three provinces. Each province is allowed to consume up to one-half of what would be the natural flow of water that arises there and of water that flows through from upstream, allowing the remainder to pass to the downstream province. There are other operational rules about minimum flows and about the equitable allocation of downstream flows within each year. The agreement establishes a board to oversee the agreement and to recommend on related matters of water management, and sets out a cost-sharing and dispute resolution mechanism. The agreement has been amended a number of times, as recently as 1999.[4]

Percy (2012) describes the political environment in which this master agreement was forged after the provinces gained responsibility for water resources in 1930. Alberta and Saskatchewan enacted legislation in 1937 that would have permitted the creation of a bi-provincial board to regulate inter-provincial boundary flows, but this plan was never implemented. In 1946, all three provinces discussed the advisability of studying water resources basin-wide on a tri-lateral basis, but were discouraged from doing so on jurisdictional grounds by a federal official (Percy, 2012, p. 10). In 1948, the three provinces plus the federal government created an advisory body, known as the Prairie Provinces Water Board (PPWB), to make recommendations about future water allocations based on the criterion of 'best use'. When Alberta and Saskatchewan presented competing demands that would have exceeded available supply, the 'best use' doctrine was not relied upon, since it asked an advisory body to recommend one province's proposal over the others. Amicable resolution of these competing claims was encouraged by the federal government, which was funding many of these provinces' new investments in irrigation and hydropower as part of post-World War II reconstruction.

The principle of 'equitable allocation' of natural flows had been used as the basis for the Colorado River Compact in the US in 1929, and was studied by the PPWB as an alternative to 'best use'. As Percy notes, by the 1950s, the concept of equitable apportionment was well established in international water law. Some version of equitable allocation would likely have been imposed by the courts if any of these water allocation disputes would have been litigated by the provinces (Percy, 2012, p. 10). There were diverse motivations for all parties to join the Canada–Alberta–Saskatchewan–Manitoba Master Agreement on Apportionment. These included continuing access to federal water funding and fear that, without an effective allocation mechanism, the federal government might seek exclusive jurisdictional control of inter-provincial rivers, or the courts might impose their own remedies.

Although the master agreement has functioned well historically, legal scholars have questioned whether its terms could be upheld in the case of a serious dispute, especially if any party were to exit the agreement (Saunders, 1988; Percy, 2012). Percy points out that this agreement has lower legal status than either the Colorado River Compact or the Murray–Darling Basin Agreement and Basin Plan in Australia. The former is authorized under the Compact Clause of the US Constitution and the latter is embedded within the Water Amendment Act (Australia). By comparison, the Canada–Alberta–Saskatchewan–Manitoba Master Agreement on Apportionment is not part of any statute. Its force ultimately may rely on the continuing goodwill of the parties.

The provinces also collaborate through the Canadian Council of Ministers of the Environment. This group includes the federal government and the governments of all provinces and territories, but excludes explicit Aboriginal participation. The work of this council is consensus-based, since the council has no specific powers or authority. The council provides a forum for governments to collaborate on common issues such as revising guidelines for safe drinking water that each participating member government could adopt with or without further amendment.

The ten provinces and three territories also participate (without federal government participation) in the Council of the Federation. In 2010, this council of elected leaders adopted a Water Charter, and in 2011 they formed a Water Stewardship Council and a Water Partner Advisory Committee. These groups promote water stewardship and conservation, environmental monitoring and information sharing.

6.5 SELECTED WATER MANAGEMENT CHALLENGES

There are a large number of threats to water security in the Saskatchewan River Basin and three specific examples are addressed next.

6.5.1 The Effectiveness of Year-to-Year Water Allocation

The current system of water allocation in each province relies upon seniority-based systems that date from the 1800s, as further regulated by the 1969 Canada–Alberta–Saskatchewan–Manitoba Master Agreement on Apportionment. These systems do not have the flexibility to address developing water shortages easily, such as by restraining wasteful uses of water.

Consider specifically the case of the upstream province of Alberta. In the relatively water-short South Saskatchewan River sub-basin, the provincial government recently discontinued issuing new water licenses. Demands had reached levels that might only be met in years of average (or greater) precipitation. Issuing any new licenses was thought to place new licensees in a position of potential hardship and frustration.

In 2001–2002, there was a regional drought of such significance that holders of licenses issued as early as 1960 were at risk of being denied any water at all. Some of these vulnerable license holders were small towns with residents and businesses to supply. Others were vegetable processors whose ability to operate would be important to the irrigators who held the senior licenses. A number of senior licensees agreed to take reductions voluntarily that would share the burden of cuts more evenly, thus avoiding the need to invoke the seniority system of allocation. The specter of having to impose the formal rationing system in future reinforced the commitment not to issue new licenses.

If new users could not apply for new licenses, the historical pattern of water allocations to specific parcels of land would have been locked in place. Alberta adopted legislative amendments that allow various forms of short-term and long-term trading among licensees and newcomers within specified sub-basins. The early experience with this revised system has occurred during years of relatively average precipitation. There have been very few long-term trades of water to new users, and relatively few short-term trades.

Historically, large senior licensees had little profit motive to conserve water, since these users paid no volumetric fees, and since pumping costs were low. More recently, the ability to lease or to sell unused water allocations has provided clearer incentives to conserve water, such as by lining irrigation canals or changing irrigation methods. In one much publicized case, the water saved by constructing a closed culvert allowed an irrigation district to make a permanent transfer of a large water allocation to a new shopping mall. In another case, a city has proposed the sale of some of its treated sanitation return flows to an industrial user. One predictable effect of these changes is that the consumptive portion of the total water withdrawals will increase, and, in general, the volume of return flows that would be available to downstream jurisdictions will decrease.

These examples suggest that the current system of water allocation in upstream Alberta is putting pressure on the province's own users and reducing flows to downstream provinces. As an issue of basin-wide governance, there may be potential unrealized gains from addressing this issue collectively. Even for the upstream province of Alberta, one can envision enhanced schemes of spatial or inter-temporal water trading (water banking) that would help address variations in water availability relative to demand. Under the current master agreement, Albertan irrigators near the Saskatchewan frontier would only be allowed to trade with upstream (Albertan) neighbors, leaving unrealized any potential trades with others downstream in Saskatchewan or Manitoba. Current governance mechanisms do not provide a direct means for individuals or for provinces to operate alternative systems of inter-provincial water sharing that deviate from the fixed limits of the 1969 master agreement.

6.5.2 Multiyear Drought

Over its relatively short history since European settlement, the Saskatchewan River Basin has experienced a number of droughts severe enough to curtail agriculture. Even earlier evidence comes from examination of the annual growth patterns recorded in tree rings. There have been droughts that have lasted for four years, eight years, or more than a decade. A comparison of the timing of low-flow years at selected points within the entire basin shows that many low-flow events are local or regional in extent. There are only three or four years in the 90-year record where the low flows were experienced across the basin in the same year (Partners, 2009, p. 159).

Since long-term droughts have not occurred in recent times, there is concern that current systems of watershed governance would not be capable of addressing them. The two upstream provinces might not be able to meet their obligations as specified in the Canada–Alberta–Saskatchewan–Manitoba Master Agreement on Apportionment, or they might incur needlessly large costs to do so.

Water shortages can invoke other adjustments that might pit one region against another. For example, a bidding war for available livestock feed supplies could cause unexpected financial harm to other customers well beyond the regional drought zone. An alternate response to drought can be a simultaneous reduction of livestock herds. This can put financial harm on neighboring jurisdictions, first as their livestock sales prices fall, then later, in the rebuilding phases, when the price of suitable breeding stock might become prohibitively high.

These examples suggest that a multiyear drought event may be a crisis with basin-scale consequences. For the moment, there is no apparent water governance mechanism that is positioned to anticipate such an event, to establish response plans in advance, or to coordinate their implementation once such situations arise.

6.5.3 Water Quality at Lake Winnipeg

The Saskatchewan River Basin is one of five large river basins that discharge into Lake Winnipeg, where nutrient loading is an increasingly serious ecological problem.[5] The lake is the 10th or 11th largest in the world as measured by surface area, and serves as a catchment for five Canadian provinces and four US states. The lake hosts a significant commercial fishery. The lake's drainage basin is 40 times larger than the lake's surface area, making it an outlier among the world's large lakes. The lake is also relatively shallow which increases its vulnerability. Having a relatively large catchment area has made this lake susceptible to receiving a large load of nutrients, especially phosphorus and nitrogen, and the lake has become eutrophic with large blooms of blue-green algae. Recent increases in phosphorus loadings have been highest from flows of the Red River to Lake Winnipeg, and these loading are attributed to increased livestock production, fertilizer usage, and urban effluent, along with increased rates of overland flooding that washes phosphorus from farm fields.

The lake's inflow volumes have been moderated and its outflows controlled, holding back water during periods of high algal growth. The Saskatchewan River once deposited relatively large sediment loads, which have been reduced under its managed flow regime. As a result, water clarity has increased in Lake Winnipeg, allowing deeper subsurface penetration of sunlight and greater algal growth. The reduction of upstream and adjoining wetlands has reduced their role in trapping the dissolved nutrients. Historically, the water in the lake has a relatively short residence time, which suggests that cleaning up inflows could show a quick response in water quality. Unfortunately, the current situation is leading to the deposition of nutrients in lake sediments, and this could cause water quality problems to persist much longer (Partners, 2009, p. 143).

There is considerable data on the sources of the nutrients overloading the lake, where phosphorus is the greatest concern. More than 50 per cent of the phosphorus comes from outside Manitoba, with about 10 per cent from the Saskatchewan River. Fifty-eight per cent comes from the Red River and much of this originates in the United States (Partners, 2009, p. 144).

Among the lake's other risks and challenges are introduced species. The threat of introducing even more such species has fueled controversy over a recent US action to direct drainage from Devils Lake, in North Dakota, into the drainage basin of Lake Winnipeg. Some of these species could move upstream into the Saskatchewan River Basin.

The water quality problems of Lake Winnipeg are related, in part, to the cumulative effects of nutrient loadings from all of its contributing watersheds. A solution to the aggregate water quality problem might involve locating and exploiting opportunities to reduce those loadings that can be achieved most effectively and with lowest cost. The Lake Winnipeg Basin Initiative, and the federal–Manitoba agreement that supports it, provide a reasonably comprehensive program to address some of these problems, but only at the local and provincial levels. Potential solutions are likely to be both inter-provincial and international in scope. Once again, this appears to be a multi-jurisdictional, basin-scale problem, but current basin-scale water governance mechanisms do not seem suited to address it.

6.6 CONCLUSION

The Saskatchewan River Basin is a river system in a federation with present threats and risks to water security. Governance processes and mandates feature overlapping provincial and federal authority to act, with an observed reluctance of the federal government to assert its own authority. There is a well-articulated model of water governance within each province, but there is a limited ability to act across provinces. At present, the operational mechanisms to address inter-jurisdictional concerns are federal–provincial agreements, such as the Canada–Alberta–Saskatchewan–Manitoba Master Agreement on Apportionment. These agreements so far have not provided an effective means for the participating governments to address fully such challenges as apportionment of increasingly scarce water, to prepare for extended droughts, or to address the cumulative effects of nutrient loadings in Lake Winnipeg.

6.7 NOTES

1. This sequence is from west to east on a map. Small areas of the basin extend into the province of British Columbia in the west and into the US state of Montana in the southwest. Montana and British Columbia exert relatively small influence on the overall hydrology of the basin, and these jurisdictions are not active in basin-scale aspects of water governance.
2. Section 35 of the *Constitution Act* (Canada, 1982) recognizes and affirms the original and treaty rights of the aboriginal peoples of Canada, who are defined in that section to be the Indian, Inuit and Métis peoples of Canada. In popular usage, the term First Nations is used to describe aboriginal people defined as Indian in this act. In operational matters respecting the supply and use of water for aboriginal peoples, including issues of water resources on First Nations lands, the federal government takes a lead regulatory role in what would otherwise be provincial jurisdiction.
3. Another smaller agreement is the 'Canada–Manitoba Memorandum of Understanding Respecting Lake Winnipeg and the Lake Winnipeg Basin', signed in September 2010. The federal and provincial governments have had a series of other agreements addressing such matters as cost-sharing and responsibilities for flood control measures, and for hydrometrics in the Lake Winnipeg Basin.
4. The 1992 amendment added targets for about 10 water quality parameters to be measured regularly in 11 specified locations. This requires the province where any limit is exceeded to address the problem, but does not impose penalties if problems persist. This amendment also commits the parties to recommend upon issues of transboundary groundwater management (Percy, 2012, p. 19).
5. Two recent video productions illustrate the nature of the problems and solutions. See: *Choking Lake Winnipeg*, http://vimeo.com/18031235 and *Save My Lake* http://www.cbc.ca/documentaries/natureofthings/video.html?ID=1867857094.

6.8 REFERENCES

Alberta Environment (2003), *Water for Life: Alberta's Strategy for Sustainability*. Edmonton, AB, Canada: Alberta Environment.

Bakker, Karen and Christina Cook (2011), 'Water governance in Canada: innovation and fragmentation', *International Journal of Water Resources Development*, **27** (2): 275–289.

Byrne, James, Stefan Kienzle and David Sauchyn (2010), 'Prairies water and climate change', in Sauchyn, David, Harry P. Diaz and Suren N. Kulshreshtha (eds), *The New Normal: The Canadian Prairies in a Changing Climate*, Regina, SK, Canada: University of Regina, Canadian Plains Research Center, pp. 61–79.

Gibbins, Roger (2012), 'Finding policy traction in a wet world', presented to the conference on 'Water in a World of Seven Billion', Calgary, AB, Canada, May 9, 2012, available at http://ww7b.org/118id (accessed September 20, 2012).

Manitoba Conservation (2003), *The Manitoba Water Strategy*, Winnipeg, MB, Canada: Manitoba Conservation.

Partners for the Saskatchewan River Basin (2009), *From the Mountains to the Sea: The State of the Saskatchewan River Basin*, Regina, SK, Canada: Partners for the Saskatchewan River Basin.

Percy, David R. (2012), 'Resolving water-use conflicts: insights from the prairie experience for the Mackenzie River Basin', *C.D. Howe Institute Commentary*, No. 341, Toronto, ON, Canada: C.D. Howe Institute.

Saskatchewan Watershed Authority (2002), *Protecting Our Water: A Watershed and Aquifer Model for Saskatchewan*, Moose Jaw, SK, Canada: Saskatchewan Watershed Authority.

Saunders, J. Owen (1988), *Interjurisdictional Issues in Canadian Water Management*, Calgary, AB, Canada: University of Calgary, Canadian Institute of Resources Law.

Saunders, J. Owen (2013), 'Managing water in a federal state: The Canadian experience', Chapter 5, this volume.

Statistics Canada (2008), 'Aboriginal identity (8), area of residence (6), age groups (12) and sex (3) for the population of Canada, provinces and territories, 2006 Census – 20% sample data', *2006 Census of Population*, Statistics Canada Catalogue no. 97-558-XCB2006006, Ottawa, ON, Canada: Statistics Canada.

Walkem, Ardith (2007), 'The land is dry: indigenous peoples, water and environmental justice', in Karen Bakker (ed.), *Eau Canada: The Future of Canada's Water*, Vancouver, BC, Canada: UBC Press, pp. 303–319.

7. Water management and ecosystems: a new framework in Mexico

Eugenio Barrios

7.1 INTRODUCTION: EVOLUTION OF WATER MANAGEMENT IN MEXICO

The history of Mexico as an independent country begins 200 years ago, but it entered its gestation as a modern nation only 100 years ago with the Mexican Revolution of 1910. The political and social evolution of Mexico during the past century is the story of the configuration of a federation, which has shaped the pathway of water management. Water, perhaps like no other resource, is debated between the camps of decentralization and concentration of powers, and the constant demand for local solutions and societal participation.

Aboites et al. (2010) identifies three major phases in the development of water management in Mexico in the last 100 years: 'local water', 'national water' and 'commercial-environmental water'. This identification helps to understand at a glance the guiding policies of the resource management. With the birth of the nation local management prevails. Water united all of the local actors both in taking care of it and in the resolution of related conflicts: some colonial water partitions were maintained for up to three centuries. Later, as of the Revolution of 1910, the management of water enters into a phase completely centralized by the federal government that, according to the view of water as a leverage for development, is aimed at responding to the big projects required by the country: in a period of 80 years, the storage capacity of the country went from 10,000 to 142 billion cubic meters. Aboites et al. (2010) also observes that toward the end of the twentieth century, the centralized model of 'national water' enters into crisis. The overexploitation of aquifers, the great waste in primary activities, and the contamination were the primary evidence. The need for a transformation in management became imminent, essentially toward decentralization and participation, in which water starts to be presented as a commercial product but also as part of the environment in the water management principles. The experiences that are analyzed in this chapter are situated in the latter years of this transformation process.

7.2 THE FEDERAL WATER ADMINISTRATION SYSTEM IN MEXICO

The United Mexican States are a representative, democratic and federal republic with a presidential political regime in which the president is the chief of state and of government and is elected by direct vote for only one six-year period. Among his powers are the appointment of the cabinet and the promulgation of the laws passed by the legislative branch. The legislative and judicial branches complement the political system: the Congress of the Union which is made up of the chamber of deputies and senators, and the judicial power lies with the Supreme Court of Justice (SCJ), the electoral court, the multi-judge and single-judge circuit courts and the district courts.

The republic is made up of 31 constituent units or states united in a federation and the Federal District. Each state consists of municipalities, a third tier of constitutionalized government which amount to a total of 2,456 in the country. An additional and important fourth tier of government, also recognized by the constitution is the common land-ownership, covering half of the country surface area (INEGI, 2010; SRA, 2008).

In the configuration of the federation, the party regime is critical (Anderson, 2008), and in the case of Mexico it has been an important driven force. The three main political parties were founded as national parties and they operate national structures. The Partido Revolucionario Institucional (PRI) governed the entire country from the end of the revolution until 2002. In 1988, after a severely questioned presidential election, it started losing power; and in 2002 it lost the presidential elections and the majority of the state governors for the first time. This redistribution of the political forces translated into considerable changes. A good example is the fiscal transfers to states, approved by the Congress: at the end of the 1990s, transfers were around 20 per cent; currently, they have increased up to 40 percent (near USD82 billion) (Herrera, 2011).

Changes in the social, political and economic balances of power had a clear influence on the control of water resources. The management of water in the hands of a central power begins to see itself questioned by the new local and regional powers. At the same time, additional pressure started appearing from the society; in the 1990s an increase in the intensity of water conflicts was observed (Torregrosa et al., 2010).

The guiding principle of water policy in Mexico is Article 27 of the Political Constitution of the United Mexican States of 1917 (CPEUM) that established water as public good, property of the nation and patrimony of the Mexicans, but subject to concession for its use.

The major transformation of water management in the country began in 1989 with the creation of the National Water Commission (CONAGUA), and the publication of the National Water Law (LAN) in 1992, its regulations in 1994, and its modification in 2004.

The LAN is the highest legal framework on the subject of continental waters under the Constitution. It regulates the use, distribution and control of water and ratifies a regime of administrative coordination strongly centralized in the federal executive exercised through CONAGUA (Gutiérrez and Emanuelli, 2010). The LAN establishes among its main principles the deconcentration, the integrated management of water resources, the participation of private sector and the involvement of social actors.

In the CPEUM and the LAN, the role of the state governments is defined for the waters of state jurisdiction, which in practice has no relevance, and aspects of coordination with the federation are only enunciated in a very general manner. On the other hand, the role of the municipal governments became clear since 1983, with the Constitution itself establishing the water and sanitation services as their exclusive responsibility.

The legal framework for water is supplemented by additional laws, regulations, and norms related with the environment, health, and fiscal issues, among others.

7.3 CURRENT WATER MANAGEMENT SITUATION: FEDERALISM AND WATER SECURITY THREATS

7.3.1 Water Governance

Water management in Mexico follows a multiple governance structure aiming to accommodate integrated water management principles and the country federalism. At the top, a central horizontal governance layer is established in Article 10 of the LAN, by a Technical Council chaired by the Ministry of Environment (Semarnat), and with the representation of the Ministries of Treasury; Social Development, Agriculture, Livestock and Rural Development; the Mexican Institute of Water Technology; and the National Forestry Commission. A vertical structure is defined by the interaction of the three levels of government, and a bottom layer is the space of stakeholders at the river basins, through the establishment of mixed spaces of coordination and consensus such as the river basin councils and auxiliary bodies.

In 2004, the new reform of the LAN strengthened the deconcentration of management creating 13 river basins agencies and increased the social participation in the river basin councils by opening spaces to civil society. At the same time, state governments started developing their own water laws, and establishing their own state water commissions and water technical and research entities. The municipal governments also strengthen their water supply and sanitation structures. These parallel processes created tensions among the governments and the new CONAGUA river basins agencies that have not been resolved yet (Torregrosa et al., 2010). In one hand, new federal arrangements with more active interaction among the federal and states governments, and in the other a more participative water management structure based on river basin organizations and the paradigm of integrated water resource management, were decided.

7.3.2 Water and Land Management Integration

According to the LAN, integrated water resource management includes the water, the land and related resources and the environment. Nevertheless, the ownership of water use rights in Mexico, inspired by the civil law doctrine, is not connected to land ownership and in the area of integrated management has had little influence on its integration. Currently, the lack of coordination between authorizations in the change of land use, a municipal matter, and extractions of groundwater, a federal matter, is one of the primary causes of the advance of the agricultural frontier, the overexploitation of aquifers, and the occupation of river beds and flood plains by people and urban areas. An example is the case of the upper Rio San Pedro Mezquital River where hydraulic capacity of flood plains have been dramatically reduced to 10 m^3/s when the legal criteria based on a five-year extreme event is 357 m^3/s (WWF, 2012).

These problems are among the most important threats to water systems from climate variability. They increase vulnerability by rising water demand from agriculture, decreasing natural water storage capacity, and severely reducing hydraulic capacity to manage extreme floods. Improving coordination among the three levels of government on these issues is an opportunity to reduce risk and build a more effective water management system.

7.3.3 Water Allocation Policy

One of the most relevant priorities, if not the most, is the allocation of water use rights. Under the current policy, the overexploitation of water sources is alarming. CONAGUA acknowledges an overexploitation of 6.5 km^3 annually, 100 overexploited aquifers, and consumption above 40 per cent of the annual availability in eight of the 13 hydrological administrative regions into which the country is divided, and where more than 75 per cent of the GDP is produced (CONAGUA, 2011, 2010). To have arrived at this situation is not by chance; it was a decision clearly made at the time, to make an exception of the sustainable extraction principle stated in the LAN, one of the key responsibilities of the federal water authority.

With the publication of the LAN in 1992, a commitment was made to regularize all of the users, estimated at 300,000, in a period of three years. This task encountered numerous obstacles typical of a highly centralized water administration, the most important the lack of information to establish reliable water balances. Around 1995, the fact that all of the applicants were granted permits without having carried out water balance studies was considered to be an 'ecological cost' that had to be paid to register all existing users for the purpose of laying the foundation for the sustainable development and management of water resources. The integration of the user registry was finally completed in 2003 after establishing a series of incentives for the users, but reducing the validity of the rights from 50 to 10 years, hoping with this period to have better information on the availability of water at the time of renewal of the rights by the users (Garduño H., 2003). Thus, by 2003, the scenario of overexploitation was already in place, turning out to have more 'paper water' than real water.

The emergence of new state powers and pressure from society represented a challenge to the federal management of water which has been reflected above all in the questioning of the officially recognized availability of water and in the demand for greater volumes of water. For Torregrosa et al. (2010), it has become clear that a political dispute for controlling water is happening as a power space, in which the main focus is on water policy allocation, rather than the lack of water in itself.

As an answer to this situation, CONAGUA focused on establishing more transparent procedures to determine water balances and by 2011 officially publish the availability of water in all of the 738 river basins and 653 aquifers in the country for the purpose of providing certainty on the quantities of water to which the users would have access and to avoid the discretion with which they had been operating, typical of the period of 'national water'. However, the environment has not been considered in the determination of availability requirements, as expressed in the LAN, alleging the lack of a reliable methodology that would risk the credibility of the process.

As a result, this policy of 'openness' in the allocation of water awards concessions up to the limit of 100 per cent of the mean annual runoff, representing a situation of elevated risk for the current and future water security of the country, since many of the years this availability will not be achieved. Thus, CONAGUA as a federal agency has lost power to enforce a sustainable water extraction.

This policy has modeled a water deficit management model in constant conflict that year after year must resolve the allocation of available water, in a power structure primarily controlled by the dominant groups in each region and with no space for adapting to future changes.

7.4 NEW MODELS OF WATER MANAGEMENT IN MEXICO

In 2004, the World Wide Fund for Nature (WWF) established an alliance with the Fundación Gonzalo Río Arronte (FGRA) to develop a project designated as Development of New Models of Water Management in Mexico. The project considers the primary question to be how to ensure water for the environment; in other words how to determine an environmental flow (eflow) with ecological significance, capable of being legally established and socially recognized for its relation with the conditions of equity, but also with the presence and development of infrastructure.

It is worth to mentioning, that the appearance of non-profit NGOs and private foundations, in the management of water turned out to be innovative, and has been relevant in the last years. Due to its independency on human and financial resources, they have been able to play a coordinating role within the multi-governance water management structure, and thus fill some of the gaps of the federal system. From 2000 to 2010, the FGRA invested USD65 million supporting 83 water projects around Mexico (FGRA, 2011); and the Alliance WWF-FGRA has carried out investments of USD14 million dollars to date, that have leveraged near USD30 million from federal and state government programs due to this coordination capacity (WWF, 2011).

For the development of the project, three pilots or model river basins were selected in three different regions of the country, with different problems and contexts, and good examples of the evolution phases of the water management in Mexico. Figure 7.1 shows the location of the river basins and Table 7.1 shows their main characteristics.

The allocation of water for the environment involves dealing directly with the process of water rights, and therefore with the water security conditions on each basin. Those related with the variability of the natural system and the unexpected changes, the manner in which the water users

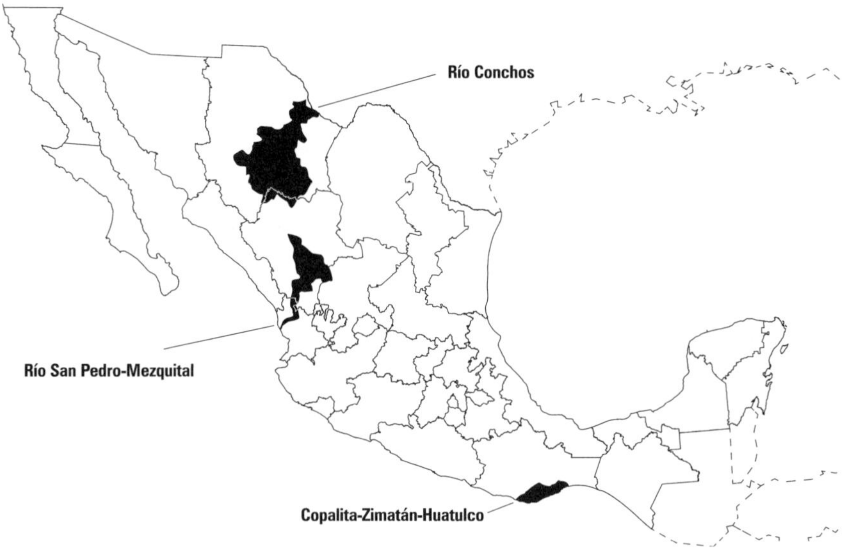

Figure 7.1 Location of the pilot river basins

relate to the authorities within a federal water management system, the current capacities of all involved parties to respond effectively, and the opportunities to reduce water risk. The situation observed in these river basins is presented below.

7.4.1 The Rio Conchos River Basin

The Rio Conchos (RC) is a key river basin for the security of the state of Chihuahua, where 47 per cent of the GDP is produced, and a sub-basin of the Rio Bravo–Rio Grande, where three states in Mexico and two in US depend on its runoff. In the past 20 years, the RC has suffered the 'death' of stationarity (Milly et al., 2008). An important drought from 1993 to 2005 reduced river inflows to 25 per cent of the historical mean for the last 60 years and revealed the limited capacity of the basin to deal with variability (Barrios et al., 2009). In 2008, extreme peak flows causes severe floods that showed the limited capacity of the watershed and the infrastructure; and in 2012 drought returned.

Over-allocation of nearly 150 per cent of mean annual water availability makes no space for any effective response to variability and is the main threat to water security of the Rio Bravo users. Over-allocation is related to the process of appropriation explained in the previous section; while

Table 7.1 Main characteristics of the pilot river basins

	Rio Conchos	Copalita-Zimatán-Huatulco	San Pedro Mezquital
Surface area (10^3 Ha)	7,000	370	2,648
Mean annual precipitation (mm)	400	1,500	643
Mean Annual Runoff (MAR) (hm^3)	6,205	1,409	3,417
Official surface water availability (hm^3)	(−467)	1,409	2,839
Water Allocation (hm^3)	3,535	4	433
Water storage capacity (hm^3)	3,528	0	339
Water rights administrative situation	Over exploitation	Surplus	Surface water extraction ban since 1955
Main water management characteristic as proposed by Aboites (2010)	Commercial-environmental water	Local water	National water
Population	1.3 million	85,000	756,249
MAR/cap (m^3/cap-yr)	4,775	15,665	4,870
Water allocation for agriculture	89%	–	58%
Irrigated surface area (ha)	245,000	–	18,179
Number of States in the main river basin	6 states in the Rio Bravo/ Grande basin	1	3
Number of Municipalities	40	20	26
% of the total surface area in the basin under common land ownership	39%	72%	74%
Estimated GDP (USD million)	13,230	Not Determined	846

over extraction is mainly caused by illegal pumping promoted by a subsidy in the power tariff for agriculture; a clear example of lack of congruency in the federal policies of water and energy.

Water management in the RC is fully dominated by the agricultural sector of the state of Chihuahua, which owns 89 per cent of the water rights in the basin, and which has rejected the recovery of an eflow to avoid any interference from actors in the management which it considers to be its own. It can be said that water in the RC basin is exclusively a productive input of agriculture, thus water security has no meaning but the impact on production of a small group of farmers: half of the irrigated land and 31 percent of total water is used to produced alfalfa, a cash-crop that needs only five workers for 100 hectares and generates four times less revenue than other crops produced in the basin (Trueba and Goicoechea, 2009). As a productive input, water in the RC has a price (USD1–2 per thousand cubic meters) in an uncontrolled market that monopolizes rights and exerts control over availability.

In contrast, for the federal government water in the RC is a national security issue. It represents one-third of the international waters of the Rio Bravo, and therefore the compliance with the 1944 Mexico–US water treaty. It is clear that due to the tensions that represent the RC and the binational conflicts that arose in the last drought, water governance operates by a de facto agreement in which the farmers and behind them the state of Chihuahua are the beneficiaries from all the available water and the investments to improve their water efficiency, providing they meet with the water quota to comply with the water deliveries under the binational treaty (Barrios et al., 2009).

The allocation of water for the environment proposal for the RC made it possible to understand that it is feasible to determine an eflow from a balance between the water uses and the ecological importance of each sub-basin; that it is, in large part, a matter of infrastructure operation and integrated management of surface and groundwater; that the creation of a right by the concession of water to an individual is the primary obstacle for its implementation; and finally, that the eflows are achieved by the contribution of all of the users, and therefore can be considered a common management objective and thus an element of transparency and accountability.

7.4.2 The Copalita–Zimatán–Huatulco river basins

It is a river basin with a very diffuse population (85,000 inhabitants in 854 towns), where only 10 percent of water availability is officially allocated, but where local communities, the fourth level of government, in 15 of the 20 municipalities of the basin, assume the legitimate common property of water and land. The basin shows acute periods of drought annually and an important expected growth planned to reach 16,000 hotel rooms by 2025, that is, 800 percent of current tourism capacity. Currently, local water consumption average 163 liters per capita, compared with a tourist consumption of up to 878 liters per capita. Tourism development is a federal project in the coastal zone in which the involvement of local communities have been severely questioned (Agroder, 2009, 2012).

The implementation of eflow in the Copalita–Zimatán–Huatulco (CZH) River Basin has focused on two areas: the recognition by the communities and municipalities, under their traditional community agreements; and, its recognition as federal water rights. In this case, in addition to awarding legal protection to the allocation of water by community agreements, it carries out the function of social allocation of water, a function that the federal administrative system has not been able to resolve. The LAN establishes that the towns of less than 2500 inhabitants can extract water freely, which in many cases creates a situation of defenselessness, since upon not requiring allocation, the volumes are not recorded in the Public Registry of Water Rights (known in Spanish as the Registro Público de Derechos de Agua) and therefore are available in the event that a bigger user may apply for them. In this case, it is evident that the mutual recognition of the federal and indigenous community right strengthens both types of management: the official federal and that of 'local water'. Currently, agreements for the vigilance and monitoring of the eflow have been discussed and recognized, and have begun to be assumed as a common goal for everyone.

7.4.3 The San Pedro Mezquital River Basin

The San Pedro Mezquital (SPM) River Basin turned out to be very important for the purpose of the project. Surface water extraction has been indefinitely forbidden since 1955 without any justification other than: 'the conservation of the prevailing hydraulic conditions to continue the studies . . . to achieve the maximum exploitation of the country's hydraulic resources' (SRH, 1955). It is a legacy of the period of 'national water' in its most centralized version. Actually, CONAGUA has kept this type of bans in 191 river basins, 175 of which are a dating back to 1920–1950 (CONAGUA, 2011a).

Currently, the SPM keeps 83 percent of its availability of surface water (2839 Hm^3). Its middle and lower sections maintain their natural hydrological regime, and it is the only river that currently crosses the Sierra Madre Occidental mountain range without dams, maintaining the connectivity of the central part of Mexico with the sea. In contrast, the upper section of the river basin, where near 80 percent of the population is situated, has been seen to be affected by this prohibition. The ban caused groundwater overexploitation currently estimated around 95 Hm^3 per year, and the subsequent increase in arsenic and fluoride concentrations and therefore the restriction as a water supply source of Durango, the most important city in the basin (SEMARNAT, 2009; WWF, 2011a).

In the lower part, the SPM basin is the main freshwater contribution to Marismas Nacionales, a coastal wetland of around 200,000 hectares protected as a Ramsar Convention wetland site since 1995 and as a biosphere reserve since 2010. The natural regime that has been preserved as a legacy of the period of 'national water' represents the sustainability of the fisheries, tourism and subsistence activities of 90,000 inhabitants, and which productivity has been estimated at USD100 million annually (Agroder, 2010).

Its condition of prohibition offered a very advantageous situation for the adoption of eflow. The complaints of the state governments and big users of the river basin to lift the ban offered the opportunity to consider a partial lifting of the prohibition, which establishes an allocation for the environment prior to initiating the procedure for the allocation of concessions.

Once a sustainable water allocation is set up, water security will be complemented by specific projects: in the upper part of the basin, a project call 'Future Water' is focusing to reduce aquifer over extraction and flood risks, and ensure potable water supply; in the lower part, a 480 MW hydropower project is being proposed that will affect natural discharge to Marismas Nacionales.

Finally, CONAGUA has decided to partially lift the ban on the river basin and decree a reserve of water for the environment, the first of its kind in the country. Water reserves are an instrument that enables eflow implementation, recognized by the LAN (article 44 and 86 bis) and are defined as: 'those specific areas of the aquifers, hydrologic basins, or hydrologic regions, in which limitations are established for the exploitation or use of a portion or the totality of the available waters, for the purpose of providing a public service, establishing a program of restoration, conservation or preservation or when the State decides to exploit such waters for the public good'.

The conditions for the allocation of water in the SPM has become a model in which the federation plays its role as administrator of the resource and establishes a sustainable limit to which the users and the development projects of the states adapt, while at the same time protecting the ecological functions of the hydrologic cycle that is a natural heritage of the country and represents the means of life of indigenous and rural communities of the river basin.

7.5 THE WATER RESERVES FOR THE ENVIRONMENT INITIATIVE

The systematization of these experiences allowed a proposal for a Mexican norm for the determination of eflows to be presented to CONAGUA, which has currently been approved. It will be applied at a national level and will allow for the allocation of water to the environment in the official water availability balances in river basins and aquifers, and thus will comply with the LAN and improve water security conditions.

Besides, drawing on the broad experiences in the realm of eflows, the Alliance WWF–FGRA and CONAGUA intend to amplify this approach, and they have conducted a scoping study that identifies potential water reserves throughout Mexico, defined as river basins with favorable conditions – high biological richness and high conservation values, availability of water and low water extraction – for ensuring allocation of water to the environment as stated under the LAN (CONAGUA, 2011a).

The study identified 189 basins where water reserves could be established and are designated to be the main target of the National Water Reserves Program. The goals of the program are to: establish a national system of water reserves; demonstrate that water reserves ensure the healthy functioning of the water cycle, as well as the environmental services they provide; and build capacity in the implementation of eflows backed by the Mexican norm throughout the country.

Nowadays, contrary to the concept of water reserves applied during the period of the 'National Water', water reserves acquired a relevant meaning as a water security tool. They are not a restrictive instrument rather they represent the implementation of a buffer capacity within the national water balances to deal with uncertainty, and thus reducing risk from water scarcity and conflicts. They imply the integrated planning and management of both groundwater and surface water; the preservation or controlled release of peak flows to prevent the interruption of river channels, invasion of river beds, and as a consequence, diminish the risk against extreme events.

In terms of ecological functions, water reserves keep connectivity of the entire basin, conserve ecosystems, and provide a buffer capacity for biodiversity adaptation. Water reserves reinforce the strategy for the conservation of the nation's most important ecosystems and their environmental benefits in 99 Natural Protected Areas, 55 Ramsar sites, and an additional 78,500 km^2 of river basins.

7.6 FINAL REMARKS

The development of the project on the three pilot river basins to allocate water for the environment has made it possible to closely understand the tensions and conflicts that water stakeholders face, most of them related to power distribution and responsibilities of a centralized federal water management system in transformation, and the coexistence of the historical models of water management. These conditions shape the different realities of water management in the country and the risks to which they are subject.

Mexico is facing a huge water risk under current circumstances of overexploitation of water sources and the disputes around water resource control. Water more than ever has become a crucial factor in the consolidation of the power spaces to control the resource (Torregrosa, 2010), rather than a factor to share power and build complementarity.

The evolution of the water allocation process presented in this chapter is a good example of the evolution of federalism and water management. Under the scope of this chapter, it can be observed that, although parallel processes, they have not had a clear congruency with each other. The advantages of having a federal system to strengthen water management have lost clarity, mainly because of the differences among regions, but also because power disputes have created confusion and mistrust among all.

Currently, CONAGUA has identified that the allocation of water to the environment as an opportunity to clearly legitimate its own role as a federal authority in charge of the natural wealth of Mexico, comply with the LAN, and adjust water balances to ensure more secure water management conditions. The initiative can be of great importance to counter the impacts of climate change.

However, it is clear that federalism will improve governance conditions and then water security only if, as stated by Watts (2002), it is understood as a pragmatic and prudential technique whose applicability in a particular situation may well depend upon the particular form in which it is adopted or adapted or even upon the development of new innovations in its application.

7.7 REFERENCES

Aboites, L., Birrichaga, D., and Garay, J. (2010), El Manejo de las Aguas Mexicanas en el Siglo XX [Management of Mexican Water in the XX Century], in Jiménez, B., Torregrosa, M., and Aboites, L. (eds) (2010), *El agua en México: cauces y encauces* [Water in Mexico: channels and guidances], México: Academia Mexicana de las Ciencias [Mexican Academy of Science], CONAGUA, Cap. 2. p. 21.

Agroder, S.C. (2009), *Propuesta de Diseño e Implementación de esquemas de incentivos compensatorios para la conservación de las cuencas Copalita, Zimatán y Huatulco Fase II* [Proposal for design and implementation of compensatory incentives schemes for Copalita, Zimatán and Huatulco watershed Conservation, Phase II], México: Alianza WWF-FGRA, Internal Report KZ89.

Agroder, S.C. (2010), Evaluación rápida del beneficio de las inundaciones en la cuenca baja del SPM [Quick assessment of the benefits of floods in the lower basin of SPM], México: Alianza WWF-FGRA, Internal Report.

Agroder, S.C. (2012), Huella Hídrica en México en el contexto de Norteamérica [The Water Footprint of Mexico in the Context of North America], Mexico: WWF, SAB Miller, p. 45.

Anderson, George (2008), *Federalism: An Introduction*, Canada: Oxford University Press, p. 88.

Barrios, E., Rodríguez, J.A., and De la Maza, M. (2009), 'Integrated river basin management in the Conchos River Basin, Mexico: a case study of freshwater climate change adaptation'. *Climate and Development* 1 (2009) 1–12.

Barrios, E., Rodríguez, J.A, González, I., Gómez, R., Reyes, J.A., Escobedo, H., Sánchez Navarro, R., and Salinas, S. (2010), *Tres propuestas de caudal ecológico en México* [Three proposals for ecological flow in Mexico], México: Alianza WWF – Fundación Gonzalo Río Arronte, p. 31.

CONAGUA (2010), *Estadísticas del Agua en México* [Water Statistics in Mexico], *edición 2010*, México: Secretaria de Medio Ambiente y Recursos Naturales [Ministry of Environment and Natural Resources], p. 249.

CONAGUA (2011), *Agenda del Agua 2030* [2030 Water Agenda], México: Secretaria de Medio Ambiente y Recursos Naturales [Ministry of Environment and Natural Resources], p. 66.

CONAGUA (2011a), *Identificación de Reservas Potenciales de Agua para el Medio Ambiente en México* [Identification of Potential Water Reserves for the Environment in Mexico], Mexico: Secretaria de Medio Ambiente y Recursos Naturales [Ministry of Environment and Natural Resources], Alianza WWF–FGRA, p. 85.

FGRA (2011), Fundación Gonzalo Rio Arronte, IAP Primera Década 2000–2010 [Gonzalo Rio Arronte Foundation, First Decade 2000–2010], p. 262.

Garduño, H. (2003), *Administración de derechos de agua. Experiencias, asuntos relevantes y lineamientos [Water rights administration. Experiences, relevant issues and guidelines]*, Roma: FAO, Estudio Legislativo [Legal Studies] 81, p. 38.

Gutiérrez Rivas, R., and Emanuelli Panico, M. (2010), 'Régimen jurídico del agua continental en México: un análisis crítico'['Legal framework of continental water in Mexico: a critical analysis'], in Jiménez, B., Torregrosa, M., and Aboites, L. (eds) (2010), *El agua en México: cauces y encauces* [Water in Mexico: channels and guidance], México: Academia Mexicana de las Ciencias [Mexican Academy of Science], CONAGUA, Chap. 23.

Herrera, Rene (2011), 'Operación de Recursos Federales Transferidos a Estados y Municipios 2011' ['Operation of Federal Resources Transferred to States and Municipalities 2011'], *Hacienda Municipal*, México: Indetec, Enero-Marzo 2011 (112), 18–31.

INEGI (2010), Censo de Población y Vivienda 2010 [Census of Population and Housing 2010], www.inegi.org.mx.

King, J.M, Tharme, R.E., and de Villeers, M.S. (eds) (2000), *Environmental Flow Assessments for Rivers: Manual for the Building Block Methodology*. South Africa: Freshwater Research Unit, University of Cape Town, Water Research Commission Report No. TT 131/00.

Milly, P.C.D., Betancourt, J., Falkenmark, M., Hirsch, R.M., Kundzewicz, Z.W., Lettenmaier, D., and Stouffer, R.J. (2008), 'Stationarity is dead: whither water management?' *Science* 1, February: 319 (5863), 573–574.

Secretaría de la Reforma Agraria [Ministry of Land Reform] (2008), *Programa Sectorial 2007–2012* [Sectorial Program 2007–2012], México: SRA.

Secretaría de Recursos Hidráulicos (SRH) [Ministry of Water Resources] (1955),

'Acuerdo que declara veda para el otorgamiento de concesiones para el aprovechamiento de aguas del río Mezquital o San Pedro o Tuxpan, en los Estados de Durango y Nayarit'['Agreement that declares a ban on granting concessions for the exploitation of the Mezquital River or San Pedro or Tuxpan, in the states of Durango and Nayarit'], *Diario Oficial de la Federación*, México: 8 February 1955.

SEMARNAT (2009), 'ACUERDO por el que se da a conocer la ubicación geográfica de 371 acuíferos del territorio nacional, sc actualiza la disponibilidad media anual de agua subterránea de 282 acuíferos, y se modifica, para su mejor precisión, la descripción geográfica de 202 acuíferos' ['Agreement that discloses the geographic location of 371 aquifers in the country, updates the average annual availability of 282 groundwater aquifers, and amends, to improve accuracy, the geographical description of 202 aquifers'], México: Diario Oficial de la Federación, 2° y 3° Sección.

Torregrosa, M.L., Paré, L., Kloster, K., and Vera, J. (2010), Administración del Agua, in Jiménez, B., Torregrosa, M., and Aboites, L. (eds) (2010), *El agua en México: cauces y* encauces [Water in Mexico: channels and guidances], México: Academia Mexicana de las Ciencias [Mexican Academy of Science], CONAGUA, Chap. 21. p. 599.

Trueba, V., and Goicoechea, J. (2009) *Escenarios del Agua 2015–2030 en la Cuenca del Río Conchos: Acciones para un Desarrollo Ambientalmente Sostenible [2015–2030 Water Scenarios in the Conchos River Basin: Actions for Environmentally Sustainable Development]*, Mexico: Alianza WWF–FGRA, 1ª revisión, p. 139.

Watts, L.R. (2002), 'Federalism Today. The Relevance Today of the Federal Idea', International Conference on Federalism, Saint Gallen, Switzerland.

WWF (2012), Evaluación Hidrológica-Hidráulica del Manejo del Agua en el Rio Tunal y de su potencial para optimizar sus efectos ambientales [Hydrologic-Hydraulic Assessment of Water Management in the Rio Tunal and its potential to optimize its environmental effects], Internal Report, Ingeniería y Gestión Hídrica S.C.

WWF (2011), *Manejo del Agua en Cuencas Hidrográficas: Desarrollo de Nuevos Modelos en México, Informe Final* [Water Management in River Basins: Development of New Models in Mexico, Final Reportl, México: WWF México, Internal Report.

WWF (2011a), Ficha Técnica San Pedro Mezquital [San Pedro Mezquital Technical Datasheet], México: WWF México, http://wwf.org.mx/sanpedromezquital/wp-content/uploads/FichaTecnica_SanPedroMezquital.pdf.

8. Main challenges and responses to federalism and water security in Brazil

Ana Carolina Coelho, Benedito Braga, Rodrigo Flecha, Nelson Freitas and Osman F. da Silva

8.1 INTRODUCTION

Water resources are among Brazil's most valuable natural resources. Indeed, water resources management was given more attention with the establishment of the National Water Resources Management System in 1997. In fact, as a federative republic with contrasting river basin systems, integrated water resources management in Brazil depends on federal states and national governments to adopt cooperative approaches, according to regional particularities. Water security risks require that cooperation and integration strategies be even more emphasized. In this sense, this chapter discusses some alternatives being applied in Brazil – in general – and at the Doce River Basin – more specifically – as mechanisms to address integrated water resources management challenges and minimize water security risks.

8.2 WATER RESOURCES MANAGING IN BRAZIL

8.2.1 Brazil and its Water Resources

Brazil is a presidentialist federative republic, constituted by the union, 26 states, 1 federal district and 5561 municipalities, each one with its own government. Although Brazil is considered to be rich in water availability, having 12 percent of the world's fresh water availability, this availability significantly differs from one region to another. Furthermore, there are important seasonal variations, and problems of water supply for different uses. These problems result, mainly, from high demands in relation to the availability and poor water quality due to pollution (ANA, 2007). Regarding water use distribution among sectors, the irrigation represents 69 percent of the total water demand (ANA, 2007). The remaining 31 percent is divided among urban supply (11 percent), animal needs (11 percent), industrial uses (7 percent) and rural supply (2 percent).

Water resources authority, according to the 1988 Federal Constitution, is divided between the union and the states. Also, 'lakes, rivers and any other streams in its lands of domain, or which flow through more than one state, or are boundaries with other countries'[1] belong to the union. Other watercourses and underground water are under state jurisdiction. As a result, multiple levels of water resources management frameworks exist in Brazil. Figure 8.1 illustrates the 12 national hydrographic regions used for macro analysis of water resources in Brazil, in dark lines, as well as state water resources planning and management regions, in light lines. The boundaries of the national regions, as expected in federations, are very different from the geo-political boundaries of the Brazilian states. This has important implications over the water resources management model adopted in Brazil.

In fact, water resources management in Brazil is compared to the management of international transboundary river basins. Therefore, all federative levels need to work cooperatively in order to overcome conflictual intergovernmental relations and achieve integrated water resources management (IWRM).

8.2.2 The National Water Resources Management System

Water resources management legislation in Brazil was introduced by the 1934 Water Code. Over more than 60 years of its application, water resources management was focused on the development of hydraulic infrastructure for hydroelectric power generation. During the 1970s, additional water users begun to demand a different set of rules based on the principle of multiple uses guarantee. The 1988 Federal Constitution recommended the creation of a national water resources management system, which was regulated in 1997. The National Water Resources Policy of Brazil (PNRH), established by the Water Law in January 1997, aims to ensure adequate water availability for present and future generations, as well as to prevent and protect against critical hydrological events, through rational and integrated water resources management. Its main principles and defining characteristics are summarized below:

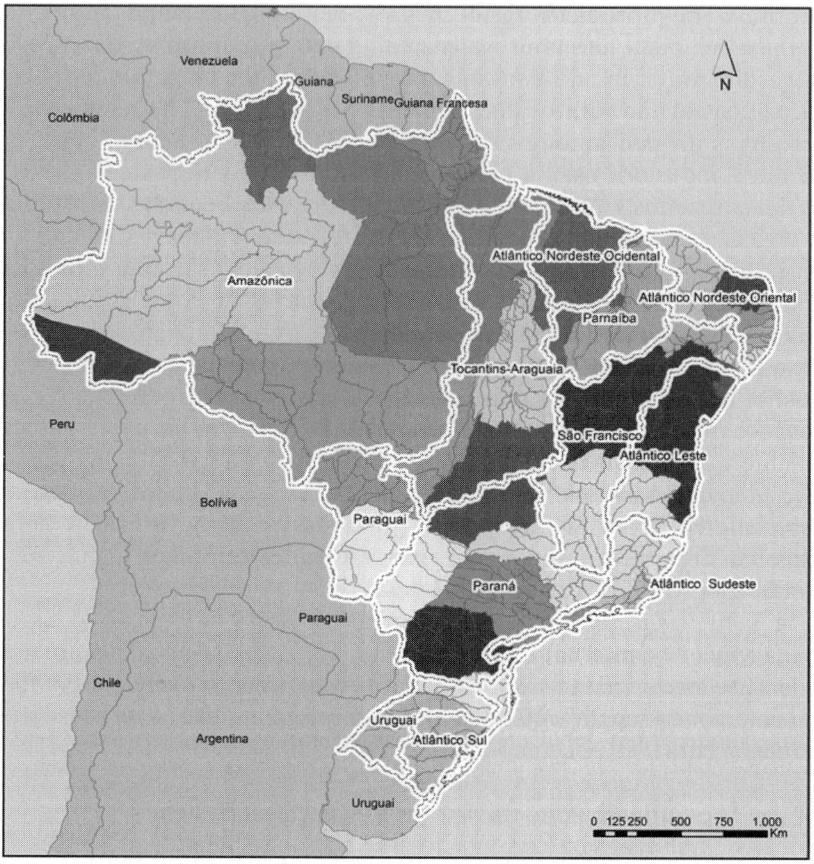

Figure 8.1 Water resources regions and state limits in Brazil

- water is recognized as a public good that has economic value;
- water management shall comprise and induce multiple uses;
- priority of water use is given to meet human's basic needs and animal watering, especially in shortage periods;
- river basin is the territorial unit for water resources planning and management;
- water management shall allow the participation of government, water users and the organized civil society in the decision-making process.

In order to implement the necessary tools and guidelines, the PNRH established the National Water Resources Management System

(SINGREH). As a modern and relatively new system, it introduced innovative principles, such as application of economic tools to water management and decentralized and participatory decision-making. The River Basin Committee, as a water parliament, was also introduced at the river basin level, including representatives of the governments, users and non-governmental organizations responsible for approving the river basin plan, among others. These committees are supposed to count with a water agency as its technical office. In addition, water resources councils at both national and state levels, responsible for resolving water use conflicts and subsidizing policy formulation, were instituted. Governmental executive organizations – the National Water Agency (ANA) and state water resources organizations – are responsible for regulating the water use and for implementing its water resources management system (Braga et al., 2006). Figure 8.2 illustrates SINGREH's institutional framework. At last, municipalities are responsible for water supply, water treatment and sanitation services, as well as the promotion of necessary integration among such policies and water resources policies.

The Brazilian Water Law has established five management instruments to support the implementation of PNRH. These are: river basin plans, water quality classification, water permits, water use charges, and information systems. In addition, monitoring, capacity building, water allocation strategies and command and control instruments, such as inspection, are also established.

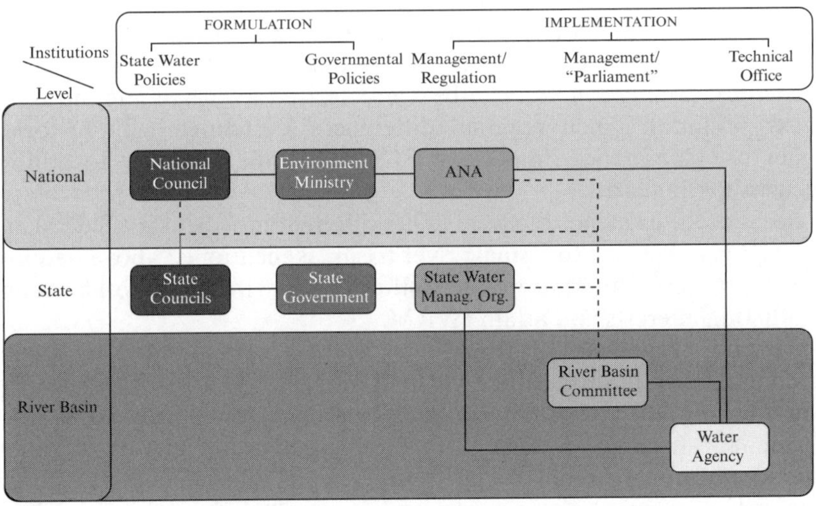

Figure 8.2 National water resources management system of Brazil

As a public policy, water resources management is unique because water is a shared resource that flows beyond borders. In a river basin area, actions performed upstream have impacts downstream, involving, in general, more than one jurisdiction. Therefore, water is essentially an integrating factor and its management depends on the cooperation among government units. PNRH recognizes these aspects and provides conflict resolution strategies as an alternative to judicial paths. Furthermore, SINGREH reflects IWRM directives, mainly regarding multiple stakeholder participation in the decision-making process, as well as the adoption of the river basin level to implement its instruments.

8.3 MAIN CHALLENGES TO IMPLEMENT SINGREH AT NATIONAL LEVEL

8.3.1 Regional Contrasts and Different Water Resources Jurisdictions

Water resources management is essentially a complex process, because it involves several different activities, usually conducted by multiple actors. As expected, SINGREH's implementation process in Brazil has to deal with significant challenges.

First, it is important to recognize that Brazil, the fifth largest country in the world, presents substantial regional contrasts. For instance, the North region has the highest water availability, smallest population and is currently under the pressure of hydroelectricity expansion. The Northeast is the driest region while the Midwest region is under the pressure of agricultural expansion. The Southeast and South regions have the largest population and can be characterized by urban and industrial water pollution. Such regional differences, combined with historical water use dominance, impose to SINGREH the necessity to address regional particularities.

Second, the existence of rivers under different jurisdiction – federal and state (Figure 8.3) – into a single river basin, as mentioned above, requires the institution of an appropriate institutional environment to harmonize conflicting interests and attain IWRM.

Around 75 percent of Brazil's territory is compounded by shared interstate river basins, as shown in Figure 8.3. Due to the federative condition, union and states are autonomous to implement water management instruments according to the respective jurisdictional authority. Therefore, different water resources management systems were established in each administrative territory, at different times (Figure 8.4). As a consequence, there can be multiple rules and procedures for

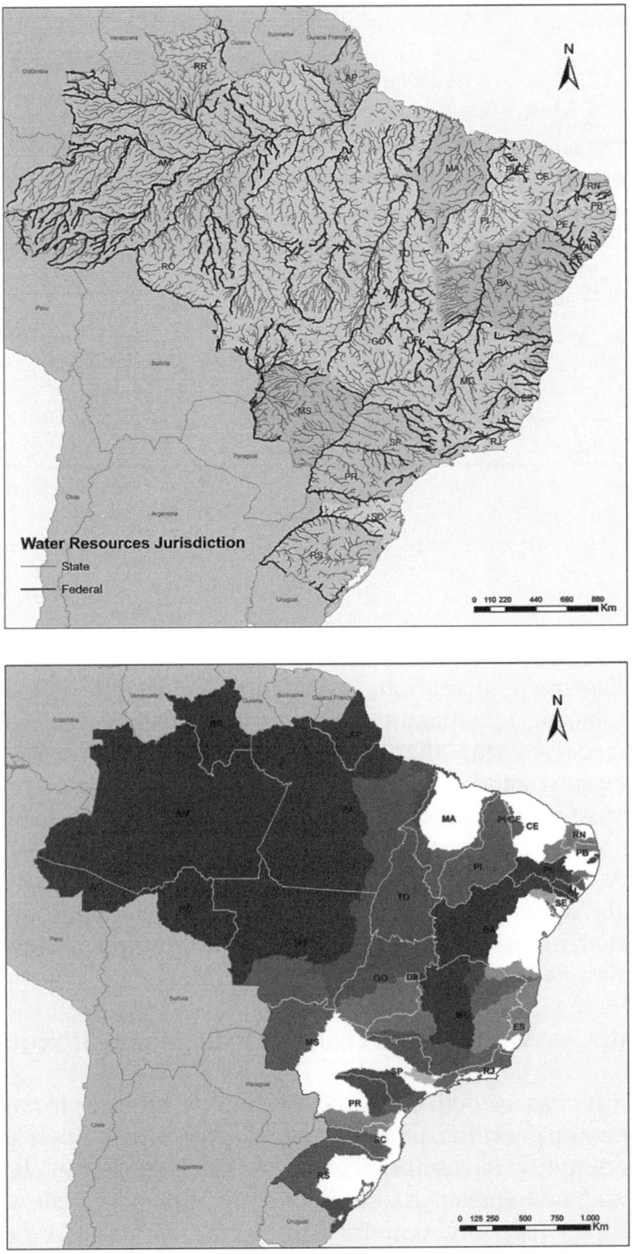

(a)

(b)

Figure 8.3 Dual jurisdiction of water bodies (a) and correspondent shared river basins (b)

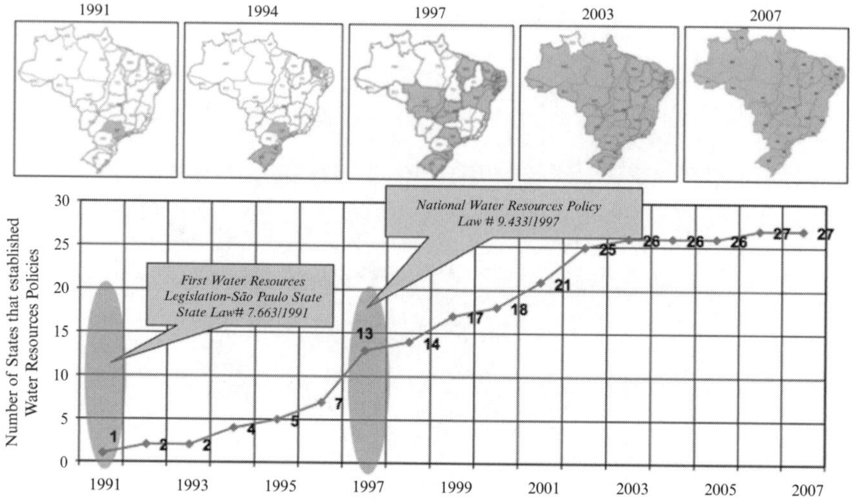

Figure 8.4 *Evolution of water resources systems establishment in each federative unit*

water management, operation and regulation, legally defined, in the same river basin. This condition is also reflected on the institutional framework, considering that river basin committees were created at multiple levels (Figure 8.5). In 2011, there were 169 river basin committees correspondent to rivers under state jurisdiction, and seven river basin committees installed at interstate river basins, on top of the state ones (ANA, 2012a). Indeed, necessary integration among governments, including the development of common objectives, common understanding of related issues and sense of unity to harmonized action (Pereira, 2003), becomes a true challenge.

8.3.2 Water Security as a New Challenge to Implement SINGREH

Water security can be defined as the availability of water for all its uses, including human development and environmental needs, at the best possible timing, quantity and quality. However, the prognostic of climate variability poses a new challenge to water resources management worldwide, and especially in federative countries, due to the fact that more stakeholders are involved and need to agree on what to do, adopting cooperative and functional approaches. In a multiparty country, like Brazil, where different interests contrast among States and between states and central

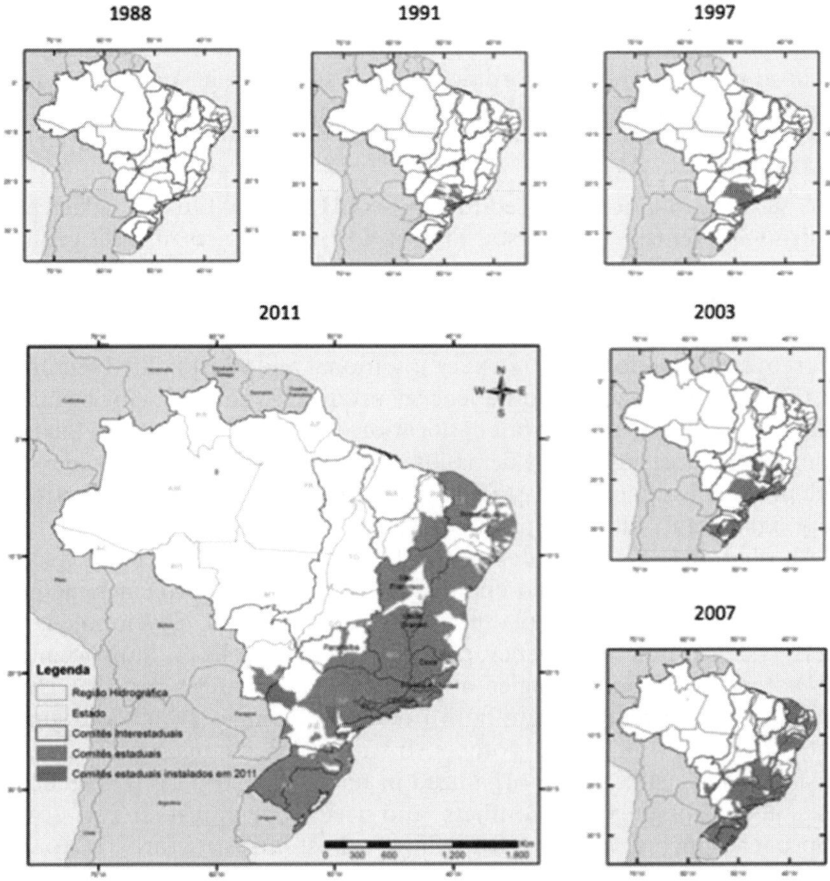

Figure 8.5 Evolution of river basin committees establishment at multiple levels

government, it is very difficult to have an integrated plan of action, increasingly required by climate change forecasts.

According to Stakhiv (2011), the most significant implication of climate change for water resources management is an increase in frequency, duration and magnitude of floods and droughts. Therefore, water security is affected, considering that it may not be available in the right time and quantity for its uses. The pressure over the challenges prior listed is increased and new conditions are imposed to implement SINGREH.

One of these conditions is the strengthening of the participation of municipalities into water resources management. For instance, flood events are closely related to land use and occupation decisions that are made at municipal level, according to Brazil's legislation. A more systemic approach should be established, at the river basin level, promoting the dialogue among key actors and including the development of alert systems and contingency plans.

A second condition imposed to SINGREH is related to the occurrence of drought events. In such cases, competition for water resources is greater and strategies to promote rational and multiple use of limited resources should be strengthened. In parallel, more robust infrastructure may become necessary, reinforcing the necessity of involving multiple interests. Moreover, hydroelectricity is a very traditional and consolidated sector in Brazil that combined with other energy production sources – for instance ethanol – requires specific water allocation strategies to guarantee the fulfillment of increasing energy demands. Although Brazil is considered to be rich in water availability, as stated before, predicted climate change effects may significantly alter this picture in some regions.

As observed, both top-down and bottom-up approaches are necessary to overcome presented challenges. Bottom up approaches include land use and occupation planning at municipal level, development of alert systems and contingency plans at river basin level, improvement of water allocation strategies and operational changes, also at river basin level, to accommodate both flood and drought events and guarantee multiple uses, among others. Institutional arrangements at river basin level are still poorly organized in terms of sharing water security risks and managing such conflicts, and need to be adjusted. Top-down approaches include the need for a more robust infrastructure, reinforcement of IWRM, involving all governmental levels and stakeholders, and strengthening of central government role as an integrating agent, among others.

In Brazil, it is still premature to assume that extreme events are directly related to climate change variability. Although the occurrence of recent extreme events – not common in the past, for instance droughts in the northern and southern parts of the country, as well as several flood events – deforestation, agricultural practices, land use modification, reservoir operational rules, among others can be the contributing factors to such events. Indeed, hydraulic data inconsistency is also possible. The question to be addressed is whether water security in Brazil is affected mostly by ineffective IWRM or possible climate change effects.

8.4 ADDRESSING INITIATIVES AT NATIONAL LEVEL

8.4.1 National Pact for Water Management

Some strategies are being developed, at the national level, in order to address discussed challenges, especially to promote integration among different water resources systems existing at federative units. The National Pact for Water Management (PNGIRH) is an initiative involving federal government and all federal states. Its objectives are: to establish commitments among federative units to overcome common challenges and lack of harmonization; to encourage multiple and sustainable use of water resources, especially in shared river basins; to promote an effective articulation between water resources management and regulation processes, at national and state levels; and, to strength water governance system in Brazil, in a integrated, decentralized and participative manner. PNGIRH is enabled based on the implementation of its Management Map.

Management Map is the first initiative in Brazil that recognizes asymmetries in water management needs across the country. It is a strategic, dynamic and evolutionary planning tool used to map different regions in Brazil, considering political versus river basin boundaries (COELHO et al., 2005). It also identifies that SINGREH should not be implemented, in the same way and same pace, in the entire territory of Brazil, requiring regional particularities to be addressed. Therefore, its results constitute a priority action plan, proposing different ways to implement SINGREH, according to different priorities existing at different parts of the territory. Among its implementation steps, there is the establishment of goals for institutional development and water management instruments implementation according to predefined typologies depending on the existing level of complexity of water management process and institutional structure. Higher complexity level demands advanced institutional structure and so on.

As indicated before, multiple pathways exist for the implementation and operation of SINGREH, especially considering that Brazil is a federation and, therefore, each state has its own water resources management system and policy choices. There is not an identical set of instruments in every system established by the federal states because of different needs and priorities in terms of water resources planning and management. The federal government should be able to support them accordingly in order to enhance interaction between different government levels and to implement a true integrated water resources management system.

8.4.2 Water Security Strategies

In addition to PNGIRH, which constitutes a strategic mechanism to enhance governability among federative units, and, indirectly increase water security in Brazil, there are more specific structural and non-structural arrangements, at national level to lower water security risks. For instance, ANA is promoting five actions to be anticipated to possible climate change effects. First, better analysis and diagnostics are being incorporated in water resources planning activities. Second, more investments are being allocated to water resources monitoring, including better quality of data, less interruptions in historic data and superior spatial distribution of monitoring stations. Third, situation rooms are being established at all federative units to improve monitoring, forecasting and warning of critical hydrologic events. Fourth, the analysis and issuing of water permits is being reinforced. Fifth, and last, social communication, stakeholder capacity building and institutional strength are prioritized (ANA, 2010).

Another issue being addressed is related to water allocation. While in some regions there is not enough water for multiple uses, in some other regions water is not being used efficiently among users. Climate change, and possible effects over water distribution, may increase water conflicts. Regulatory agreements constitute a negotiation mechanism, legally binding, to promote reasonable interstate water allocation among multiple uses, including water transfers. Such agreements can be implemented in specific regions, at very local levels, or to an entire river basin to overcome the existence of different criteria to issue water use permits among states and federal governments. Regulatory agreements are basically a three-step process, comprised of: (a) confirmation of current water uses; (b) elaboration of hydrological studies and integrated water permitting strategies among states and federal governments, and (c) issuing of water permits and associated priorities, for instance during drought events.

In terms of structural strategies, the discussion about storage infrastructure is emphasized. It is possible that the effects of climate change, considering the increase in water variability, can lead to the necessity of bigger and more expensive water infrastructure (ANA, 2010). Gomide (2012) affirms that it is necessary to invest in water infrastructure, including water storage, to reach United Nations millennium development goals. Livino (2012) suggests that reservoirs are appropriate water security mechanisms to adapt to climate changes, storing excessive flows during flood events, as well as providing water supply during drought events. In Brazil, according to the author, it is important to observe that reservoirs are also sources of clean energy, through hydroelectricity generation, supporting the expansion of intermittent energy sources such as solar and winds. Stakhiv (2011) advises, however, that it is still very difficult to appropriately reproduce the hydrologic consequences of uncertain climate scenarios into future water infrastructure projects. In addition, according to Kundzewicz and Stakhiv (2010), climate change models are still not capable of addressing water resources planning and management needs at river basin scale. 'A paradigm shift' from stationary climate to multicriteria, non-deterministic and less efficiency centered view, including social, environmental and regional economic aspects, is necessary (Stakhiv, 2011).

8.5 CASE STUDY: DOCE RIVER BASIN

8.5.1 Doce and its Water Resources

The Doce River Basin is located in the Southeast region of Brazil, as illustrated at Figure 8.6. It comprises an area of 86,700 km^2, 86 percent of those being in Minas Gerais State (MG) and 14 percent in Espirito Santo State (ES), containing 229 municipalities and being home to 3.5 million

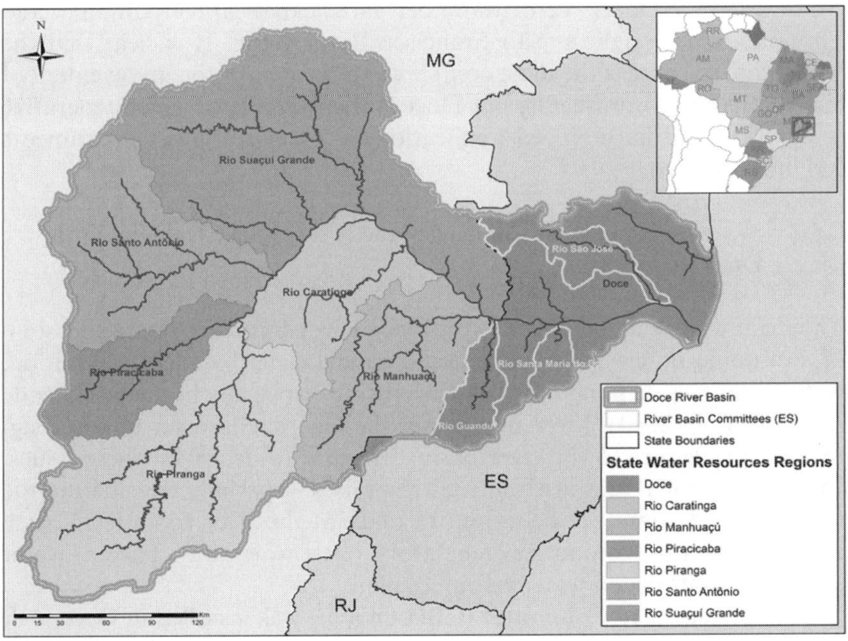

Figure 8.6 The Doce River Basin

people. The hydrological system is formed by six sub-basins located at MG and three sub-basins located at ES. From October to March the precipitation varies from 800 to 1300 mm. During the dry season, from April to September, it varies from 150 to 250 mm. The major economic activities in the river basin are industry (steel/mining), hydroelectricity, agriculture and livestock. The Doce River Basin has the biggest steel complex of Latin America, including USIMINAS, ACESITA, Belgo Mineira and Vale do Rio Doce. The region is also characterized by coffee production and exportation.

Doce is among the interstate river basins in which river basin committees are installed, as presented in Figure 8.5. In this picture, the lines show the delineation of seven interstate river basin committees: Paraíba do Sul, Verde Grande, Piracicaba-Capivari-Jundiaí, Piranhas-Açu, São Francisco, Doce and Paranaíba. Doce was selected as a case study among these because it is the river basin where water resources management is most advanced in terms of implementing integrated strategies between federative units.

All interstate river basins have to deal with the challenge of integrating water resources policies to promote IWRM, beyond politico-administrative boundaries. The number of federative units involved varies from three at Doce, Verde Grande, Piracicaba-Capivari-Jundiaí and Piranhas-Açu to eight at São Francisco River Basin. It is clear that the São Francisco case is the most complex one because of the involvement of eight different governments, but Doce is the example in which federative integration and harmonized application of water resources instruments has most evolved.

8.5.2 Water Management Challenges and Addressing Strategies at the Doce River Basin

The main water related issues in the Doce River Basin are floods, erosion, lack of domestic sewage treatment and critical water balance. In addition, water resources demand is rapidly increasing, especially in the last decade (CBH DOCE, 2012). Most important, the federal character imposes significant challenges to the river basin. In order to deal with these issues, some institutional and instrumental responses, especially considering the federative situation, are being put in place at the river basin level, such as an integration committee, regulatory pact, water use charges, water agency and integrated water resources plan.

Doce River Basin Committee (CBH DOCE) was installed in December 2002 at interstate level, with the mission to promote rational use of water resources and support the implementation of recovery actions in the river basin (CBH DOCE, 2012). In addition, it has the challenge of integrating nine river basin committees established at sub-basin level, correspondent to the rivers under MG and ES jurisdiction. Considering that water from sub-basin committees flows into the upper level river basin committee, actions should be planned and decisions should be made with the necessary integration (Abers and Keck, 2006). Representatives at both levels should work together in order to enhance public participation and promote IWRM.

In this sense, its Integration Technical Chamber is proposing the establishment of an integration river basin committee, similar to the strategy adopted at Piracicaba-Capivari-Jundiaí River Basin. According to the ongoing discussions, state river basin committees would be responsible for approving water resources action plans (PARHs), resolving water use conflicts and determining the financial value for water use charges into their territory, as well as selecting representatives for the integration committee at Doce River Basin. The integration committee, in turn, would be responsible for approving the Integrated Water Resources Plan of Doce River Basin (PIRH), resolving interstate water use conflicts and providing general guidelines for water use charge, water use priorities and allocation. Its assembly would be composed by representatives from state river basin committees, having a common nucleus that represent multiple assemblies and facilitate the decision-making process. Meetings and deliberations could be jointly taken by the ten assemblies.

In fact, Doce Integration Committee, by itself, is not sufficient to guarantee the necessary cooperation among federative units, and other stakeholders, at the river basin level. Therefore, a second institutional strategy was initiated at the basin: the Integration Water Resources Management Pact at Doce River Basin. It was agreed between all ten river basin committees, MG, ES and federal governments. The integration pact includes the provisions of integrating river basin committees' agenda, establishing minimal conditions to implement water management instruments (river basin plans, water permitting, monitoring, information systems, command–control and regulation practices), harmonizing water charging criteria, implementing a single water agency at the river basin level, integrating the application of financial resources, and facilitating communication and social mobilization. It also sets up the obligation for parties to jointly work to define milestones for water resources management; to establish responsibilities and commitments between river basin committees, states and ANA; and, to create a systematic strategy for continuous monitoring of the goals established at river basin level. A cooperation pact can also be negotiated with the federal government, in case the states need to reinforce their institutional framework and technical instruments to cope with the challenge of decentralized and participative management. Furthermore, management contracts among federal or state governments and the water agency allow that financial resources collected by the Union or the States through water use charging to be transferred to the river basin.

The water agency and water use charges also integrate the governance system that is being put in place at Doce River Basin. According to Brazilian legislation, these strategies should be jointly developed, in order to enable water decentralization at the river basin level. Therefore, both the institution of the water agency and the implementation of water use charges were approved by the National Water Resources Council (CNRH) in 2011. The non-profit organization – Instituto BioAtlântica (IBIO) – received the authorization to exercise water agency functions at Doce River Basin, with the approval of all river basin committees; resulting in a single water agency for the entire basin. IBIO is responsible for the elaboration and/or supervision of all technical studies and the application of all financial resources collected by ANA and the states through water use charges. A management contract was also agreed between ANA, MG/ES states and ten river basin committees as a result-driven contract. Its goals include information, planning, and management, application of the water resources charges revenue, leverage of new financial resources and social control. Annual assessment of the results is performed using specific indicators. The management contract also guarantees that the financial resources are invested by the water agency, in a more efficient way, according to a multi-year implementation plan.

Regarding water use charges, CNRH Resolution #123 approved the mechanisms and the prices to be charged for water use at Rio Doce River Basin. For the first time in Brazilian history, the implementation process of water use charges occurred simultaneously at all government levels and was approved by all ten river basin committees.[2] PIRH was incorporated in the decision-making process, and the integration pact subsidized the necessary articulation between river basin committees and state and federal water resources institutions to implement water use charges. A progressive unitary price was established, conditional on meeting the per centage agreed for the disbursement of the funds, as stated in the management contract, and developing new studies about water use charges at the basin. The administrative costs of the water agency must be covered by the water charges, limited to 7.5 percent of the total collected. The potential revenue for the four initial years is estimated in U$57 million to be allocated at the river basin, representing 25 percent of PIRH in ten years of collection. Since the water use charging was effectively initiated at the end of 2011, such results are still not consolidated.

Doce is the fourth river basin to implement water use charges and its water agency in Brazil. Based on past experiences, Doce has evolved to a more solid and integrated agreement involving all river basin committees. The integrity of the negotiation process between all significant stakeholders increases the chances of success.

The last strategy in course at Doce River Basin is related to planning process. Again, in an effort to overcome the challenges of the federative situation at the river basin, the three water resources governmental organizations (MG: IGAM, ES: IEMA and ANA) and respective river basin committees agreed to develop one single river basin plan: PIRH. A technical and managerial group (GTA), composed by 26 representatives of river basin committees and governmental organizations, was established to supervise the elaboration of the river basin plan.

PIRH was developed jointly with PARHs, which represent the action plans determined at sub-basin level. The strategy PIRH–PARH addresses the necessity to apply the financial resources collected in each sub-basin into their territory. At last, a multi-year implementation plan was approved by all ten river basin committees to implement the goals established at integration pact, management contract and PIRH–PARH, in articulation with sectorial policies. Its main programs are: flood protection, water supply and sanitation, effluent control, water rational use in agriculture, payments for environmental services, and restoration of riparian vegetation, river basin committees strengthening, environmental education, social communication, training and capacity building.

Regarding some water security strategies, an alert system was put in place in 1997 (CPRM, 2012) at Doce River Basin in order to minimize flood impacts. The Brazil Geologic Survey (CPRM), in partnership with ANA and Minas Gerais Water Resources Management Institute (IGAM), works to collect and analyze hydrologic data, elaborate hydrological forecasts, and transmit information to the civil defense, military police and 16 city halls. In addition, the incorporation of structural measures to mitigate flood effects is also being studied, along the lines of the Paraíba do Sul River Basin experience (ANA, 2012b). Because responsibilities among stakeholders are still not clearly defined, the next step should be the development of a contingency plan to establish who is responsible for what and how, at multiple levels.

8.6 CONCLUSIONS

With the implementation of SINGREH, a sophisticated and modern water resources management system, Brazil counts with a governance arrangement that enables decentralization at river basin level, participation of users and the organized civil society in the decision-making process, and increase in water use efficiency. In fact, IWRM requires two action fronts of parallel and mutual interaction: *governance* for management and *governability* over water resources issues, which are both fundamental (Brazil Section, 2011).

Considering the governability front, a very positive river basin experience has been reported in this chapter, at Doce River Basin. As described, an integrated river basin plan was approved for the interstate river basin, combined with action plans for the sub-basins at state level; water use charging is being successfully implemented jointly with the establishment of a single water agency for the entire basin; the integration water resources management pact was agreed between all ten river basin committees, MG, ES and ANA; and Doce Integration Committee complement the institutional strategies build up over the years to overcome federative challenges at the river basin. The governance system that is in place has been strengthening IWRM, including the participation of various actors involved in decision-making at decentralized level. However, two additional measures are still necessary: first, the incorporation of actions established at PIRH–PARH into sectorial policies (water supply and sanitation, agriculture, industry, energy, etc.); and second, the establishment of hydroenvironmental performance indicators to monitor the PIRH–PARH implementation process.

At the national level, the lack of articulation with sectoral policies is also true. Water issues are not very compelling or well articulated with other sectors; therefore, water is still not comprehensively treated among Brazil's priorities. Agencies responsible for energy, sanitation and irrigation, among others, plan their water policies and pursue its goals independently (Abers and Keck, 2006). A more systemic vision of water sector needs to be incorporated into the political agenda. Sectoral integration depends on water sector articulation on planning stages. Moving in this direction, in 2011, ANA was invited to participate, for the first time, in the Federal Government General Multi-year Implementation Plan.

Furthermore, the National Pact for Water Management, as described, is capable of minimizing presented challenges to implement SINGREH, especially due to the federative condition of Brazil. The strengthening of ongoing strategies should also diminish the impacts of administrative discontinuities, which weaken IWRM advancement. Coordination, without violating the autonomy of the states, needs to be incorporated effectively into water policy.

Finally, it is important to stress that global changes, especially climate variability, will bring new and important challenges to water resources management around the world in general, and in Brazil in particular. Federations can have increased difficulty adopting cooperative approaches in situations of conflicts due to hydroclimatic variability. More extensive droughts and intense floods forecast by the Intergovernmental Panel on Climate Change (IPCC) should have direct consequences in the definition of a more resilient hydrologic infrastructure; socially responsible and environmentally sustainable, and the adoption of more efficient and effective consumption standards. It constitutes a great challenge that needs to be addressed by the world of hydrologic science, so that long-term decisions can be made and water security risks reduced.

8.7 NOTES

1. Art. 20, § III, 1988 Federal Constitution of Brazil.
2. At Espirito Santo State, the effective implementation of water use charges still depends on specific regulations. In this sense, the state legislative assembly is currently analysing some amendments to the State Water Resources Policy (Law 5.818/98).

8.8 REFERENCES

Abers, R.N. and Keck, M. Muddy Waters: The Political Construction of Deliberative Water Basin Governance in Brazil, *International Journal of Urban and Regional Research* 30: 3 (September 2006), 601–622.

ANA Agência Nacional de Águas. Diagnóstico de Recursos Hídricos (*Water Resources Assessment*), 2004.

ANA Agência Nacional de Águas. Disponibilidade e Demandas de Recursos Hídricos no Brasil (*Water Resources Availability and Demands in Brazil*), Brasília, 2007.

ANA Agência Nacional de Águas. Nota Técnica: Os Efeitos das Mudanças Climáticas sobre os Recursos Hídricos: Desafios para a Gestão (*Technical Report: Climate Change Effects over Water Resources: Challenges for Management*), Brasília, 2010.

ANA Agência Nacional de Águas. Relatório de Conjuntura dos Recursos Hídricos- Informe 2011 (*Water Resources Overview – 2011 Report*), Brasília, 2011a.

ANA Agência Nacional de Águas. Pacto Nacional pela Gestão das Águas: Construindo uma Visão Nacional. Documento Base – Volumes I e II (*National Water Resources Management Pact: Building a National Vision. Basis Document – Volume I and II*), Brasília, 2011b.

ANA Agência Nacional de Águas. Relatório de Conjuntura dos Recursos Hídricos- Informe 2012 (*Water Resources Overview – 2012 Report*), Brasília, 2012a.

ANA Agência Nacional de Águas. Estudo e concepção de um Sistema de Previsão de Eventos Críticos para a bacia hidrográfica do rio Paraíba do Sul e Sistema de Intervenções Estruturais para a Mitigação dos Efeitos de Cheias nas Bacias dos Rios Muriaé e Pomba (*Study and Conception of an Extreme Events Early Warning System to Mitigate Flood Impacts at Muriaé and Pomba River Basins*), Brasília, 2012b.

Braga, B. et al. (ed.). Águas Doces no Brasil: Capital Ecológico, Uso e Conservação (*Fresh Water in Brazil: Ecological Capital, Use and Conservation*), Escrituras Editora, 3rd ed., São Paulo, 2006.

Brazil Section. Good Governance For Water Resources Integrated Management. 6th World Water Forum – Time for Solutions. Brazil's reference document submitted to the Americas' Regional Process. 2011.

CBH DOCE. Comitê da Bacia do Rio Doce (*Doce River Basin Committee*). Available at: http://www.riodoce.cbh.gov.br/, accessed April 2012.

Coelho, A.C.P., Gontijo, W.C., and Cardoso, A. Unidades de Planejamento e Gestão de Recursos Hídricos: uma proposta metodológica (*Water Resources Planning and Management Units: a suggested approach*). In: Anais 7o Simpósio de Hidráulica e Recursos Hídricos dos Países de Língua Oficial Portuguesa – Silusba, Portugal. 2005.

CPRM. Sistema de Alerta Contra Cheias da Bacia do Rio Doce (*Floods Early Warning System at Doce River Basin*). Available at: http://www.cprm.gov.br/alerta/site/index.html, accessed April 2012.

Gomide, J.L. Sobre Reservatórios e Segurança Hídrica (*Reservoirs and Water Security*). Agência Nacional de Águas. Brasília, 2012.

Kundzewicz, Z.W. and Stakhiv, E.Z. Are climate models "ready for prime time" in water resources management applications, or is more research needed?, *Hydrological Sciences Journal* (2010) 55: 7, 1085–1089.

Livino, A. Reservatórios de Regularização de Usinas Hidrelétricas: Contribuição para uma Matriz Energética Mais Limpa (*Regularization Dams for Hydroelectric Power Stations: Contribution for Cleaner Energy Sources*). Exame de Qualificação. COPPE, Universidade Federal do Rio de Janeiro, Rio de Janeiro, 2012.

Pereira, D. (ed.). Governabilidade dos Recursos Hídricos no Brasil: a Implementação dos Instrumentos de Gestão na Bacia do Rio Paraíba do Sul (*Water Resources Governability in Brazil: Implementation of Water Resources Management Tools at Paraíba do Sul River Basin*), Agência Nacional de Águas, Brasília, 2003.

PIRH. Plano Integrado de Recursos Hídricos da Bacia do Rio Doce (*Doce River Basin Integrated Water Resources Plan*). Available at: http://www.pirhdoce.com.br/, accessed April 2012.

Stakhiv, E.Z. Pragmatic Approaches for Water Management Under Climate Change Uncertainty. *Journal of the American Water Resources Association JAWRA* (2011) 47: 6, 1183–1196.

PART III

EUROPE – THE SPANISH EXPERIENCE

9. River basin governance and water policies in Spain

José Albiac, Elena Calvo and Encarna Esteban

9.1 SUMMARY

The chapter describes the water institutions and water policies in Spain. A distinctive feature of water management in the country is the key role of basin authorities, where stakeholders are involved in all management decisions. Water policies entail very large investments in water technologies to ensure water security, with different degrees of success in terms of human provision and ecosystem protection. The evolving interplay between the federal and state water governance levels is illustrated through the main water conflicts in recent years. This strong institutional and policy setting seems powerless to offset the worsening water scarcity and degradation in basins.

9.2 INTRODUCTION

Spain is located in the Iberian Peninsula covering a territory of 500,000 km^2 and with a population of 46 million inhabitants. The climate in the Iberian Peninsula is very diverse, with Atlantic, Continental and Mediterranean climates ranging between northwest and eastern regions. The diversity in climate, orography and soils supports a large biodiversity, sheltering more than half of species found in Europe.[1] The Doñana and Daimiel wetlands, the Ebro River delta and L'Albufera are important environmental assets supporting aquatic ecosystems in Spain. All of them are seriously threatened from the pressure of human activities on water resources.

The average annual precipitation is 680 millimeters (340,000 million cubic meters, Mm^3), but the variability in precipitation across time and space is very large. The average flow in rivers amounts to 110,000 Mm^3 and the groundwater renewable resources are around 30,000 Mm^3. The large time and space variability of water regimes in rivers has led to the

Table 9.1 Water extractions by sector in Spain (Mm^3)

Year	1960	1990	2009
Irrigation	12,800	23,000	25,000
Urban & Industrial	2,700	5,800	7,000
Total	15,500	28,800	32,000

Sources: INE (2010), MIMAM (2000), MOPT (1993).

construction of a considerable number of dams with a large storage capacity.

A strong process of economic development has taken place in Spain during the last 50 years. The consequence has been a growing pressure of economic activities on water resources, and the ensuing water scarcity and water quality degradation. This degradation is a common threat to human water security and environmental biodiversity across the world, which is compensated with huge investments in rich countries that ensure human security (Vörösmarty et al. 2010).

Water scarcity in Spain is linked to the enormous development of irrigation, while quality degradation is linked to pollution from urban, industrial and agricultural sources. Water extractions for consumptive uses were 15,500 Mm^3 per year in 1960. These water extractions have doubled at present by the expansion of irrigation acreage and the growth in urban and industrial demand (Table 9.1). Urban and industrial uses represent less than a fourth of extractions, but their pollution loads are the main causes of water quality degradation and the resulting environmental damages.

The more severe problems of water scarcity and quality degradation are located in the southern and eastern basins, with pressures coming from intensive agriculture, urban sprawling and tourism on the Mediterranean coast. In the inland basins of the Tajo, Ebro and Duero rivers, irrigated crops are cultivated under extensive production systems with surface water. Surface water resources are under the effective control of basin authorities, while groundwater resources are beyond their control. The surge in aquifer pumping by farmers that occurred in recent decades in the southern and eastern basins of the Júcar, Segura, Sur, Guadiana and Guadalquivir rivers has triggered a severe degradation of water resources in these basins (Figure 9.1).

These threats from water scarcity over human water security have triggered a process of reform in the management of water resources in Spain. This process has taken place during the last 15 years and involves multi-

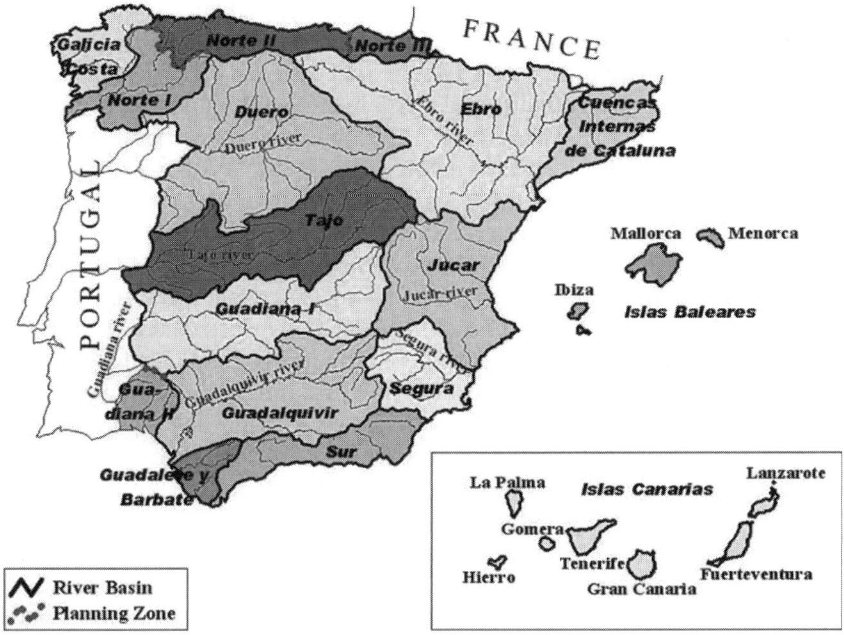

Figure 9.1 River basins in Spain

billion investments in water technologies, in the form of water transfers, water storage and conservation, advanced irrigation systems, wastewater treatment plants, seawater desalination, and water reuse. A sizable part of these investments in water technologies has been financed with development funds from the European Union, especially for wastewater treatment plants, advanced irrigation technologies and seawater desalination. This process of improving water management though large investments in water systems has relied on the strong institutional approach characterizing water management in Spain, while water management in other developed countries relies mostly on judicial processes and economic instruments.

The main institutional organizations in Spain are the river basin authorities, which manage water and deal with the problems of water scarcity and droughts. The distinctive feature of this institutional arrangement is the key role played by stakeholders in basin authorities, with stakeholders involved in all their governing and management bodies. However, important water policies such as the National Irrigation Plan or the National Hydrological Plan have been unable to reverse or dampen down the progressive water scarcity in basins. This is a common problem in arid and semiarid basins across the world, where expansion of water supply is more important than protection of water resources and ecosystems.

Spain is a federal country with 17 states called *Comunidades Autónomas*, having their own governments and parliaments, and managing at present almost half of public expenditures. The federal organization of Spain derives historically from the medieval kingdoms of Castilla-León, Navarra, Aragón and Valencia, which had their own governments and parliaments under the Habsburg dynasty until the eighteenth century. The modern federal principle gained support during the nineteenth and twentieth centuries in Spain, leading to the proclamation of the Federal Republic of 1873 and the Second Republic of 1931–1939. After the dictatorship, the federal constitution of the Second Republic inspired the current Spanish Constitution of 1978. States have large political and economic powers, and matters such as education, health and justice are under their responsibility.

In relation to water resources, states are in charge of urban water supply and sanitation, agriculture and forests, environment and land use planning. This devolution of competences to states implies that coordination between the federal and state levels becomes a key issue for water management. The participation of states in basin authorities has been included by creating a new body (called public authorities committees) in each basin authority.

These public authorities committees are only consultative bodies, and the main thrust for water management in the basin remains in the hands of stakeholders using the water. Therefore states influence water management indirectly, through their decisions on urban supply and sanitation and their agricultural, environmental and land use policies. During recent years, states are trying to expand their control over water resources in their territories. The consequence has been a series of water conflicts among states, with the federal government yielding to the political clout of states. The conflicts are politically driven to gain electoral support, and they frequently undermine sound water governance in basins. These conflicts show that states are trying to gain additional power by pressuring the federal government in detriment of other states in the basin or other basins. In fact, better management in basins requires cooperation among states which is hindered by these confrontations.

Another layer of water governance is the Water Framework Directive of the European Union (WFD). This European regulation sets the objective of achieving the 'good ecological status' for all water bodies, giving a large flexibility to member countries to achieve the objective. The emphasis of the WFD is on water quality, which is the main problem in northern and central Europe, while the main problem in southern countries is water scarcity. Another governance issue is that the European WFD advances economic instruments for water management, but this is at odds with the Spanish approach based on institutional instruments. Therefore, the flexibility of the WFD is important for southern countries such as Spain.

9.3 RIVER BASIN GOVERNANCE WITH STAKEHOLDER PARTICIPATION

The institutional organization of water management is based on the administration bodies, the legal system, and the water policies. The basic law regulating water management is the 1985 Water Act, adapted in 2003 to the European WFD. Water management is also affected by related legislation on water supply and sanitation, land use and water pollution, which are matters managed mostly by states. All continental water, either surface or groundwater, is public domain. There are no water rights but rather temporary public concessions that grant the holder the right to use a given amount of water. The Spanish legislation grants an important role to water authorities in each basin, in charge of implementing the basin plan. The planning objectives are to satisfy demand, attain an equilibrated and harmonious water sector, and further regional development. Since 2003, the objective of achieving good ecological status has been added to hydrological planning, in order to comply with the European water legislation.

The institutional organization of water management results from the interaction between the European and the Spanish federal and state levels of government (Figure 9.2 from Varela et al. 2010). The European government establishes general water policies developed by member countries. The water administration in Spain is divided between the federal and the state governments, with federal administration ruling over river basins covering more than one state, and states ruling over basins which are within their state boundaries. The basins under state administration are Galicia-Costa, Cuencas Internas de Cataluña, Sur, Guadalete, Baleares and Canarias, and the rest are federal basins (Figure 9.1).

The Ministry of Agriculture is the federal authority on water resources, and its main tasks are the design of water policies, the National Hydrological Plan, and managing federal basin authorities. The National Water Council is an advisory board giving recommendations on water policies and projects. The state governments have the responsibilities on urban supply and wastewater treatment, agriculture, land planning, environment protection, and state river basins.

An important feature of the Spanish water institutions is the key role given to water users, especially to those in agriculture. Agricultural users

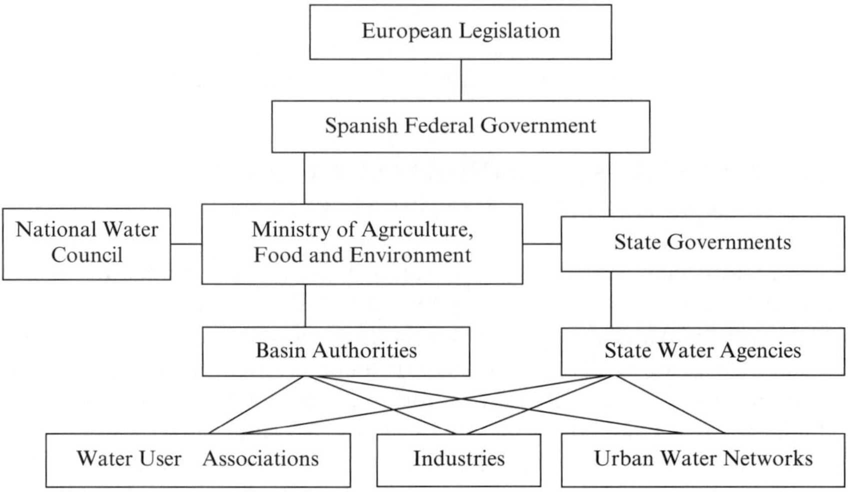

Figure 9.2 Institutional organization of water management

must create a water users' association, with their own equity and legal status. These associations are in charge of organizing irrigation turns, controlling water allocation and consumption, collecting water fees, and investing in the modernization of irrigation systems.

The basin authorities are the main administrative bodies largely responsible for water management. Each basin authority is organized around the governing boards, the stakeholder boards, and the management services. The governing boards include the governing council, the water council and the public authorities committee. The governing council approves investments, management plans and user associations, while the water council approves the basin plan. The public authorities committee is a coordination body between federal, state and local public authorities (Figure 9.3).

The stakeholder boards are the user assembly, the watershed boards and the dam release commission. The user assembly coordinates waterworks and water resources, and the watershed boards are in charge of management in each watershed of the basin. The dam release commission deals with water release regimes. The administrative services include the water commission, the technical directorate and the planning office. The water commission is in charge of approval of permits for water use and discharges, registry of public and private waters, water monitoring, enforcement, and data collection. The technical directorate is in charge of waterworks management. The planning office is in charge of designing and implementing the basin plan. The basin authorities collect four levies from water users, and these levies are payments for the use of the public

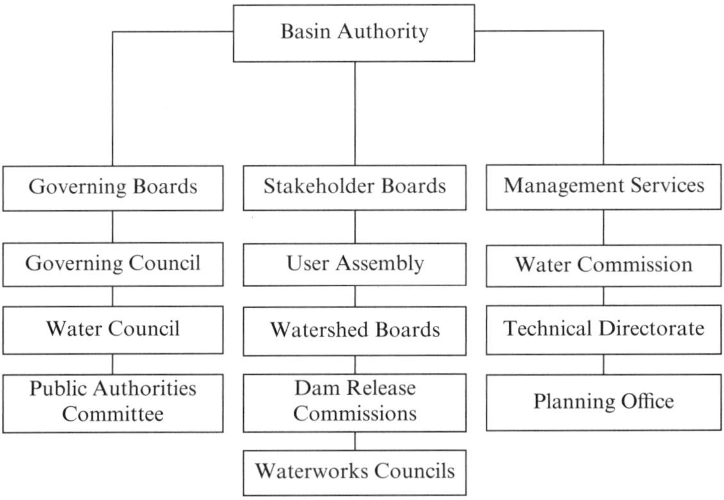

Figure 9.3 Organization of basin authorities

domain, for discharges of effluent loads, and for investments in water supply facilities.

An important feature of basin authorities is the key role of stakeholders, as indicated by their original name that was 'hydrografic union confederation'. This involvement of stakeholders has been a permanent characteristic since the creation of basin authorities in the 1920s. The stakeholders' representatives are present in all governing and participation bodies at basin scale, and run the watershed boards at local scale. Therefore, the stakeholders are involved at all levels of decision making: planning, financing, waterworks, measures design, enforcement, and water management at basin and watershed levels. The advantage of having the stakeholders inside the basin authority taking decisions is that the implementation and enforcement of decisions is carried on smoothly.

9.4 WATER POLICIES: ACHIEVEMENTS AND FAILURES

The responses and adaptation to water scarcity and water quality degradation in Spain during the last twenty years have been shaped by the national water policies and also by the European legislation, with large investment projects costing billions of Euros. The main water policies have been the National Hydrological Plan proposal of 1993, the National

Hydrological Plan of 2001 modified with the AGUA Project of 2005, the National Irrigation Plan of 2002, the Upper Guadiana Plan of 2008, the First and Second Sanitation Plans of 1995 and 2008, and the European Water Framework Directive enacted in 2003 in Spain.

Large-scale aquifer depletion in southern and eastern Spain led to the proposal of water interbasin transfers in both National Hydrological Plans of 1993 and 2001, with investments amounting to 28 and 19 billion euros, respectively. These large interbasin transfers were met by the opposition of political parties, groups of interest, and donating territories. Finally, these water transfers have been abandoned and substituted by the so-called AGUA project, which involves large investments in seawater desalination.

The investments of water policy planning in Spain are considerable with the main policies summing up almost 50 billion euros: 19 billion euros for the National Hydrological Plan, 6 billion euros or the National Irrigation Plan, 3 billion euros for the Upper Guadiana Plan, and 20 billion euros for the Second Sanitation Plan. The National Irrigation Plan and the Second Sanitation Plan seem well designed to improve water quality by reducing pollution loads. But other policies such as the Upper Guadiana Plan and the AGUA project seem poorly designed or even misguided.

The National Irrigation Plan has a good potential of saving water and curbing pollution through investments in advanced irrigation technologies. These investments do not guarantee the solution to all problems, but it is obvious that technical innovations in irrigation systems facilitate the private and public control of water quantity and quality. Realizing the potential of the National Irrigation Plan will requires strong coordination between water authorities and irrigation water user associations.

Aquifer depletion in southern and eastern basins sums up to 700 Mm^3, despite the fact that groundwater was declared public domain in 1985. Registration of both concessions and private rights of groundwater is far from completed, and the number of non-registered wells could be above half million. In contrast, water scarcity and degradation is rather moderate in inland Spain because irrigation is based on collective systems: basin authorities control concessions, river flows and dam reserves, while irrigation user associations manage irrigation districts.

There are some examples of unconvincing water policies such as the Plan of the Upper Guadiana, approved in 2008 and canceled in 2012. The plan aimed at curbing overdraft in the Western La-Mancha Aquifer and recovering the Tablas de Daimiel, one of the main wetlands in the country. Previous efforts to sanction illegal abstractions were turned down by the federal government yielding to state pressures. Instead of curtailing abstractions, the plan anticipated investments of 3 billion euros to eliminate 220 Mm^3 of overdraft. What is surprising in this plan is that no lessons have been learned from the neighboring Eastern La-Mancha aquifer, where the solution for sustainable extractions is based on the collective action of farmers rather than on huge investments. The Plan of the Upper Guadiana was a misguided policy to recover the aquifer, because stakeholders' cooperation requires serious commitments to manage the aquifer, and cannot be bribed for by side payments (Esteban and Albiac 2011, 2012).

A second example of a questionable water policy is the current AGUA project. The AGUA project includes investments of 4 billion euros to build desalination plants and expand supply by 600 Mm^3, of which 300 Mm^3 are for irrigation in the coastal fringe. Although there is a potential irrigation demand in the area from greenhouses and other high-profit crops, the pumping costs are much lower than desalination costs, and farmers will not buy desalinated water. Public investments in desalination are only justified if basin authorities are able to strictly enforce a ban on aquifer overdraft, forcing farmers to buy desalinated water. But the solution found by the government is to subsidize desalinated water up to the level farmers are willing to pay (pumping costs).

9.5 WATER CONFLICTS AND FEDERAL AND STATE GOVERNANCE

There has been a series of water conflicts in Spain during recent years involving territories and economic sectors. These conflicts show the evolving relationship between the federal and state governments in relation to water resources. Two examples of water management cooperation at different governance levels are also presented: the Albufeira agreement between Portugal and Spain, and the application in Spain of the Water Framework Directive that sets a common water policy for the whole European Union.

9.5.1 The Challenge to Water Federal Governance by State Governments

The first two large water disputes involving conflicts among territories arose in both the 1993 and 2001 National Hydrological Plans. More recent conflicts appeared after 2005 with the reform of the state bills, called autonomy statutes. The states have been trying to expand their control over the water flowing through their territories, undermining in fact the federal basin authorities.

The National Hydrological Plan of 1993 was intended to interconnect the main basins of Spain with huge investments. The size of investments and the large volume of transfers caused controversy and were met with distrust by social and political groups, and by territories (MIMAM 2000). Because the plan was only a proposal that never was approved by the parliament, there were no severe conflicts among territories, and the state governments were not involved in fights to change the plan.

The National Hydrological Plan of 2001 was a new water planning effort to face the escalating and acute water scarcity, but this time based on a unique water transfer of 1100 Mm^3 from the Ebro to the Júcar, Segura, and Sur river basins. The Ebro transfer faced strong opposition from social and environmental organizations, and from some states located in the donating Ebro basin. The Ebro transfer caused a serious fight between the receiving Valencia and Murcia states, and the donating Aragon and Cataluña states (Figure 9.4). There were large street demonstrations and significant political impacts in the elections of the four states. Finally, the Ebro water transfer was cancelled in 2005 replaced by the AGUA project of seawater desalination.

Several water conflicts have been developing in recent years between most states, which are trying to take over water resources threatening the

Figure 9.4 States in Spain

existence of federal basin authorities. One conflict is between Cataluña and the rest of the states in the Ebro basin. The fall in environmental flows in the Ebro delta was a prominent argument against the Ebro water transfer, and when the Ebro transfer was canceled the federal government decided to elaborate a special plan for protecting the Ebro delta. The special plan would have been a bilateral agreement between the federal and Cataluña governments, including environmental flows requirements in the Ebro delta. The other states in the basin were against setting these flows, because they would jeopardize the water uses in the whole Ebro basin. Cataluña asked for very large environmental flows and finally the special plan was not approved. Other conflicts on the Ebro basin derive from the fact that Cataluña wants to set water policy measures in the lower Ebro and take control of its aquifers, overriding the federal Ebro basin authority.

Several states have reformed their state laws or autonomy statutes in recent years, and this has been a source of water conflicts. Valencia state calls for water transfers from 'surplus' basins (the Ebro) in its reformed statute. To the contrary, the Aragon and Cataluña reformed statutes require the approval for any water transfer from the Ebro. Also, the reformed Aragon statute establishes a reserve of 6000 Mm^3 over the Ebro in order to prevent water transfers to other basins.

There is another water conflict between Castilla-La Mancha and Valencia for the Jucar basin authority. The reason of the conflict is the severe scarcity in the Jucar and the resulting dwindling flows in the lower Jucar, which Valencia attributes to the expansion of irrigation in the upper Jucar in Castilla-La Mancha. Castilla-La Mancha wants that some Valencia watersheds are taken out of the Jucar basin authority and given to the Valencia state government. This will increase the power of Castilla-La Mancha representatives in the Jucar basin authority. The federal government approved the separation of the Valencia watersheds from the basin, but then Valencia took the case to the Supreme Court and won, and therefore the Jucar basin authority boundaries will remain unchanged.

Castilla-La Mancha is also fighting with Murcia and Valencia for the existing water transfer between the Tajo basin in Castilla, and the Segura basin in Murcia and Valencia. Castilla-La Mancha included the abolishment of the Tajo-Segura transfer in its reformed statute, but finally the statute was not approved in the Spanish parliament because of the opposition of Murcia and Valencia, two very populated states with strong political clout.

Andalucía reformed its statute including the control of the Guadalquivir basin as the key issue, based on the argument that most of the basin is in

Andalucía, and then the federal government transferred the Guadalquivir basin authority to the Andalucía government in 2008. The Castilla-Leon state followed Andalucía and included the control of the Duero basin in its reformed statute. But then the Extremadura and Castilla-La Mancha governments with territory in the Guadalquivir basin took the cases of the Guadalquivir and Duero basins to the constitutional court and to the Supreme Court. Both the constitutional and supreme courts ruled that the federal basin authorities cannot be transferred to the states. The Guadalquivir basin authority has been returned to the federal government in 2011, and the Duero basin authority will remain federal after the sentences.

9.5.2 Examples of Water Management Cooperation with Portugal and the European Union

Spain and Portugal share the Duero, Tajo and Guadiana basins, which are important rivers in the Iberian Peninsula (Figure 9.1). Portugal is the downstream country in these rivers and a series of treaties were signed during the twentieth century dealing with the use of water for hydropower and the sharing of water in international river tracts. The Spanish National Hydrological Plan of 1993 was a significant threat to Portugal because the large interbasin transfers in Spain would reduce the downstream flows of the main Portuguese rivers. The planned water net exports per year were 1050 Mm^3 from the Duero and 50 Mm^3 from the Tajo, threatening the hydropower generation in the Duero. After Spain canceled the plan in 1993, a negotiation process started in 1994 leading to the Albufeira Convention signed in 1998. The Albufeira agreement was the result of the expanding cooperation between the two countries, with underlying principles that show the influence of the European Water Directive that was being drafted at the time (Garrido et al. 2010).

The Albufeira agreement establishes the sustainable management of shared water bodies to achieve a good ecological status, with coordinated measures for floods and droughts. The agreement included minimum annual flow regimes depending of climate conditions. The severe drought in the Iberian Peninsula during 2005 was a test for the agreement. Spain called upon the drought clause for not complying with the minimum annual flow in the Duero. Also, there was a large irregularity of the Tajo river flows entering Portugal, intensified by the Spanish extractions of Tajo water to supply the Tajo-Segura water transfer. Portugal complained that the water transfers to the Segura basin were larger than the water left in the river to enter Portugal. These drought problems and the need for further coordination lead to the modification of the Albufeira agreement in 2008. New minimum quarterly and weekly flow regimes have been introduced, and the two countries are working to integrate the management of water resources into a unique plan for each shared river basin.

Another example in Spain of water management coordination among different levels of government is the European WFD. This legislation creates a common water policy to protect water resources and achieve the 'good ecological status' of water bodies. The directive introduces the principle that water prices should be close to full recovery costs, in order to improve the efficiency in the use of water. Recovery costs include the abstraction, distribution and treatment costs, and also the environmental costs and the resource value. Water pricing is a feasible measure in the industrial and urban sectors, since there is a water demand response to prices in these sectors. But water in irrigation and the environment is a common pool resource and the use of pure economic instruments such as water pricing is questionable.[2] Other policy measures such as institutional and command and control instruments seem more appropriate when water is a common pool resource.

The European WFD does not address water scarcity and droughts, which are the main issues in Spain, but rather deals with water quality and the attainment of good ecological status. However, the Spanish water authorities are going beyond the requirements of the directive by strengthening the minimum environmental flows that existed already in previous basin plans.

The implementation of the WFD in Spain based on the water pricing instrument is a quite difficult task, because the main use of water is irrigation. Water management is based on institutional instruments, where basin authorities are run by stakeholders in their governing, implementation and enforcement bodies. This management organization fails for groundwater, and this is the reason for the severe problems of water resources degradation occurring in the southern and eastern basins. The solution is to integrate groundwater users in basin authorities, following the example of the Eastern La Mancha aquifer described in Esteban and Albiac (2012).

The basin authorities in southern and eastern Spain do not control the number of wells or the volume of individual extractions from aquifers, and hence they cannot impose the recovery costs principle of the directive. Furthermore, the required price level to curb demand in some of these areas is around 3 €/m^3, which is far above urban and industrial prices and therefore politically unfeasible (Albiac et al. 2006). Basin authorities may try to impose water pricing in the areas of inland Spain based in low profitable crops, because they have the effective control of collective irrigation systems. This would require the support of stakeholders, but the question is then the following: why they should play around allocating water through water pricing when they are making direct and wise water allocations with the current institutional setting?

The WFD requires generating knowledge and information on the ecological status of water bodies, in order to support the analysis of measures to be taken for water quality improvement. This has resulted in the improvement of the water quality control networks in Spanish basins. Spain is lagging behind the directive deadline of 2009 for presentation and approval of basin plans, despite its experience in the elaboration of the current basin plans in the 1990s. The reason is the challenge to federal water governance resulting from the conflicts among states described above. The Ebro plan is delayed by the discussion among states on the level of environmental flows. The Jucar, Tajo and Segura plans are delayed because of the fight between Castilla-La Mancha, Valencia and Murcia states, involving the allocation of water in the Jucar, and the extractions of water in the Tajo to supply the Tajo-Segura interbasin transfer.

9.6 PROSPECTS FOR WATER SECURITY AND ENVIRONMENTAL PROTECTION

The economic development of Spain in recent decades has resulted in acute pressures that are degrading water resources and producing large environmental damages. The water resources scarcity and quality problems are more severe in the eastern and southern basins, which are under strong pressures from irrigation, sprawling and tourism. The resulting threats to human water security triggered a process of change in water management based on multi-billion investments in water systems and technologies, taking advantage of the strong water institutions that facilitate the involvement of stakeholders.

These institutions are based on administrative, legal and policy components managed at European and Spanish federal and state governance levels. The distinctive features of the institutional setting in Spain are that the involvement of public administrations in water management, and the key role played by stakeholders in these administrative bodies. The main water institutions in Spain are the Ministry of Agriculture, the basin authorities and the state governments. Basin authorities are the main organizations responsible for water management in each basin, with stakeholders' representatives in all governing and participation bodies.

Water policies in Spain during recent years have been addressing the progressive water degradation in basins. These policies are the AGUA project, the National Irrigation Plan, the Upper Guadiana Plan, and the

First and Second Sanitation Plans. The investments involved in these plans are very large, but the outcomes are mixed. These plans are not curtailing the growing water extractions, and the water quality improvements are insufficient. Regarding water quality, the irrigation and sanitation plans have reduced the pollution loads in basins. But the plans addressing water quantity, the AGUA project and the Upper Guadiana Plan (now canceled), are poorly designed and seem misguided to solve water scarcity.

Water institutions and water policies in Spain are being affected by a series of water conflicts among states, which are challenging the principle of federal water governance. States are trying to take over water resources in their territories by capturing the water management responsibilities from the federal governance level, and threatening in fact the existence of federal basin authorities. The major challenges have occurred in the Guadalquivir, Duero and Jucar basins, where the federal government yielded to state pressures. The Guadalquivir basin actually passed to state responsibility, while the Duero basin undertook the same process. The Jucar basin district was divided into two parts to modify the power of states in the basin. Finally, the conflicts were settled by the sentences of the constitutional and supreme courts, which returned the three basins to federal governance.

There are also examples of international water management cooperation with Portugal and the European Union. The Albufera Convention is an agreement to achieve the sustainable management of shared basins between Spain and Portugal. The agreement guarantees minimum flows entering Portugal, trying to reduce conflicts during severe drought periods. The Water Framework Directive is another example of international cooperation between Spain and the European Union, but this directive is not so well adapted to water scarcity and droughts which are the main problems in Spain. The directive promotes water pricing, which is an economic instrument appropriate for urban and industrial demand, but questionable for irrigation and environmental flows. Water management in Spain relies on institutions supporting stakeholders' cooperation. At present, this management organization fails for groundwater systems, but the solution is to integrate groundwater users in basin authorities, rather than pushing water pricing while ignoring the need for cooperation in common pool resources such as irrigation and environmental water.

The institutional organization of water management in Spain is now at a crossroads with states challenging the federal governance of basins, and the European water legislation calling for economic instruments that could undermine the strong institutional approach that sustains collective action. Water conflicts among states are politically driven to gain electoral support, at the cost of threatening to break down federal basins. If states succeed in taking over federal basins, the cooperation among basin stakeholders would be much more difficult to achieve. The implication is that one necessary condition for good water governance is that federal river basins are under federal authority. Despite the strong institutional organization in Spain, the situation of the river basins is similar to most major river basins in arid and semiarid regions worldwide, which are in a process of progressive water scarcity that seems unstoppable. The case of Spain illustrates the difficulties of achieving a more sustainable management of water resources.

Since the Dublin statement in 1992 declaring water an economic good, economic instruments such as water pricing or water markets have been advocated by many experts and decision makers as key ingredients to solve water problems. But there cannot be simple and wide-ranging policy solutions based on economic instruments or any other instrument. The water management experiences in Spain seem to suggest that policy efforts should be focused on nurturing stakeholders' collective action, and on providing the necessary institutional setting.

9.7 ACKNOWLEDGMENTS

This study was made possible by the financial support of the project INIA RTA2010-00109-C04 from the Spanish Ministry of Economy and Competitiveness.

9.8 NOTES

1. Spain is home to the highest number of vertebrate species (600) and vascular plants (8000), and harbors a third of all endemic species in Europe.
2. Cornish and Perry (2003) and Bosworth el al. (2002) show compelling results on the impossibility of using water prices to allocate water in irrigation.

9.9 REFERENCES

Albiac, J., Y. Martínez and J. Tapia (2006), Water quantity and quality issues in Mediterranean agriculture, in: OECD (ed.) *Water and Agriculture: Sustainability, Markets and Policies*, Paris: Organisation for Economic Co-operation and Development.

Bosworth, B., G. Cornish, C. Perry and F. Van Steenbergen (2002), *Water Charging in Irrigated Agriculture. Lessons from the Literature*, Report OD 145, Wallingford, UK: HR Wallingford.

Confederación Hidrográfica del Ebro (2008), *Los Organismos de Cuenca en el Siglo XXI*, Zaragoza: Confederación Hidrográfica del Ebro.

Cornish, G. and C. Perry (2003), *Water Charging in Irrigated Agriculture. Lessons from the Field*, Report OD 150, Wallingford, UK: HR Wallingford.
Esteban, E. and J. Albiac (2011), 'Groundwater and ecosystems damages: Questioning the Gisser-Sánchez effect', *Ecological Economics*, **70**, 2062–2069.
Esteban, E. and J. Albiac (2012), 'The problem of sustainable groundwater management: the case of La Mancha aquifers, Spain', *Hydrogeology Journal*, 20(5), 851–863.
Garrido, A., A. Barreira, S. Dinar and E. Luque (2010), 'The Spanish and Portuguese cooperation over their transboundary basins', in A. Garrido and R. Llamas (eds) *Water Policy in Spain*, Abingdon, UK: CRC Press.
Instituto Nacional de Estadística (2010), *Cuentas satélite del agua en España*, Madrid: Instituto Nacional de Estadísitca.
Ministerio de Medio Ambiente (MIMAM) (2000), *Libro blanco del agua en España*, Dirección General de Obras Hidraúlicas y Calidad de las Aguas. Secretaría de Estado de Aguas y Costas. Madrid: Ministerio de Medio Ambiente.
Ministerio de Obras Públicas y Transportes (1993), *Plan Hidrológico Nacional. Memoria y Anteproyecto de Ley*, Madrid: Ministerio de Obras Públicas y Transportes.
Varela, C. and N. Hernández (2010), 'Institutions and institutional reform in the Spanish water sector: a historical perspective', in A. Garrido and R. Llamas (eds) *Water Policy in Spain*, Abingdon, UK: CRC Press.
Vörösmarty, C., P. McIntyre, M. Gessner, D. Dudgeon, A. Prusevich, P. Green, S. Glidden, S. Bunn, C. Sullivan, C. Liermann and P. Davies (2010), 'Global threats to human water security and river biodiversity', *Nature*, 467, 555–561.

10. Between a rock and a hard place: redefining water security under decentralization in Spain

Elena López-Gunn and Lucia De Stefano

10.1 INTRODUCTION

Spain is recognized worldwide for a long tradition and history of water management. The country is nestled in the Mediterranean basin, a cradle of civilization and an area famous for its warm but dry climate. In this geographic location, the need to guarantee water security due to natural water scarcity has triggered innovation throughout history, both in institutional terms and through the development of increasingly sophisticated (and expensive) water infrastructure. This chapter however is not focused on the history of addressing past challenges, but rather on the way the ongoing decentralization process in Spain is interacting with the institutional capacity to define water security. This means addressing not just the climate variability inherent to the Mediterranean climate but also climate change impacts. In order to do so, the chapter focuses on two aspects: how water security is defined and pursued over time across different scales; and how an ongoing decentralization process and the Europeanization of water management (after the approval of the European Water Framework Directive (WFD) in 2000) is reshaping national water policy paradigms.

As it will be seen below, at the national level the emphasis has partly shifted towards demand management and a lukewarm compliance with European water policy principles and guidelines which now give center stage to achieving an environmental good state. In contrast, many states in Spain are still focused on securing access to water resources, and/or increased supply to meet the existing demand from agriculture and the growing demand from non-agricultural sectors. In this context, the central federal government is reluctant or finds it politically difficult to re-allocate water resources from the agricultural sector through command and control regulatory approaches and is moving towards market based approaches (even if only as thin markets). The main challenge is how to trigger change and facilitate the task of reducing pressure on existing resources in fully or over-allocated basins. Added to these ongoing processes, it is also important to assess how this is played out in the political economy of water. In a highly complex system the ongoing parallel processes related, on the one hand to political decentralization and on the other to addressing resource pressures, can lead to non-linear unpredictable outcomes. This highlights the importance of identifying gaps and room for potential institutional strengthening and reform to address existing challenges. A possible *Pacto del Agua* (federal government water pact) among the main political parties and players in the water sector would provide a focus and litmus test, since the failure to achieve this elusive goal is a good indicator for the current ongoing decentralization process in Spain.

The chapter is structured as follows: the first section briefly outlines the constitutional organization of Spain, particularly in relation to water resources management. The second section outlines particular issues that have arisen in Spain in relation to the ongoing decentralization process and discusses how this is played out in relation to river basin management with reference to four case studies and two time periods. The final section reflects on the complexity of a three level game in Spanish federal water politics, because of the superimposition of the European Union Water Framework Directive.

10.2 SPAIN´S WATER RESOURCES AND CONSTITUTIONAL SET UP FOR WATER MANAGEMENT

Spain is a quasi-federal country of 500,000 km^2, a population of 45 million inhabitants, a mean rainfall of about 670 mm/year, which disguises a wide difference between the so-called 'wet' north, the 'dry' Spain in the interior, with a harsh continental weather, and the Mediterranean coast and the archipelagos, where much of the population is concentrated. Water resources are evaluated at 114 km^3 of which 47 km^3 are used (level of abstractions). If the most recent climate change projections for Spain materialize, by 2040 the mean annual temperature could increase between +1.4 to +1.9 °C and annual precipitation could decrease between 5 and 6 percent (CEDEX, 2011). This would translate into a potential reduction of between 7 percent and 14 percent in stream flows. It is less clear what the impact would be on groundwater recharge, due to the possible increase in short, flash floods.

In a context of water security, Spain is in theory well positioned to deal with its inherent climate uncertainty and variability, due to a long tradition of a focus on supply-based policies. This offers both challenges and opportunities. First, there is an advantage in the range of options to play with and the different types of water resources both conventional (surface, ground, soil), and non-conventional (artificial recharge, reclaimed, and desalinated). Second, European Union public policies requiring good water status represent a challenge because they require a reconsideration of the impact that highly regulated rivers and intensively used aquifers have on weakening the overall resilience of the physical resource base (Pittock and Finlayson, 2011). Finally, due to globalization, the role of virtual water is becoming more central, with debates and discussions about how imported embedded water could relieve pressure from domestic stressed water systems (Table 10.1). However this also raises concerns on the potential unintended consequences that these imports of 'cheap' virtual water from countries like Argentina or Brazil as animal feed crops might be externalizing the environmental impacts of Spanish consumption and production patterns (Willaarts et al., 2011).

In terms of water management Spain (together with the US) pioneered the catchment management approach in the last century. The creation of the Ebro River Basin authority in 1926, was followed by ten other river basin organizations (RBOs), as well as island water administrations for the Balearic and the Canary archipelagos, covering the whole of the country. Spain has a quasi-federal political system to acknowledge a tapestry of strong regional and nationalistic identities. The very nature of its quasi-federal system results in frequent debates and tensions because when the two main national parties are not able to get a majority in the federal government parliament, often regional parties are essential for coalitions needed to form a government. After the approval of the 1978 Spanish Constitution, the existing RBOs model had to be adapted to the newly created *Estado de las Autonomias* (federal government with autonomous regional governments or states as referred to in this chapter). Thus the Constitution established that water had to be managed at federal government level for the inter-community basins (those shared by two or more states), whereas for intra-regional basins (that is, those located within a single state) water is managed through the creation of regional water agencies (see Figure 10.3). River basin organizations are the executive arm of the central administration and are responsible for inter-sectorial allocation, water quantity and quality monitoring and enforcement, the authorization of water and discharge permits and water pricing, for example in agriculture. Similar organizations with the same responsibilities exist in the intra-regional basins. As a result, currently there are 11 RBOs (including Ceuta and Melilla) that depend on the central government and 12 regional

Table 10.1 Summary of water resources in Spain

Type of resources			Water resources magnitudes
CONVENTIONAL	SURFACE WATER	Surface water resources	22.000 million m^3
		Storage capacity[a]	55.590 million m^3 in more than 1,300 reservoirs
		Estimated annual demand[c]	31,500 million m^3/yr
	GROUNDWATER	Estimated storage volume till 100–200 m of depth	150,000–300,000 million m^3
		Available groundwater resources for human activities[c]	22,000 million m^3/yr
		Estimated recharge rate[b]	30,000 million m^3/yr
		Estimated annual demand[c]	6,900 million m^3/yr
NON-CONVENTIONAL	DESALINATION	Desalination capacity	111.5 million m^3/yr (57 desalination plants planned, with 21 functioning, 16 under construction)
	RE-USED WATER	Water reuse	430 million m^3/yr (2009)
	VIRTUAL WATER	Average net virtual water imports (1997-2006)[d]	12,800 million m^3/yr

Sources: [a] MIMAM 2000 [b] Molinero et al. 2011 [c] De Stefano et al. 2012; [d] Garrido et al. 2010.

water agencies. In this context, land use management is the responsibility of regional governments. This creates a latent tension in water management because of the need to liaise with agricultural (and irrigation) policies decided upon by the states, in a country where irrigation accounts for 70 per cent of national water use and holds an important symbolic power (Swyngedouw, 1999, 2007; López-Gunn, 2009).

10.3 FEDERAL RIVERS AND THE UNRAVELING DECENTRALIZATION PROCESS

The decentralization process sanctioned by the 1978 Constitution is still in flux and this has also been reflected in the water administration. It can be observed that 'constitutions reflect their societies' and the functioning of constitutions will change over time as society changes. Water in Spain is a major issue politically. Taking into consideration that federal systems are designed to share power, and are not static, water is part of a tug of war that takes place between specific states and the central federal government.

Decentralization has adopted a number of formats: de-concentration, political, administrative, and fiscal and market decentralization (Turner and Hulme, 1997). In the case of Spain all these aspects of decentralization apply to water. In the case of fiscal decentralization, this is becoming more central because of the current economic and financial crisis and the need to raise revenues directly, for example through water user fees.

In fact, Spain is now bucking the trend of most countries by partly reversing a process where rivers were managed on a catchment basis. To an extent disbanding RBOs and re-organizing water administration on a regional basis. This is because of the parallel political decentralization process which uses water as one of the arenas in a tug of war between the federal government and the states testing how deep the process will go. Therefore the most substantive and radical change in relation to water has been a consequence of political decentralization – or devolution – with directly elected sub-national governments asserting their interests. This is reflected in the ability of regional governments in Spain to issue their own laws and also in the development of parallel bureaucratic structures like regional water agencies. The next section analyzes the evolution of the concept of water security in Spain to show how both processes (i.e., ongoing changes in defining water security and political decentralization) have become entangled with unpredictable outcomes.

10.4 DEFINING WATER SECURITY AT DIFFERENT SCALES

This section will explore the different definitions of the concept of water security (Cook and Bakker, 2012), and the tension this creates across different scales. Traditionally, the water security concept has been approached from a physical perspective linked to national security, a concern over potential 'green wars' (Tuchman Mathews, 1989) and to cases where water could be a potential cause of conflict between neighboring countries (López-Gunn et al., 2012a). Therefore, securing sufficient water resources and guaranteeing access has been seen – and in some cases still is – strategic for national security (UNESCO-IHE, 2010; Houdret, 2004). More recently however, attention has shifted to domestic conflicts, between groups or users, since in the case of water, geographic scale and intensity of conflict are inversely related (Wolf, 1998). Lately, the term has also been associated with virtual water as a means for a government to ensure 'water security' through food imports (Allan, 2011; Guodong, 2003). In the case of Spain, water security has in many ways focused on reducing vulnerability to climate variability by developing storage infrastructure (Figure 10.1), with Spain being the fifth country in the world after China, the US, India and Japan in regulation capacity with 1300 reservoirs that can store up to 55.59 km^3 (Estrela and Vargas, 2012).

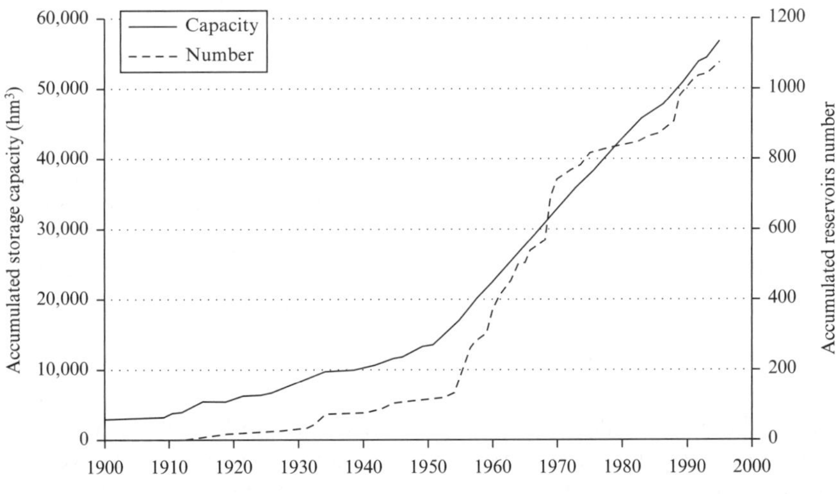

Figure 10.1 Accumulated evolution from 1900 to 2000 on the number of reservoirs built in Spain and storage capacity (modified from MIMAM, 2000)

In recent times, this has been complemented by initiatives to increase desalination capacity, water reuse and artificial recharge (Downward, 2007). A focus on increasing access and storage and increased supply, was gradually upgraded and complemented by a risk-based approach with a provision under water planning to include protection against extreme events, mainly floods and droughts (Estrela and Vargas, 2012). However, a more recent approach is starting to emerge, which adopts a complex systems frame, where stationarity is dead (Milly et al., 2008). It links with a more subtle and integrated view on the way to achieve water security, associated with the 'ecosystems services' concept, where the 'good functioning' of the hydrological cycle and associated ecosystems brings system resilience and security (Willaarts et al., 2011; Grey and Sadoff, 2007). This approach is reflected in the 2000 EU Water Framework Directive, which acknowledges the need to protect all types of water (and associated ecosystems) to contribute to the long term provision of good quality water. Thus in the most recent definition water security is now encompassed within the wider concepts of human security, framed by environmental security (López-Gunn et al., 2012b).

As a quasi-federal country, Spain operates on the premise of relinquishing part of its sovereignty, upwards towards the European Union and downwards or across to the states. However, it is precisely the different understandings of water security at different scales that get entangled with tensions over centralizing/decentralizing trends.

The European Union at the supranational level has introduced *and* pushed water security towards a conceptualization framed by environmental security, and the ultimate good status of the system (see Figure 10.2). Thus, the characteristics of the Spanish federal system when looked through the lens of water security, have to be understood as a polycentric system, with tensions across scales on how to define water security, where the Spanish federal government is caught up between the states' preference for the 'old' water security paradigm, focused on securing access to resources (internal and external if necessary and/ or possible) and the 'new' water security paradigm focused on ecosystem resilience.

10.4.1 A Three Level Game

One of the ongoing debates is whether the European Union is a loose federation or a unique organization that has similarities to a federal system. In reality for EU federal member states like Spain, this has meant an additional third layer in terms of supranational relations. That is, the policies originating from the EU under EU law are supreme to its member states, who have voluntarily relinquished part of their sovereignty.

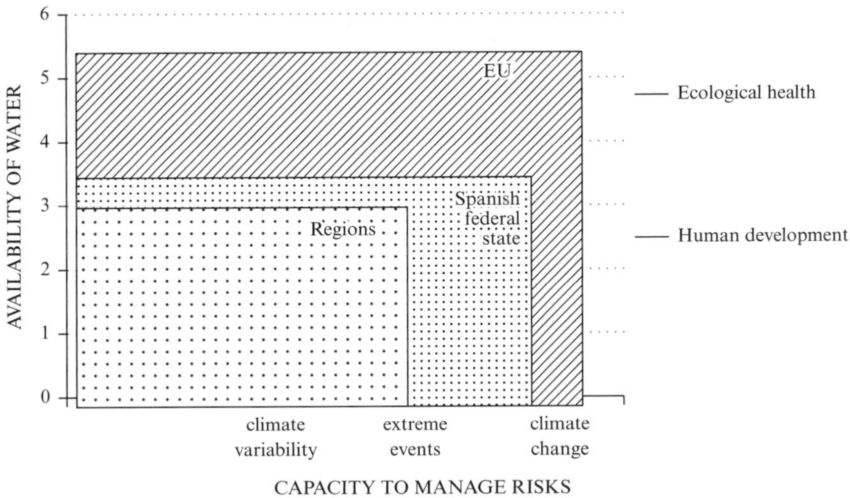

Figure 10.2 Definitions of water security across scales

The Water Framework Directive addresses water security from a resource perspective, and a critical milestone in the implementation of the WFD is the approval for each river basin district (RBD) of a river basin management plan (RBMP), designed to achieve the WFD objectives within six-year planning cycles. However, the deadline for approving new plans expired in December 2009 and by December 2012 only five RBDs had completed the full process leading to a RBMP. In most basins, water planning got 'caught up' in political negotiations, quagmires and legal disputes. The federal government has been unable or unwilling to successfully address tensions between states. Consequences of this generalized delay include not only legal actions initiated by the European Commission against Spain (in 2010, 2011 and 2012), but also and more importantly, reductions in the time available in most of the river basins to implement the RBMPs. The tight link between water and land use also puts the focus on the need to plan and make decisions from the perspective of integrated water and land use management. A well-targeted management of land uses is the most effective way of addressing water security problems from a complementary perspective (relative to water resources management) and is the missing piece in the puzzle of Spanish water policy. Thus ultimate success in water security, understood in its acception of ecological integrity, necessarily depends on coordination and cooperation between the federal government and the states. Yet at the moment there is limited or no coordination.

In the Spanish context, regional governments are in charge of defining, funding and implementing most of the sectorial policies that determine land use. Thus, states play a key role in the achievement of water security based on the achievement and maintenance of good water status. At the same time, states are not held responsible before the EU for the achievement (or failure to achieve) the WFD requirements, as the ultimate responsibility for compliance with EU legislation lies with the Spanish central federal government. To facilitate the integration of water and land management, a new coordinating body has been established in each RBD, the Committee of Competent Authorities, that includes representatives from different administrative levels (national, regional and local) and sectors (agriculture, industry, etc.), as well as ports and coastal management. However, these new RBD bodies still have to prove to be an effective way of conciliating often apparently diverging interests.

10.4.2 Twists and Turns: How to 'Secure' Water at Federal Level

As outlined at the beginning of this section, the concept of water security is defined differently through time and across scalar levels (Cook and Bakker, 2012). At national level, in line with the push–pull model argued by Börzel (2001), compared to the EU which was focused on environmental security, the focus for the federal government and the states is centered on water security defined as guaranteeing access to physical water. Yet within this conceptualization, two different approaches can be identified: the first associated to hard-engineering solutions which involves states claiming water from rivers outside their regional boundaries through water transfer schemes. A second approach, more recent and with complex political and institutional implications, is that states claim 'sovereignty' and control over the resources of rivers within their borders but shared with other states, while also constructing non-conventional water resources infrastructure. The first model worked to an extent in the twentieth century, under a highly centralized government and in a moment where Spain needed to harness its water resources to spur socioeconomic development. This model however has gradually been questioned and eroded by a civil society movement spearheaded by 'New Water Culture' paradigms (Martinez-Gil, 1997; Tabara and Ilhan, 2008). This coincides with a parallel claim for the local and the re-localization of resources, helped by the fact that states are no longer willing to give away water resources for the sake of 'hydro-solidarity' (Falkenmark, 2005; Llamas and Pérez-Picazo, 2001). The role of the central federal government, states and the EU in relation to water disputes and decentralization tensions, has followed two distinct stages.

- The first stage covers the period 1991–2004. During this period conflict was played out largely at the federal level, with the presentation in 1993, and 2001 of two national water plans, where the centerpiece was based on interbasin transfers between states. States largely responded reactively, through contestation via the media and through large public demonstrations in favor or against the planned transfers (depending on whether they were being asked to export or import give). They also actively attempted to inflate water demands in the catchment river basin plans to 'demonstrate' lack of surplus water resources in the case of potentially exporting basins. In the case of possible importing basins they sought to demonstrate structural water deficits. These disputes played out through the development of a series of national water plans (see MIMAM, 2000).
- The second stage extends from 2004 to 2011. The conflict in this period has been downscaled to the level of the states. The federal government has taken a backstage position. The main aim was to claim full sovereignty over regional water resources while expanding the resource base through non-conventional water resources. These conflicts have been largely resolved through court rulings, with the federal government in many ways retreating from playing a mediation role. This has resulted in failure to submit river basin plans for the WFD. Thus the constituent units have not been able to resolve some key issues among themselves while the federal government – which had the opportunity to intervene and help resolve the issues – failed to do so. In terms of power asymmetries some states which are not as powerful economically, have been able to gain political power at national level through water. They have focused on securing access to 'regional' water resources in a 'water war' against (wealthier) neighboring states.

In terms of conflict resolution, the judiciary has played an increasingly more central role in settling disputes regarding jurisdictional competences over water. It seems at this stage (although additional research and analysis is needed) that the judiciary has generally developed or enforced a doctrine regarding water rights and management which gives precedence to the federal level in shared basins. Meanwhile judicary it is also expecting states to fully exercise their constitutional assigned duties in relation to water when relevant.

10.4.3 Downscaling Water Security at State Level

Spain's states in general address water security issues with special emphasis on securing water resources for regional socioeconomic development.

That is secure resources should be sustained also during droughts, and, to a far lesser extent, for ecosystem maintenance. To get an overview on the evolution of water security as a concept and how it is entangled with decentralization, it is illustrative to look at a set of four emblematic cases of disputes arising from the search for water security in Spain's federal context (Box 10.1) which looks at the existence or functioning of the Ebro and Tagus-Segura water transfer schemes; and the disputed jurisdiction over the Guadalquivir and Jucar basins. The downscaling of water security to state level is focused on securing access to water, in the first period as policy takers from the federal government which pursued the goal of physically transferring water between basins. This hard path contrasts with the institutional way of the second period, which has become much more centrifugal. During this time states have made claims to sovereignty for the resources within their borders, to claim back water already forfeited in existing and operating transfers like the Tagus-Segura scheme. The Tagus-Segura entered a phase of retrospective analysis and discussion about its continuity in the future. The Ebro transfer so far is only a plan, officially approved in 2001 and subsequently revoked by royal decree in 2004. Its history exemplifies the shift of focus from the federal level which stepped back and gave more room to states on water issues. The Guadalquivir and the Jucar basins are examples where states have engaged into a fierce legal battle with the Spanish federal government and other neighboring states to get a higher or exclusive control over water resources within their regions.

These four cases under discussion cover a territory which includes around half of Spain's population and economic production from agriculture and tourism (see Table 10.2). They are also located in naturally water scarce areas (see Box 10.1). These areas represent undoubtedly a daunting bottleneck for Spain's water planning under the WFD requirements. These cases show that there are very few dispute resolving mechanisms and arenas available, which means it is a win/lose situation where many disputes are resolved through the courts. Thus there is plenty of room and scope for strengthening and developing collaborative mechanisms to mediate at the different scale boundaries (federal/state; state/state; federal/local). Questions remain over how we define water security: At what level? Is water security an evolving concept? Do different levels pursue different strategies rooted in different water security concepts? What happens if these strategies are mutually exclusive or incompatible?

BOX 10.1 CASE STUDIES ON WATER SECURITY AND DECENTRALIZATION

Tajo-Segura transfer

The Tagus-Segura transfer was completed in 1979, to bring around 1000 Million m^3/yr of water surplus from the Tagus headwaters to the Segura basin in the Mediterranean coast. In reality the scheme has never carried more than 600 Million m^3/yr and on average transfers have been around 320 Million m^3/yr (i.e. original plan) due to lack of water in the donor area. The transfer is managed by the Central federal Government, with the Council of Minister´s approval for each water delivery. In 2004–2007 a prolonged droughts – which affected both basins heated conflicts over the transferred volumes and also sparked the mobilization of a local social platform in the upper Tagus. Between 2004 and 2010 the Castilla La Mancha parliament, the state in the upper tagus basin, discussed several drafts of its Statute, which earmarked between 4000 and 6000 Million m^3/yr for the state and required cancelling the transfer by 2015. The main recipient state, Murcia, in the Segura basin, fiercely defends the continuity of the transfer, arguing it is essential for sustaining high value agriculture and tourism. Both states have repeatedly ended up before the Supreme Court to argue their case and the battle is now played in the new Tagus and Segura River Basin Management Plans. Both Plans have to be issued by the Central federal Government via the Tagus and Segura RBOs and the RBMPs will have to clearly state which of the two basins can include the transferable volumes in its accounting of 'available resources'.

Ebro transfer

The 2001 Spanish National Hydrological Plan (SNHP) – approved by law by the Central Government – established the construction of a 914-km-long water transfer 1,000 Million m^3/year from the lower Ebro river, to the different states along the Mediterranean coast. The main donor states, Aragon and Catalonia, and a coalition of environmental and social movements fiercely opposed the project, and the derogation or continuation of the transfer was a political hot issue in the 2004 national elections campaign. The infrastructure was going to be partly funded by the EU, which, seeing the unresolved dispute over the transfer, delayed the final funding approval. Eventually in March 2004 a new government was elected and, in order to keep its electoral pledge, cancelled the transfer scheme. After this, both donor and recipient states have used the reform of their state Statutes to claim the Ebro waters for themselves. The 2006 Catalonia statute included a clause establishing the power to veto any water transfer affecting the state, and the 2007 Aragon Statute earmarked 6550 Million m^3 for its own development, which in practice is a veto to any transfer. On the recipient side, the 2006 Valencia Statute claimed the right of Valencian citizens to receive surplus water from other river basins.

Jucar

The Jucar RBO was created in 1934 to manage – on behalf of the Spanish State – several sub-basins including the Júcar that ends in the Mediterranean sea. In 2007, as part of the WFD planning process, the central federal government defined the boundaries of the Jucar River Basin District, and decided to exclude the intra-regional sub-basins from the newly established RBD, thus transferring authority over them to the Valencia state. However, the Valencia state filed a complaint before the Supreme Court, asking to include the intra-regional sub-basins in the RBD. The main interest of Valencia to maintain the status quo and not having to manage its intra-regional sub-basins is that the Jucar watershed and the regional sub-basins are highly interconnected through water schemes and having two different jurisdictions over water – regional and federal – would highly complicate water supply to the intra-regional basins. At the same time the Castilla-La Mancha state, in the upper part of the Jucar's main stream, asked the Supreme Court to ensure that the new territorial boundaries were fully enforced – as this would mean Castilla-La Mancha would have a larger control of the Jucar waters. In 2011 a Supreme Court ruling rejected Castilla La Mancha´s complaint, and accepted Valencia's one and established that the central federal government should revise the RBD boundaries to all the original sub-basins.

Gualdaquivir

Historically the Guadalquivir basin has been managed by the Central federal government, through the Guadalquivir RBO created back in 1927. The 1978 Constitution did not change this set up, since it established that rivers shared by more than one state – like the Guadalquivir – are under the jurisdiction of the Central federal government. In 2007 the Andalusian parliament, argued that a state which has 90 per cent of its territory in the Guadalquivir watershed, should have exclusive jurisdiction over the waters of the Guadalquivir river, thus it reformed its state Statute of Autonomy. In 2009 the federal government, accepted Andalusia's request, and transferred to the state most of the water competences, including the Guadalquivir River Basin Authority. However, the Extremadura state, which has a small share of the Guadalquivir river contested the Andalusian Statute, claiming that the clause relative to the jurisdiction over water was unconstitutional. In 2011 the Constitutional Court ruled that Andalusia had to return the transferred authority to the federal government, which means that the situation is back to the pre-2008 status quo.

Figure 10.3 *Map highlighting (a) basins managed by the federal government through RBOs and basins managed by state water agencies (dark gray); and (b) case study areas*

Table 10.2 Typology of regional disputes over water

	Basin Area (Km^2)	Basin Population (thousands inhabitants)	Average rainfall (mm/yr)	Mean 1980–2006 annual flow (million m^3/yr)	Federal governments sharing the river (% area of the river basin)	Authority responsible for the river/ water transfer management	Approach of federal governments to water security	Focus of the dispute
SPAIN	500,000 (total Spain's territory)	46.160 (total Spain's population)	670	NA	17 Autonomous Communities	Central Government in inter-regional basins and Regional Governments in intra-regional basins	NA	NA
Guadalquivir	57,527	4,100	538	7,000	Andalusia (90.2);Castilla-La Mancha (7.1), Extremadura (2.5, Murcia (0.2)	Central Government	Increase the control of a river flowing within the region	Jurisdiction over the Guadalquivir basin. Andalusia, who owns 90% of its territory, claims full control over the river
Jucar	42,832	5,100	500	3,250	Valencia (49.6) Castilla-La Mancha (36.6, Aragón (13.2), Catalonia (0.6)	Central Government	Increase the control of a river flowing within the region	Dispute over the boundaries of the RBD, because their delimitation changes power balance of the states sharing the basin

Ebro/Med. Coast transfer scheme	Ebro: 85,534	Ebro: 3,000	Ebro: 620	Ebro: 18,200 (period 1940–86) Planned transfer: 1,100	Ebro: Aragón (49.5), Catalonia (17.6), Navarra (11), Castilla Leon (9,6); La Rioja (5.9), Pais Vasco (3.2), Valencia (1), Cantabria (0.9) Med. Coast: Valencia, Murcia and Andalusia	Central Government	Securing water resources from external basins/ Blocking water transfers leaving the region	Dispute over the construction of an inter-regional transfer
Tagus/Segura transfer scheme	(Spanish) Tagus: 28,033 Segura: 18,870	Tagus: 7,800 Segura: 2,000	Tagus: 648 Segura: 365	Tagus: 8,273 Segura: 871 Average transferred water: 320	Tagus: Castilla-La Mancha (48.2), Extremadura (14.4), Madrid (14.4), Castilla Leon (7.1), Aragon (0.4) Segura: Murcia (59), Castilla-La Mancha (25) Andalusia (9), Valencia (7)	Central Government	Securing water resources from external basins/ Blocking water transfers leaving the region	Dispute over the volumes transferred and the future continuity of a transfer. The states in the Tagus basin now want to control and use the resources normally transferred to Segura

10.5 CONCLUSIONS

During the past decade, the Spanish national government has swung from a highly controlled and centralized approach to Spain's water policy – crafted in the 2001 National Hydrological Plan and the Ebro transfer plan – to a de-centralized and accommodating approach that was highly responsive to state demands – exemplified by the transfer of the Guadalquivir water competence to Andalusia in 2008. Up to now, however, none of these approaches has been able to deliver a long-term satisfactory solution and national water policy remains a controversial issue. The first approach fails because it seems to deny acknowledging and dealing with ineluctable – no matter if legitimate or not – regional aspirations and disputes triggered by the states' need to secure more and better resources for their socioeconomic systems. The second approach also fails because it does not treat the watershed as a physical unit and seems to forget that any long term plan over shared resources needs to be based on a common understanding of problems and needs.

The process of Europeanization (outwards) and decentralization (downwards) through the *Estado de las Autonomias* has changed the politics of water in Spain. Questions are still pending whether this will eventually be reflected in water security in its most recent and complex definition, focused on social-ecological positive outcomes. Possibly knowing the economic value of water uses – through transparent water accounting systems – helps dilute the idea that water abundance is a prerequisite to thriving societies. It opens the possibility to shift water debates from rights ('water is mine for historical reasons') to needs ('I need water to cover these needs') and from needs to sharing benefits ('how can I cover these needs'). This would mean moving away from nested positions on water, towards the achievement of an agreement – a water pact – among the main political parties, listening to Spanish society aspirations – based on a shared vision of the future of water in Spain. In this context the role of the federal government is pivotal. Not just for funding infrastructure. Its involvement is essential for an effective response to new challenges of potentially a different order of magnitude like climate change risk and adaptation, international trade changes, or the knock on effect from EU agricultural policy reforms.

10.6 REFERENCES

Allan, T. (2011), *Virtual Water: Tackling the threat to our Planet's most precious resource*. London and New York: I.B. Tauris, p. 384.

Börzel, T.A. (2001), 'Non-compliance in the European Union: Pathology or Statistical Artefact?', *Journal of European Public Policy*, **8**(5) (October): 803–24.

CEDEX (2011), *Evaluación del impacto del cambio climático en los recursos hídricos en régimen natural*. Madrid, Spain: CEDEX.

Cook, C. and K. Bakker (2012), 'Water Security: Debating an Emerging Paradigm', *Global Environmental Change*, **22**: 94–102.

De Stefano, L., Martínez-Cortina, L. and D. Chico (2012), 'Chapter 7: An Overview of Groundwater Resources in Spain' in Lucia De Stefano and M. Ramón Llamas (eds), *Water, Agriculture and the Environment in Spain: Can we square the circle?*, Leiden: Taylor & Francis.

Downward, S.R. and R. Taylor (2007), 'An Assessment of Spain's Programa AGUA and its Implications for Sustainable Water Management in the Province of Almería, Southeast Spain', *Journal of Environmental Management*, **82**: 277–289.

Estrela, T. and E. Vargas (2012), 'Drought Management Plans in the European Union. The Case of Spain', *Water Resources Management*, **26**(6): 1537–1553.

Falkenmark, M. (2005), *Towards Hydrosolidarity: Ample opportunities for human ingenuity*. Fifteen-year Message from the Stockholm Water Symposia SIWI Stockholm, Sweden.

Garrido, A., Ramón Llamas, M., Varela-Ortega, C., Novo, P., Rodríguez-Casado, R. and M. Aldaya (2010), *Water Footprint and Virtual Water Trade in Spain: Policy implications*. New York: Springer.

Grey, D. and C. Sadoff (2007), 'Sink or Swim? Water Security for Growth and Development', *Water Policy*, **9**(6): 545–571.

Guodong, C. (2003), 'Virtual Water – A Strategic Instrument to Achieve Water Security', *Bulletin of the Chinese Academy of Sciences*. Issue 4. Available at: http://en.cnki.com.cn/Article_en/CJFDTOTAL-KYYX200304005.htm.

Houdret, A. (2004), 'Water as a Security Concern – Conflict or Cooperation?', *5th Pan-European Conference of International Relations*. September 8–11, 2004. The Hague, Netherlands, available at: http://www.afes-press.de/pdf/Hague/Houdret_Water_Security.pdf (accessed 14 November 2011).

Llamas, R. and T. Pérez-Picazo (2001), 'The Segura Catchment Management and the Debate on Hydrosolidarity in Spain', *Water Security for Cities, Food and Environment – Towards Catchment Hydrosolidarity*. Proceedings of the 2001 SIWI Seminar. Stockholm, Sweden: SIWI.

López-Gunn, E. (2009), 'Agua para todos: A new regionalist hydraulic paradigm in Spain', *Water Alternatives*, **2**(3): 370–394.

López-Gunn, E., Willaarts, B., Dumont, A., Niemeyer, I. and P. Martínez-Santos (2012a), 'Chapter 2: The Concept of Water and Food Security in Spain' in Lucia De Stefano and M. Ramón Llamas (eds), *Water, Agriculture and the Environment in Spain: Can we square the circle?*, Leiden: Taylor & Francis.

López-Gunn, E., De Stefano, L. and M.R. Llamas (2012b), 'The Role of Ethics in Water and Food Security: Balancing utilitarian and intangible values', *Water Policy*, 14, Supplement 1: 89–105.

Martinez-Gil, F.J. (1997), *La nueva cultura del agua en España* [The New Water Culture in Spain]. Bilbao, Spain: Bakeaz.

Milly, P., Betancourt, J., Falkenmark, M., Hirsch, R., Kundzewicz, Z.,

Lettenmaier, D. and R. Stouffer (2008), 'Stationarity is Dead: Whither water management?' *Science*, **319**: 573–574.
Ministerio de Medio Ambiente (MIMAM) (2000), *Libro Blanco del Agua en España* [Spain's White Water Book]. Madrid, Spain.
Molinero, J., Custodio, E., Sahuquillo, A. and M. Ramón Llamas (2011), 'Groundwater in Spain: Legal framework and management issues', in Angelos N. Findikakis and Kuniaki Sato (eds), *Groundwater Management Practices*. Leiden, Netherlands: CRC Press.
Pittock, J. and C.M. Finlayson (2011), 'Australia's Murray Darling Basin: Freshwater ecosystem conservation options in an era of climate change', *Marine and Freshwater Research*, **62**, 232–243.
Swyngedouw, E. (1999), 'Modernity and Hybridity: *Regeneracionismo*, the Production of Nature and the Spanish Waterscape, 1890–1930', *Annals of the Association of American Geographers*, **89**(3): 443–465.
Swyngedouw, E. (2007), 'Technonatural Revolutions: The scalar politics of Franco's hydrosocial dream for Spain 1939–1975', *Transactions Institute British Geographers*, **32**: 9–28.
Tabara, D. and A. Ilhan (2008), 'Culture as Trigger for Sustainability Transition in the Water Domain: The case of the Spanish water policy and the Ebro River Basin', *Regional Environmental Change*, **8**, 59–71.
Tuchman Mathews, J. (1989), 'Redefining Security', *Foreign Affairs*, Spring, Council on Foreign Relations.
Turner, M. and D. Hulme (1997), *Governance, Administration and Development: Making the State Work*. London: Macmillan.
UNESCO-IHE (2010), *Water security*, available at: http://www.unesco-ihe.org/Research/Research-Themes/Water-security (accessed 16 November 2011).
Willaarts, B., Niemeyer, I. and A. Garrido (2011), 'Land and Water Requirements for Soybean Cultivation in Brazil: Environmental consequences of food production and trade', *Conference Proceedings prepared for the International Water Resources Association* (IWRA), Brazil.
Wolf, A.T. (1998), 'Conflict and Cooperation Along International Waterways', *Water Policy*, **1**(2): 251–265.

PART IV

SOUTHERN AFRICA

11. Allocating powers and functions in a federal design: the experience of South Africa

Mike Muller

11.1 INTRODUCTION AND BACKGROUND

This chapter reports on the case of South Africa where, by virtue of major political reform occurring more or less contemporaneously with the need for a fundamental review of water resource management, a systematic process of institutional design for water management was undertaken. The chapter provides the background, describes both the political and the technical processes that were undertaken and how they interfaced, considers the initial design of the system that was introduced and reflects on progress with its establishment and the outcomes derived from it.

Interventions during the negotiation of a new Constitution avoided the political federalization of river management, which would have resulted from the allocation of water resource management functions to sub-sovereign levels. This technical outcome was achieved in part because the interventions that addressed the management of water resources were located in the context of improving water supply and sanitation services, which were the political priority for most of the negotiators. This highlights the importance of understanding the complex nature of water management and the interaction between resource management and service provision at both a political and a technical level. The immediate outcome was to maintain the historic trend towards greater centralization of management and to insulate water resource management from "political federalism", the division of powers and functions between different levels in a federal system.

The approach to river management in subsequent legislation also avoided introducing "basin federalism", the allocation of powers and functions to administrative units based on basin boundaries. While provision was made for some resource management functions to be undertaken on a catchment basis, generic water management areas rather than river basins themselves were made the primary unit of management and the national water authority retained overall authority over planning, allocation (including inter-basin transfers) and trans-boundary management issues as well as retaining oversight powers over water administration within basins. South Africa has thus avoided the pitfalls inherent in allocating water resource management powers and functions to sub-sovereign political structures whose boundaries do not coincide with river basins or "balkanizing" its management at the level of the physical river basin.

11.2 NEGOTIATING CONSTITUTIONAL CHANGE

South Africa's democratic revolution in 1994 transformed many of the country's public institutions. The Constitution that was adopted in 1996 (Government of South Africa 1996) inaugurated a constitutional democracy with a Bill of Rights (including rights to a healthy environment as well as to sufficient food and water) in which the powers of executive, legislature and judiciary were formally balanced. It also established new provincial and local political jurisdictions, described, together with the national as "spheres" rather than "levels", in recognition of the original powers and functions which they were given.

This structure, seen by some commentators as "quasi-federal" (Simeon 1998) was described rather as a system "of cooperative government in which powers in a number of important functional areas are allocated concurrently to the national and the provincial levels of government" (Constitutional Court 1996). It specifically distinguished this from the system of "competitive federalism" which some political parties had favoured.

Nonetheless, the final Constitution contains important elements of federalism in that it reserves certain powers and functions as the exclusive competence of a particular sphere although allowing the central government certain rights of oversight and intervention. A feature of the constitutional development process, described in more detail below, was that specific criteria were put in place for the allocation of powers and functions.

11.3 WATER REFORM IMPERATIVES AND CONSTITUTIONAL NEGOTIATIONS

Far-reaching political reform coincided with another critical moment in South Africa's life, the pressing need for a new approach to be adapted to the management of the country's scarce water resources. South Africa is a relatively water scarce country, ranked 29th (out of 193) (UNESCO-WWAP, 2006) in terms of available water resources per capita, on the border of widely used definitions of water stress. While the relevance of total water availability per person has been questioned, a more obvious indicator of stress is the intensity of water use, expressed as the water exploitation index, the proportion of available water actually used. At 31 per cent, South Africa is, again, among the most intensive water users in the world (32nd out of 163). Higher intensities are found mainly among Gulf and Mediterranean petro-states, although Spain, India and Pakistan are also intensive users; the controversies and challenges in those countries illustrating the value of "water use intensity" as a measure of "water stress".

South Africa was approaching the limits of its easily exploitable water resources, as was evident from the extent to which deficits in major basins were being met by transfers from basins still in surplus. A review of management approaches was thus becoming urgent and there was increasing emphasis on "soft" management options to optimize the use of the available resource rather than "hard" infrastructure development options to increase supply.

The managed nature of the political transformation created the conditions in which a structured process of policy review could be undertaken. In particular, the underlying political dynamic meant that powerful interest groups, such as the commercial farmers and large industrialists who had previously enjoyed hegemony over the minority regime's water policy, could no longer dominate the debates, which now had to include the formerly disenfranchised majority.

This was emphasized by the fact that many of the reforms that were introduced were built on a foundation that had already been laid through a policy review, conducted 20 years before (Commission of Enquiry 1970) but whose recommendations had not been put fully into effect. As was pointed out by then Minister of Water Affairs Kader Asmal, they had to wait until a democratic government was in place (Asmal 1998).

The transition period between 1990 and 1994 thus provided fertile ground for a structured process of policy analysis and design within a political framework that made it possible for "rational policy" approaches to have some influence. The consequence was that powers and functions relating to water management were distributed between the different spheres of government in accordance with technical recommendations rather than as a result of interest group lobbying.

Politics did play a part in the process to the extent that its initial focus was not on water resources but on the water supply and sanitation services derived from them. This focus ensured that the recommendations had political legitimacy since the concern, at the time of transition, was to redress the huge imbalances between black and white communities among which access to basic services was one of the more egregious. To the extent that "water" was on the political agenda, it was water for human consumption that informed the strategies of different political parties as was demonstrated by opinion surveys conducted at the time. The fact that recommendations for water resource management were presented as part of a package of the measures needed to respond to the challenges of water supply and sanitation services meant that they were more likely to be accepted.

11.4 SOUTH AFRICA'S WATER MANAGEMENT HISTORY

As in many countries, the history of water management in South Africa reflected the country's broader socio-economic development (these issues are addressed in more detail in Muller 2012). During the first few centuries of colonial development, water was essentially managed at local level with investment often private or cooperative, with agriculture by far the largest user. With colonial expansion, the development of major mining industry and urbanization, that picture changed. Increasingly, water demands could no longer be met without major investment that exceeded the capabilities of private users or the capacity of local sources. State involvement thus began to dominate as the needs of major consumers required the development of distant sources, increasingly often in different catchment areas involving expensive schemes and investments on a decadal scale.

This process had already begun when, in 1910, the Union of South Africa was established, uniting two British colonies and two Boer republics (while formally disenfranchising for the next 80 years the majority native population). The water supply needs of the rapidly expanding gold industry were being met by Rand Water, initially a user-owned utility, established in 1906 and still supplying bulk potable water to the country's economic heartland. The establishment of the union was closely followed by the establishment in 1912 of the Department of Irrigation, the precursor of today's national Department of Water Affairs. The department focused initially on the expansion of irrigation as a driver of both economic growth and the expansion of settlements into the interior of the country.

In 1956 the Department of Irrigation was transformed into a Department of Water Affairs, under new legislation, in recognition of the growing importance of access to reliable water supplies for other sectors of the economy and society. That legislation also entrenched the role of the central government in the administration of what had up until then, been largely a local matter:

> The centralized planned management and control of the country's water resources was made possible through fundamental amendments to our legislation passed in 1956, whereby full control over the allocation and use of water in public streams was vested in the State. These amendments placed in the hands of the Government the instrument for so apportioning the available water resources among various sectors of the economy so that they could make maximum contribution to the development of the country as a whole.
>
> In the Republic, development, management and administration of water affairs are now strongly centralized. Only in providing for urban needs and in the distribution of water do local and regional authorities play an important but subsidiary role. (Commission of Enquiry 1970)

And, the commission concluded:

> It is desirable for various reasons that the Republic's water resources should continue to be developed to the benefit of the country as a whole under the full control of the Central Government. (Commission of Enquiry, 1970: 12)

This architecture remained intact until 1994 although some half-hearted efforts were made in the 1980s to introduce a degree of autonomy for the so-called "homelands", self-governing territories established as part of the policy of racial separation. Legislative provision for these were revoked as part of the 1994 political settlement but they did serve to put federal management rather than centralization on the table as an option for shared rivers. This was reinforced by the fact that black politicians from the then "homeland" of Kwazulu together with the white provincial politicians of Natal had long been interested in promoting a TVA (Tennessee Valley Authority) style hydropower development on the Thukela River, which falls almost entirely within what is now the province of Kwazulu-Natal.

11.4.1 The Nature of Water Management

Consideration of the evolution of "water management" in South Africa helps to clarify a term which is widely used but poorly understood. "Water management" covers a multitude of activities, at scales ranging from household to continental. A simple popular definition is that "water management is the activity of planning, developing, distributing and managing the optimum use of water resources" (Wikipedia 2012). However, such definitions do not explicitly address some key elements of the activity, notably the management of water-bodies for transport, recreation, the protection of communities from floods as well as the protection of the natural environment. It also conflates the use of water inside a private premises (home, factory or farm) for private purposes with the management of the water resource in the natural cycle and its distribution to users through public infrastructure.

A further useful distinction, frequently drawn, is between the management of the natural water resource and the management of water services, usually provided at a local government or regional utility level. Water services are normally considered as the supply of water that has been treated to meet users' requirements and is delivered to an agreed location with a managed degree of reliability as well as the removal and treatment of wastewater, which is often an integral part of water service management. In agriculture, the equivalent distinction is between the bulk supply of water to a farm (or farming area) boundary and the distribution and in-field management of that water by the farmers.

The lack of definitional precision derives in part from large differences in context. In countries with plentiful water resources, the focus of water management may be on environmental protection and flood management. On the other hand, in countries in which water use has intensified and demand for services expanded, there is a growing focus on the parallel challenges of ensuring reliable bulk supplies and maintaining water quality while also providing services.

Water stressed South Africa is an example of a country where more explicit attention is required to all dimensions of water management and in which the distinction between the management of the resource and the management of services is clearly drawn. For both technical reasons and to comply with the constitutional allocation of powers and functions, it has been necessary to develop a structured interface between resource management and service provision activities.

11.4.2 Constitutional Negotiations and the Role of Forums

South Africa's new constitutional dispensation was negotiated between 1990 and 1996. Agreement was reached with regard to the macro-structure of the state, an agreement set out in the Constitutional Principles, adopted in 1993 (Government of South Africa 1993). These determined that there would be national, provincial and local government levels; that the powers and functions of the national government and provincial governments and the boundaries of the provinces would be defined in the Constitution together with a framework for local government powers, functions and structures.

The criteria to be applied in the allocation of powers were explicitly set out, based on the level at which decisions could most effectively be taken which would be conducive to financial viability and effective public administration:

> The level at which decisions can be taken most effectively in respect of the quality and rendering of services, shall be the level responsible and accountable for the quality and the rendering of the services, and such level shall accordingly be empowered by the Constitution to do so.
>
> Where it is necessary for the maintenance of essential national standards, for the establishment of minimum standards required for the rendering of services, the maintenance of economic unity, the maintenance of national security or the prevention of unreasonable action taken by one province which is prejudicial to the interests of another province or the country as a whole, the Constitution shall empower the national government to intervene through legislation or such other steps as may be defined in the Constitution. (Government of South Africa 1993)

These criteria had relevance to the water sector.

As the formal high level political negotiations proceeded between all political parties, divided primarily into the white government and the democratic opposition groups, there was much wider engagement at a more technical level. In addition to the economic sectors, negotiating forums were established in a variety of sectors, notably local government, health, education, housing, electricity and environment. The objective of these forums was to develop a common understanding of the policy challenges and preliminary strategies to address them. In addition, they served important practical purposes of:

- building contact and confidence between parties;
- enabling the previously disenfranchised majority parties to gain greater knowledge of the sectors concerned; and
- helping the different parties to narrow their areas of difference about the way forward.

In turn, through their processes, the forums informed the high level negotiations since there was extensive interaction between the forums and the negotiators and positions agreed in forums were often adopted in the political negotiating process.

The contribution of these forums, at both national and local level, to the early policy of the national government was recognized in the government's post-1994 Reconstruction and Development Progamme and it was hoped by some that they would become long term participatory structures:

> A variety of sectoral negotiating forums has developed a participatory approach to policy formulation. National line function departments will be encouraged,

> where appropriate, to continue ongoing policy interaction with sectoral forums, which comprise key sectoral stakeholders and technical experts. Forums will advise Ministers either on request or proactively. (Government of South Africa 1995)

However, when the formal institutions of government were put in place, many of the activists who had participated in forum processes moved into government or related organizations, the role of the forums was greatly diminished and most disbanded.

11.4.3 A Forum for Water

A forum was established for water although it deliberately avoided the focus on negotiation, since it was agreed that the real need was to consider the technical challenges facing the sector and to advise the political players accordingly, rather than to negotiate distinct policy positions. According to the forum's documents:

> The Standing Committee on Water Supply and Sanitation (SCOWSAS) was established in January 1992 to address the water supply and sanitation needs of all South Africans. Its intention is to bring together as many as possible of the key stakeholders in the sector to develop policy options designed to achieve the goal of ensuring that all South Africans have access to adequate water supplies and sanitation facilities.
>
> Membership of the Standing Committee includes representation from the Democratic Movement, Central Government, Local and Regional Government, water industry parastatals, the organised professions, research organisations, development funding organisations and the non-government service organisations.

The expansion of membership to include both professional and political organizations contributed to the technical quality of its work as well as to the credibility and legitimacy of its recommendations.

11.4.4 SCOWSAS' Technical Recommendations

In 1993, as attention turned to the drafting of the final Constitution, the work of the forums took on new relevance. Decisions had to be made on the distribution of powers and functions and one obvious source of advice, given the fractured nature of the society, were the forums. To the extent that they succeeded in reaching a consensus on technical issues between a wide range of interests, they were in a strong position to provide advice.

Among the key conclusions of SCOWSAS' institutional review were that "There was an urgent need for a national agency to take overall responsibility for water supply and sanitation" and that "For second tier regional water institutions to be effective, the committee was firmly of the opinion that their boundaries should be based on natural physical boundaries rather than political divisions" (SCOWSAS 1993a).

It was in this context that the Standing Committee on Water Supply and Sanitation (SCOWSAS) engaged with the main negotiating process. As explained in their submission the negotiators:

> We are particularly concerned that decisions regarding certain issues which could affect the effective functioning of water supply and sanitation may be taken without adequate consultation with the sector. These include:
>
> - the definition of boundaries for second tier (regional) government which may, because they do not coincide with rational water management boundaries, give rise to inefficient provision of services;
> - the allocation of specific powers and responsibilities with regard to the water and sanitation sector between the different levels of government and utilities such as water boards . . .. (SCOWSAS 1993b)

Based on its consultations and research, SCOWSAS made the following recommendations:

> 8. The recommendation of the Standing Committee is that certain functions be retained at the first tier or national level, namely:
>
> - national water and sanitation legislation;
> - operation of multi-regional and international water schemes;
> - international water relations;
> - inter-regional water resource planning and development;
> - inter-regional conflict mediation;
> - long-term national water resource planning; and
> - national water information system and research.

However, the submission also addressed key concerns about water supply and sanitation services, in particular recognizing that in a highly unequal country, with substantial service deficits in poor communities, national involvement in water supply and sanitation services, previously a predominantly local service, would be required:

> In addition, new functions have been identified including:
>
> - establishment of national water supply and sanitation standards
> - promotion of the establishment of water and sanitation institutions
> - assumption of residual responsibility for water supply and sanitation management

- administration of state funding assistance
- national performance auditing . . ..

It also considered the role of regional and local organizations where the interface between water resources and services needed to be addressed on a practical level:

> 9. At the second tier level, organisations would be required to perform other functions, including:
>
> - management of water resource systems to assure bulk water supplies within a region
> - abstraction, treatment and bulk distribution of water
> - identification and control of waste water discharges into river systems and management of water quality
> - bulk collection and treatment of wastewater
> - support to local authorities in the direct provision of water supply and sanitation systems.
>
> 10. At third tier level, it is envisaged that local government organisations will provide the reticulation and sanitation services required.

These recommendations, notably those for the "second tier", addressed areas in which there were clear grounds for considering provincial jurisdiction. The concern that there would be a disjuncture between the political and the technical boundaries was explicitly addressed:

> 11. From this description it will be seen that the second tier agencies have a key role to play in the overall management of the water resources within their given region. The view of the Standing Committee is that their effectiveness in the performance of these duties will be substantially affected by the boundaries within which they work.
> 12. Many of the functions which second tier agencies will be called upon to perform are related to actions within naturally defined river catchment systems. Any division of major catchments whether longitudinally (by placing boundaries along rivers) or horizontally (by drawing boundaries across rivers) will mitigate against effective resource management by a second tier agency. This is because the agency will have no control over the actions of other agencies across the river or upstream of it.

Rather than simply argue for the powers to be held at a national level, SCOWSAS offered the alternative of drawing boundaries based on rivers catchments rather than other criteria:

> 13. It is thus the view of the Standing Committee that regional boundaries be set with proper regard to the location of catchment boundaries and the impact

> on water usage and water quality should boundaries be set that contravene these criteria.
> 14. Should this not be possible, the Standing Committee believes it is important to ensure that the allocation of powers and functions to second tier water management agencies should be separate from the attribution of powers and functions to regional government.

The submission concluded by asking the commission to address the allocation of powers and functions relating to water management:

> 15. The Standing Committee is mindful of the limited terms of reference of the Commission. It nonetheless believes that it is within the mandate of the Commission to consider the requirements of water management when making its recommendations and regional demarcation.
> 16. Further, within the context of criteria 1.2 (administrative considerations), 1.3 (the need to rationalise existing structures) and 1.4 (the necessity of limiting financial and other costs) the Standing Committee believes that it would be appropriate for the Commission to draw attention of the MNP [multi-party negotiating process] to the special needs of the water and sanitation sector when powers and functions are allocated between central and regional governments.

11.5 THE OUTCOME

11.5.1 The Constitutional Position

The SCOWSAS intervention was successful insofar as the final decisions taken reflected its recommendations. Water and its management was explicitly addressed in the final Constitution in a number of different contexts. A "right to water" was established, which in terms of the general provisions of the Bill of Rights, placed a duty on national government and the other spheres to take all reasonable legislative and other measures to achieve its progressive realization (Government of South Africa 1996: s.27) confirming, the responsibility of national government to legislate and set standards for water supply and sanitation services.

In the attribution of powers and functions (Government of South Africa 1996: Schedule 4), water and sanitation services (limited to potable water supply systems and domestic waste-water and sewage disposal and matters related thereto) were defined as local government functions for which national and provincial government would have concurrent legislative responsibility. While this gave local government the responsibility of providing water services, it provided for national oversight of their performance, regulation and intervention.

Water resource management issues were dealt with by omission, since any functions not specifically mentioned in the constitutional schedules remained national competences. One indicator of the success of the SCOWSAS intervention was that the environmental function, about which similar arguments could have been presented, was divided between national, provincial and local spheres. The suggestion that "environment" might be construed to include water resources was dispelled when legislation on water resources was passed as a matter of national competence. This formalized the position adopted in the White Paper on a National Water Policy (1997) that South Africa's water resource is an "indivisible national asset" of which national government is the custodian.

> The Constitution provides national Government with the sole responsibility for the management of water resources, although responsibility for certain water use sectors (such as agricultural and municipal use) and certain conservation and pollution control functions are concurrent provincial competencies. (DWAF 1997)

11.5.2 Implementation and its Limitations

The progress of implementing the new arrangements provides further insight into the issues around the management of federal rivers. The first step was the development of policy and legislation to give effect to the direction provided under the Constitution. A policy process followed that produced a White Paper on a National Water Policy for South Africa, which guided a fundamental revision of the water legislation (DWAF 1997).

The National Water Act (Government of South Africa 1998), promulgated on the basis of the white paper, made provision for the administration of certain water management functions to be undertaken in defined water management areas (WMAs) which did not follow either political or river basin boundaries. The major rivers were divided between a number of different WMAs (bringing together stakeholders served by common systems or in reasonable proximity to each other, with common interests) while smaller rivers were clustered into single WMAs (to avoid a proliferation of agencies). While powers can be delegated to catchment management agencies (CMAs) in these areas, this is at the discretion of the national minister who retains the power to intervene in delegated functions.

The establishment of CMAs has been a protracted process. It has been difficult to develop governance arrangements that achieve a balance between very different interest groups, from large, capable corporates to poor, largely illiterate rural communities and still enables effective and efficient as well as fair and equitable operations. In response, it is currently proposed to reduce the number of CMAs.

The establishment of a national water resources planning framework has however worked reasonably well. Each CMA (or national government, where CMAs do not exist) has a duty to prepare a management plan for their area, guided by an over-arching national water resource strategy which establishes some general parameters such as provision for inter-basin transfers. Either CMAs or, in their absence, the national department have produced management plans which have informed both for sectoral and inter-sectoral planning and decision-making. This has helped to identify and prioritize major investments and other interventions required to maintain water security. Meanwhile, at a local level, municipal water services development plans identify the resource requirements for water supply and wastewater services – at least in those municipalities which have the planning capacity.

11.6 DISCUSSION AND CONCLUSIONS

The South African example illustrates some of the challenges involved in governing and managing federal rivers. The relatively structured way in which water management matters were addressed during South Africa's political transformation has highlighted the importance of understanding the many different activities inherent to the sector. The application of formal criteria to the allocation of powers and functions between national and provincial governments brought some rigour to the process.

The generic challenge for water management is to ensure an effective and efficient interface between the technical arrangements for managing the resource and the relevant political institutions, which invariably follow different geographies. These arrangements must also ensure that the technicians remain politically accountable.

In South Africa, careful analysis of the functions, starting with the broad distinction between resource and service management and then analysing each in more detail provided a basis for allocation of powers and functions in a manner which best met the country's overall objectives. The outcome was not just a rejection of "political federalism" but also of "basin federalism", in which water management powers and functions are placed in the basin.

This recognized the specific local complexities of the resource and its management but also the dynamic nature of water resource management in a developing society. The approach taken was informed by a historical experience which had seen water supply to the economic centre of the country evolve, over 100 years, from a few private farm springs to an extensive inter-basin, trans-boundary system, linking many of the country's main rivers and crossing the borders of five of its nine provinces. Clearly, in a dynamic society whose activity is already constrained by water availability, institutional arrangements for water management and their geographical scope must be able to cope with this dynamism.

This was an important reason for keeping many of the resource management functions at national level, allowing some administrative functions to be undertaken at a regional/basin level while most service provision functions were retained at local level. It allows both for the effective and efficient performance of functions as well as facilitating the interface with political governance and enabling adjustments to be made over time.

While challenges that have arisen in the implementation of the framework, these do not appear to be specifically related to the institutional architecture that was adopted. Indeed, recent research into the potential for stakeholders to initiate local water resource management institutions found that most believed that only national intervention could provide the framework within which contested local water resource management issues could be addressed (Muller et al., 2011).

One outcome of South Africa's negotiations for a democratic transition was to ensure that water resource management could be undertaken on a national basis. In a country which is already water stressed and dependent on transfers of water between catchments and across local and provincial political boundaries, this has avoided many of the challenges that have arisen in other jurisdictions.

More generally, the process explicitly addressed the challenges of locating the water resource management function in the framework of a new constitutional democracy that was established in an emerging federal system. It has thus provided some insight into the issues that need to be addressed in the management of federal rivers, the nature of the functions that need to be considered and the distribution of powers required to enable them to be performed. This experience can contribute to the development of a more generic framework of analysis to support other processes of water resource management reform both within and beyond federal systems.

One important lesson from South Africa is that while the river basin is an important unit of management, it may not be the dominant spatial dimension once water use intensity increases. For this reason, the approach adopted in South Africa not only rejected "political federalism" in water resource management but also turned its back on "basin federalism", in which river basin authorities are given the primary powers to manage the resource. Under basin federalism, the transfer of water from a surplus basin to a deficit basin becomes a political decision between two separate populations, encouraging rent-seeking behaviour rather than finding optimal outcomes.

If political federalism poses serious challenges to the effective management of water (as evidenced by the experience of Australia, the western US and India) so too can basin federalism create difficulties, specifically, conflicts between legitimate political institutions of general governance and the more specialized agencies of basin management. While the implementation of its water management framework remains challenging, at least South Africa has avoided the unnecessary challenges that federalism can bring.

11.7 REFERENCES

Asmal, K. (1998) Speech to the National Assembly by Prof Kader Asmal, MP, Minister of Water Affairs & Forestry, on the second reading of the National Water Bill, Tuesday 9 June 1998, available at: http://www.info.gov.za/speeches/1998/98916_4179811002.htm/.

Commission of Enquiry into Water Matters (1970) Report of the Commission of Enquiry into Water Matters, Pretoria: Government Printers.

Constitutional Court (1996) Certification of the Constitution of the Republic of South Africa, 1996 Case CCT 23/96, Johannesburg.

DWAF (1997) White Paper on a National Water Policy for South Africa, Pretoria: Department of Water Affairs and Forestry.

Government of South Africa (1993) Constitution of the Republic of South Africa Act 200 of 1993 (accessed 15 August 2012 at: http://www.sahistory.org.za/archive/document-35-constitutional-principles-schedule-4-interim-constitution).

Government of South Africa (1995) Reconstruction and Development Programme White Paper: Sectoral forums (accessed 15 August 2012 at http://www.info.gov.za/view/DownloadFileAction?id=70427).

Government of South Africa (1996) Constitution of the Republic of South Africa Act, 108 of 1996.

Government of South Africa (1998) National Water Act, Act 36 of 1998.

Muller M. (2012) Lessons from South Africa about the management and development of water resources for inclusive and sustainable growth (background paper for) European Development Report, ODI/DIE, Brussels.

Muller, M., Hollingworth, B. and Mdluli, M. (2011) Prospects and processes for the establishment of stakeholder-initiated catchment management agencies, Final report WRC Project K5-1972, Pretoria: Water Research Commission.

SCOWSAS (1993a) Institutional Options for Water Supply and Sanitation, mimeo, Johannesburg.

SCOWSAS (1993b) Submission to the Multi-party Negotiating Process Commission on the Demarcation of Regions, Kempton Park.

Simeon, R. (1998) Considerations on the design of federations: the South

African constitution in comparative context, SBR Simeon, Queens University Working Paper 1998 (accessed 15 August 2012 at http://www.queensu.ca/iigr/WorkingPapers/Archive/1998/1998-2RichardSimeon.pdf).
UNESCO-WWAP (2006) *World Water Development Report No. 2*, Paris.
Wikipedia (2012) water management (accessed 15 August 2012 at http://en.wikipedia.org/wiki/Water_management).

12. Meeting the challenges of equity and sustainability in complex and uncertain worlds: the emergence of integrated water resources management in the eastern rivers of South Africa

Sharon Pollard and Derick du Toit

12.1 INTRODUCTION

South Africa is widely acclaimed for statutory water reforms following the first democratic elections in 1994. Flowing from the national constitution, the institutional reorientation – embraced largely in the National Water Act (1998) and the Water Services Act (1997) – aimed to address issues of racially-based inequities in access to water together with long-term sustainability which together constituted the cornerstone principles for change. Not only did the transformative policies highlight the socio-economic impacts of these historical disparities on the vast majority of the country's people, but they also recognised that with 13 out of 19 basins either in – or approaching – water deficit, more holistic approaches to water security[1] and long-term sustainability were needed. The Department of Water Affairs (DWA) was thus charged with finding a balance between the urgent need to address economic development while ensuring sustainability and inter-generational rights.

Central to water resources management was the abolition of privately-held riparian rights, the adoption of catchments as the basis for integrated water resources management (IWRM) and a commitment to the long-term sustainability of water resources through establishing environmental water requirements (EWRs), known in South Africa as the Reserve. Interestingly in addition to the *Ecological Reserve*, the Reserve includes water to meet the *Basic Human Needs* (termed the BHN Reserve) which aims to sustain basic human requirements, such as for drinking and domestic uses. The implication for water resources management is that no water can be allocated prior to setting the Reserve for a water resource and equally, that licence conditions need to be cognisant of the Reserve requirements.

Achieving these far-reaching goals has required a fundamental restructuring of institutional and fiscal arrangements focusing strongly on the need for devolution (of both authority and responsibility), as well as stakeholder participation and cooperative governance. With equity and sustainability as key principles and in recognition of water's pivotal role in socio-economic development, cooperative and holistic approaches such as integrated water resources management (IWRM) have become the hallmark. In this regard, the NWA establishes the minister as the custodian of water resources, supported by regional offices with a largely oversight function. The NWA provides for the establishment of 19 mega-basins – or water management areas (WMA) – to be managed by catchment management agencies or CMAs (Figure 12.1). These have recently been collapsed into nine WMAs, an issue we return to later. The CMAs ensure representative participation through water user associations (a transformed version of the former irrigation boards which comprised largely white, commercial farmers) and catchment management forums

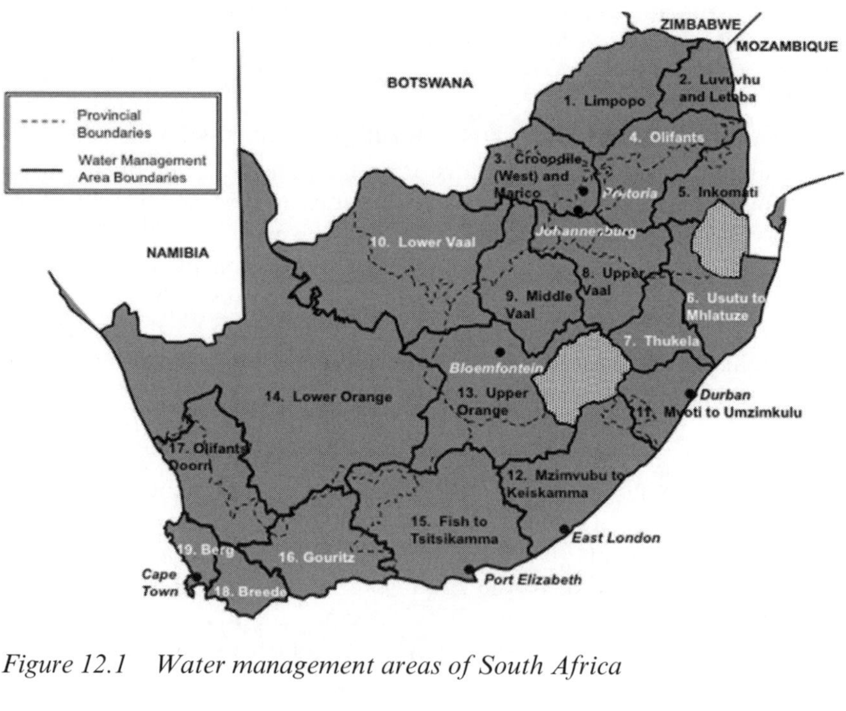

Figure 12.1 Water management areas of South Africa

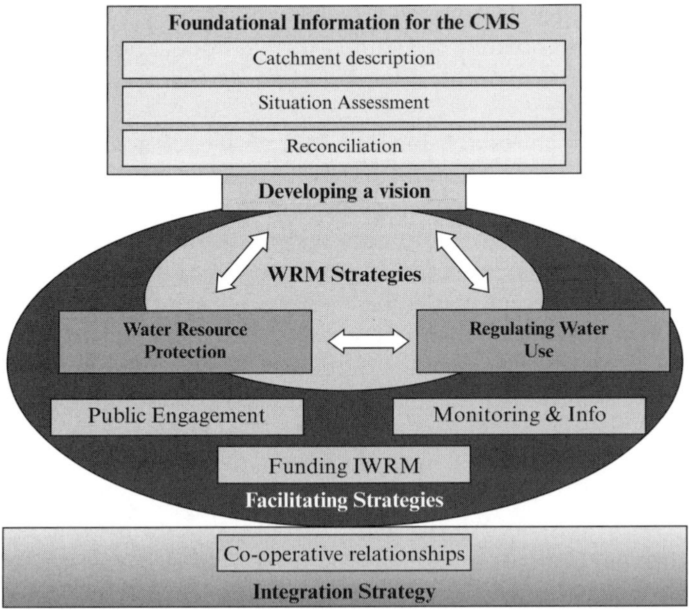

Figure 12.2 Conceptual framework

which includes all stakeholders. Guiding the IWRM work of the CMAs is a catchment management strategy central to which is a stakeholder-based vision (Pollard et al., 2007). From the vision a number of key strategies are developed including those dealing with water resources protection, water allocation and use, information management, stakeholder participation, monitoring and enforcement, and cooperative governance (DWA, 2007) (Pollard et al., 2007) (Figure 12.2). Water supply on the other hand is guided by the WSA and is managed through local government together with water boards. Clearly with stakeholder participation being central, cooperative arrangements – with water-based institutions – and between departments and sectors is key (Schreiner and Hassan, 2010). An enabling policy environment is derived constitutionally and is provided for through the Intergovernmental Relations Framework Act (2005, IRFA but see later comments) while other cooperative mechanisms and platforms are also in place.

While much has been said about the policy environment in South Africa, particularly with respect to IWRM, implementation is still in the early stages. For example, the establishment of CMAs is not on track and only three WMAs have been established nationally (Pollard and du Toit, 2011). Nonetheless, in other areas such as in determining the Reserve requirements (water quantity and quality), much progress has been made and South Africa is widely recognised for methodological sophistication. At smaller scales some progress has been seen and these cases are discussed later in the chapter.

The constitutional establishment of three tiers of government characterises South Africa as a federal system but this is not a commonly used term in this country. Nonetheless the important point for the purposes of this assessment is to explore how such institutional transformation is unfolding across different scales of governance and what enabling or constraining factors are emerging. Such reflections are an important part of adaptive management of complex systems – such as catchments – in providing the basis for learning and adaptation. This chapter aims to explore such issues based on the outcomes of a research programme on the three WMAs (or basins) of the eastern region of South Africa known as the lowveld, which is transected by the Luvuvhu, Letaba, Olifants, Sabie-Sand and Crocodile-Komati catchments (Figure 12.1) all of which are in water deficit (DWAF, 2004). All the major rivers flow into Mozambique, and Swaziland in some cases, and so while sovereign (federal) arrangements are important nationally, the impacts of institutional reform reverberate beyond the borders of South Africa to these downstream riparian states.

12.2 FRAMING THE STUDY

Over a decade on since the promulgation of the NWA there are increasing perceptions among some stakeholders that the condition of the rivers of the lowveld is declining, raising questions about implementation and the impacts of policy. Much of this was catalysed by flow cessation of the Olifants River on a number of occasions since 2005 and increasing water quality problems associated with failing waste water treatment works and acid mine drainage due to mining in many of the lowveld rivers. As noted above, the eastern region comprises three WMAs: the Luvuvhu/ Letaba, the Olifants and the Inkomati which share many similar biophysical and socio-economic characteristics. Although all of the six rivers originate in the Drakensberg escarpment (1800 m asl) and Soutpansberg with a rainfall of 1200 mm a^{-1}, they rapidly descend into a vast plain known in South Africa as the lowveld (600 m asl) which extends to coastal floodplains and estuaries of Mozambique where rainfall is less than 450 mm a^{-1}. The landscape includes a rich variety of landforms, climate and vegetation, as well as cultures and land uses. The climate is tropical to subtropical, with drought being endemic to the region (Tyson, 1986). At least 2 million people are settled within the lowveld representing a diversity of cultures who were segregated under the apartheid government into the bantustans from the late 1960s until the first democratic government of 1994 (Pollard et al., 2003). As a pervasive legacy of apartheid, these densely-populated areas (as high as 300 people km^{-2}) are characterised by high levels of poverty and unemployment which persist today (Pollard et al., 1998). Wage remittances and grants are a key livelihood source with subsistence farming providing limited food security. Basic services are improving but many people still live without access to regular and safe supplies of water. In contrast the adjacent sparsely settled, former white areas (5–20 people km^{-2}) are used mainly for commercial agriculture and more recently wildlife tourism, and in general people enjoy a generally good standard of living. In the upper Olifants and more recently the Inkomati, mining is a major economic activity and is increasing rapidly. Here acid mine drainage and unlawful mining is of particular concern for long-term water security. During the apartheid era water was considered only insofar as it was needed for consumptive purposes, and in keeping with global norms, water deficits were dealt with through increased infrastructure such as dams and inter-basin transfers (Pearce, 1992). Little concern was given to issues of sustainability or equity, and indeed this period of development was seemingly unconstrained by water resources.

Given the aforementioned concerns regarding the increasingly precarious water-security situation, the Shared Rivers Initiative was established to examine progress in implementation – particularly with respect to measures aimed at ensuring sustainability. Fortunately in South Africa the Reserve offers a critical benchmark against which to track the progress in meeting such a commitment to sustainability. Importantly while the focus is on the Reserve, the findings are of far wider application. This is because – as is elaborated later – meeting the Reserve (or any other policy objective) is only possible through the collective contribution and synergies of a number of strategies, plans and practices that make up IWRM. It is these factors that are explored in this chapter.

It is also important to remember that this is a complex and dynamic environment best suited, we contend, to governance arrangements that are framed by a systemic understanding and flexibly and adaptive responses (see Pollard and Du Toit, 2008 for a more detailed review of systems thinking and IWRM). Indeed, there is evidence of growing discomfort with governance and management based on linear cause-and-effect paradigms, often supported only by reductionist science (Meadows et al., 1972; Forrester, 1992; Levin, 1999; Holland, 1994). Alternatives to these orientations are emerging in many sectors from natural science (Gunderson et al., 1995; Folke et al., 2002; Meadows et al., 1999; Walker and Salt, 2006; Ison et al., 2007), to business (Snowden, 2003), education (Forrester 1992;

Ison et al., 2007; Wals, 2007), service delivery by government (Radzicki and Taylor, 1997) and in disaster relief (Ramalingam and Jones, 2008). These concerns are supported by philosophical and epistemological critiques by scholars such as Edgar Morin, Paul Cilliers and others (see for example Heylighen et al., 2007).

12.3 EXPLORING THE PRAXIS OF IWRM IN FEDERAL RIVERS

The first step was to ascertain whether or not the Reserve was being met and what the trends had been since 1998 which involved understanding the status of compliance with the quantity component of the Ecological Reserve (see Pollard et al., 2011). The second, which aimed to understand why the status of compliance was as it was, involved a dialogical approach based on semi-structured interviews with a range of stakeholders. Key conceptual frameworks provided the methodological orientation, while an organisational analysis for each catchment identified key role players involved in water resources management or use who were then engaged through semi-structured interviews. Participants were categorised into five broad categories: (1) regulators; (2) water users; (3) operations and maintenance; (4) researchers and (5) interested and affected parties. The interview process was guided by the overarching framework for IWRM (the catchment management strategies) in South Africa (see above). Thus questions focused on water resources protection, authorisation, monitoring, enforcement, financing, stakeholder participation and cooperative governance. Data were then analysed thematically as follows:

- Development of an integrated, systems view as the basis for planning and action (supporting IWRM) – especially between WRM and water supply.
- Adaptive capacity: Support for self-organisation and robust, multi-scale feedbacks in integrated, adaptive action and management.
- Strong, flexible leadership and governance for transformation and sustained action.
- The presence and function of participatory and representative platforms for collective action and learning.
- Capacity and skills.
- Unlawfulness and the regulation of unlawful use including self-regulation.
- Lags in the implementation of the EWRs and emergence of sustainability discourse as the basis for planning and action.

- Governance: polycentric governance and a recognition of legal pluralism.

12.3.1 Compliance with the Ecological Reserve as a Benchmark for Sustainability

When examined for compliance, none of the flow regimes of the eight rivers met the Reserve requirements, with the incidence of non-compliance varying between 15 and 85 per cent (Pollard et al., 2011). Moreover, with the exception of the Sabie River this situation appeared to have deteriorated since the promulgation of the NWA in 1998 (Pollard et al., 2011). In many cases water quality was reportedly also deteriorating but this requires verification.

A number of factors (discussed below) underlie this and despite the catchment-specific nuances, common themes emerged across all catchment. These are illuminating in that they highlight some of the challenges that face attempts to integrate across political and administrative boundaries – that is, in federal systems. Equally, while this might present a somewhat pejorative picture of the progressive realisation of basin-wide governance, it is only a snapshot along a path of institutional reform. For example, the incidence of non-compliance is likely to decrease substantially in the Inkomati WMA, certainly in the Crocodile River, as new integrated WRM approaches come on line.

12.3.2 Operationalising the Reserve Based on an Integrated, Catchment-based Approach (Supporting IWRM)

The central tenet of this chapter is that implementing or operationalising the Reserve – or any other policy commitment such as equity – moves beyond a focus on single strategic areas such as water protection or water re-allocation. Rather, it is dependent on water reform and the growth of IWRM as a whole and hence is predicated on the collective contribution and synergies of a number of strategies, plans and practices (Figure 12.2). Central to this is an integrated, catchment-based vision and approach, based on the principles of sustainability, equity and stakeholder participation (Pollard and du Toit, 2008). As noted this requires cooperative arrangements within the water sector and across different government departments and across different administrative boundaries signifying a major shift in the nature of governance (Wester et al., 2003, Pollard and du Toit, 2005). The implication is that to understand progressive realisation of any specific policy intent – be it equity or sustainability – one must turn to the 'bigger picture' to understand enabling and constraining factors.

Nationally an integrated approach to water resources management is still in its infancy in South Africa. In this regard the national objectives – which are to establish and to devolve governance responsibilities and administrative and financial support to CMAs – are not on track. For example, the Olifants CMA was to be established in 2005 (DWA, 2007, App. 1). As mentioned of the three WMAs represented in the study area, only the Inkomati CMA (ICMA) has been established (and only three have been promulgated nationally). The ICMA has developed a first draft of a catchment management strategy (ICMA, 2010) but has been hamstrung in many respects due to the lack of any substantive delegation of powers and the associated budget (Pejan, 2013). For example, the ICMA can currently only regulate water quality infringements since water quantity infringements (as specified within a licence) still fall under the remit of the regional Department of Water Affairs office. While the separation of functions between organisations or departments can theoretically be dealt with through cooperative governance, this requires strong leadership and significant resources making this particularly challenging in reality.

In this regard, the almost total lack of integration between water supply and water resources management – a notable feature of the former apartheid system – continues to pose a significant barrier to achieving sustainable IWRM. For instance, some municipalities have developed detailed plans for inter-basin transfers with little recognition of the fact that the water source is in a different catchment with different priorities. In the case of the Olifants River, which still does not have a CMA in place, water resources management and supply fall under completely different DWA regional offices nearly 600 km apart; seeking cooperation in such circumstances has proved less than tenable. The continued issuing of prospecting licences to mines in the upper Olifants, with significant water use/groundwater de-watering implications, despite the existing and growing water resources issues has exacerbated the tensions and further highlighted the need for leadership, cooperative and integrated praxis especially with respect to water allocation and regulation.

A significant – and largely unanticipated issue – has been the hiatus caused by debates to collapse the 19 WMAs into fewer, and larger, entities. The 19 WMAs have now been collapsed into nine (DWA, 2012), a process which has stalled devolution and created further uncertainty. Nonetheless – and despite the institutional uncertainties – there are substantive aspects of IWRM and systemic approaches that are underway at more localised scales. Within the Inkomati catchment, the ICMA is developing and testing a near-real time *system for integrated planning and operations of river systems* on the Crocodile River which is the most stressed of all the WMA's rivers. This system integrates information on water resources and water use and involves users in river management. The dialogue that this has fostered between stakeholders has led to a marked decrease in tension and increased willingness to cooperate (Mr B. Jackson, Acting CEO, ICMA, 22 October 2012, personal communication). Equally in the Groot Letaba River (part of the Olifants WMA) a system developed by the local DWA manager involves stakeholders in managing their water use patterns below Tzaneen Dam in a way that illustrates the importance of self-organisation and self-regulation (Pollard and du Toit 2011). In both cases leadership, innovation and a developing interface between management, research and practice are key.

The preceding discussions have highlighted a number of key elements that merit attention, namely leadership and the role of self-organisation, self-regulation and feedbacks which support adaptability, flexibility and learning in complex environments. In many cases leadership is weak and severely constrained by the lack of a coherent governance system such as in the aforementioned case of the Olifants WMA where WRM and water services are dealt with by different DWA regional offices with little integrated planning. The coordination of WRM falls to a single individual who carries additional, multiple responsibilities, severely constraining his ability to provide substantive leadership. With a catchment as large, complex and vulnerable as the Olifants, dedicated leadership, such as through a CMA, is essential. However, even in the case where the CMA has been established, the potential scope of leadership in the Inkomati CMA is severely constrained by the lack of assigned functions. Also, ensuring integration requires support from leadership in other sectors and in various institutions outside of water resources which has yet to gain momentum.

Even given weak leadership at higher scales (for example at the basin scale), localised leadership and stakeholder support and buy-in can bring about significant transformations. Such cases include the Letaba and Crocodile rivers, both of which are instructive in terms of emerging positive outcomes, namely sustainable use of the resource through stakeholder buy-in. We have suggested elsewhere that functional, responsive multi-scale feedbacks are essential for management in complex systems like catchments since they provide the basis for learning, reflection and response to an evolving context (Pollard and du Toit, 2011). However, the existence of these is *variable* from non-existent to emergent. These have emerged organically as a necessary component of a viable system but have yet to be explicitly recognised as an important component of water governance (for example, Snowden, 2003). In a number of cases, such as the Groot Letaba, Crocodile and Komati rivers there is a need to be supported through strengthening links to the wider system (for example, to national and transboundary processes). In all cases leadership and the ability to self-organise appear to be central. In other cases such as the Sabie River, despite good efforts to self-organise around water quality problems, little progress is being made. A critical issue is that the feedbacks are of a limited scale and constituency (the formerly white commercial farmers) which cannot bring about change. Wider-scale linkages and accountability are needed. One of the underlying causes here relates to the lack of integrated approaches and the tendency of various governmental departments and directorates to not have clear lines of communication or simply poor communication in general (Pollard and du Toit 2011).

Currently there is little accountability for – or consequences of – not planning within the context of water resource constraints. Among other reasons, this situation is unlikely to change without leadership as well as appropriate platforms and mechanisms for integrated planning together with stakeholders. Transformation towards a collective, catchment-based vision (such as would be developed through a catchment management strategy) can only be achieved through a collective understanding of context and a consensual approach. One reason is that the *collaborative actions* of individual stakeholders are needed; in other words even a single actor can undermine the efforts of others by acting unlawfully. Arguably overall compliance could be secured through autocratic means and stringent punitive measures but the spirit of the NWA talks to an alternative approach built on understanding and consensus (of water resources in this case) as a means to secure compliance. Although still in its early stages, collective action towards IWRM is weak in general and requires attention. While local platforms do exist, these often reflect single-sector interests or focus on specific sections of the river. Furthermore, the research suggests that different sectors within the same WMA continue to see very different priorities for managing the shared water resource. Nonetheless this is changing in the Crocodile and the longer history of cooperative management in the Letaba also points to positive potential. With the stagnation of CMA establishment, the existing platforms are bedeviled by a sense of inaction. Stakeholder platforms are not the answer on their own; participants need *a focus around which they act* – such as in the drafting and implementation of their catchment management strategies.

A key ingredient of transformation is 'learning', especially social learning that confronts the diverse understandings and meanings of the different sectors. This needs to be supported. A case in point is the current understanding and hence uptake of the Ecological Reserve in practice. While familiarity with the concept of the ER varied considerably it was generally better in the Inkomati WMA, partly reflecting the explicit acknowledgement by the ICMA to meeting the Reserve (Pollard and du Toit, 2011).

Even so, the pervasive view was that it would benefit others and that the sector in question would carry all the risks. The concept of sustainability so as to ensure water for people and inter-generational rights was rarely understood as a guiding principle. Exacerbating this negative view is the fact that water managers expressed frustration in interpreting and operationalising outputs from a Reserve determination study.

Finally mention must be made of the issue of unlawful use.[2] Water security and sustainable water resources (including EWRs) cannot be achieved without a compliant or lawful catchment-based system such that water use is authorised, regulated and monitored against the Reserve requirements. This requires cooperative arrangements between departments, adequate skills and resources for compliance monitoring and enforcement and stakeholder involvement in monitoring, reporting and rectification of transgressions (Pejan, 2013) as well as in self-regulation (such as in the Letaba sub-catchment). The perception of 'illegal (consumptive) use' as a major contributor to dwindling water resource is pervasive among stakeholders and while largely unsubstantiated, a number of cases of unlawful use were evident. For example, all of the catchments are bedevilled by major issues with regard to municipal expansion and the lack of effluent control (DWAF, 2010). Many stakeholders questioned the ability to regulate and enforce if the government itself (i.e. the municipalities) were acting unlawfully. Equally the expansion of mining in the Komati, Crocodile and the Olifants with little consideration for impacts on the water resources – in particular water quality – was seen as further evidence of inaction on the part of government. While agreeing stakeholder perceptions, regulatory officials in contrast pointed to a lack of support (especially judicial support) and under-staffing as major constraints and even pointed to political interference in monitoring and licensing in some cases. Moreover, bearing in mind that the South African NWA recognises 11 water uses – and not just consumptive practices alone – the remit of compliance monitoring and enforcement is very expansive and requires strong cooperation with other departments, another issue that the DWA regulators reported as extremely challenging. Importantly here, the very policy designed to promote inter-departmental cooperation (IRFA see earlier) is being used as a constraint to taking regulatory action: that is one state department is widely discouraged from taking legal action against another department that acts unlawfully such as in the case of municipalities and mining.

Finally, lags (the time between policy intent and realisation through strategic and operational actions) are a natural consequence of changes in policy, law and the administrative procedure. Policies are meaningless, however, if the lag is excessive. Meeting the Reserve is subject to progressive implementation and in some cases, such as the Crocodile catchment, steps are being put in place. Determining 'reasonableness' – especially from a legal perspective – is challenging especially in the absence of benchmarks (Pejan et al., 2013).

12.4 LESSONS FOR FUTURE DIRECTIONS

Meeting transformative policy commitments such as to EWRs or water for the poor (such as espoused in the Millennium Development Goals), is predicated on the *collective contribution and synergies of a number of strategies, plans and practices* (Figure 12.2). In the eastern WMAs of South Africa, cases where system resilience is strengthening offer useful lessons and frameworks for weaker situations. We have discussed in some length the importance of cooperative, polycentric governance and the role of strong adaptive leadership, multi-scale feedbacks, self-organisation and self-regulation especially through collective action. If a people-centred approach that is guided by sustainability is to be sought, then we also need to find new ways of understanding, collaboratively, the benefits associated with water resources protection measures (such as the Reserve and classification). Such thinking needs to extend across boundaries – be they upstream-downstream, sectoral or international. This is because we need to find ways of sharing responsibilities for our scarce freshwater resources collectively. While some progress has been made in this regard through the Southern African Development Community (SADC) and with the establishing of committees aimed at governance of shared water courses (The Tripartite Technical Committee for the Crocodile River and The Limpopo Commission – LIMCOM – for the Limpopo Basin) much needs to be done in practical terms of governance (discussed earlier under self-regulation, feedback, communication and learning through adaptive management).

Establishing a discourse of sustainability and entrenching this in practice warrants transformative ways of working and not one-off awareness-raising campaigns so often associated with 'stakeholder engagement'. Consideration needs to be given to the nature of learning; here we advocate for social learning approaches that confront the diverse understandings and meanings of the different sectors and provides platforms for the co-construction of meaning towards a common vision. Without multiple stakeholder platforms at which the status quo of the catchment is discussed and informs the collaborative vision from which sustained programmes emerge, this is unlikely. Currently also, despite the paper commitments to and incentives for cooperation, there is little consequence for different stakeholders or sectors that do not participate or act cooperatively – an issue which needs to be addressed not just in South Africa but globally.

The fact that the state may be acting unlawfully and that one state department is discouraged from taking action against another is a thorny legal and governance issue that requires further scholarship and discourse.

The role of self-regulation is central in the management within complex systems because, due to their openness and unpredictable nature, complex systems cannot be managed only from the outside. This is particularly true in situations where formal regulatory skills are limited, placing severe constraints on the ability of the state to regulate (Murphree, 2004). We suggest that active strategic and practical support for self-organisation and self-regulation in complex, multi-stakeholder situations is essential.

Finally, it is important to remember that all the major rivers of eastern South Africa are shared with other states, most of which are downstream neighbours. Thus not only are sovereign (federal) arrangements important nationally for water security, but the impacts of institutional reform have implications far beyond the borders of South Africa.

12.5 NOTES

1. *Water security*, although not specifically defined in the NWA is based on the definition provided by Norman et al. (2010) as 'sustainable access, on a watershed basis, to adequate quantities of water of acceptable quality, to ensure human and ecosystem health'.
2. Broadly defined in the NWA (1998) to include: taking and storing water, stream-flow reduction activities, waste discharges and disposals, controlled activities (activities which impact detrimentally on a water resource), altering a watercourse, removing water found underground for certain purposes, and recreation.

12.6 REFERENCES

DWA (2007) Guidelines for the Development of Catchment Management Strategies: Towards equity, efficiency and sustainability in water resources management. By POLLARD, S.R., DU TOIT, D., REDDY, J. and TLOU, T., Department of Water Affairs and Forestry, Pretoria, South Africa.

DWA (2012) National Water Resources Strategy. Second Edition. Pretoria, Department of Water Affairs and Forestry.

DWAF (2004) National Water Resources Strategy. First Edition. Pretoria, Department of Water Affairs and Forestry.

DWAF (2010) Blue Drop Report. South African Drinking Water Quality Management Performance.

FOLKE, C., CARPENTER, S., ELMQVIST, T., GUNDERSON, L., HOLLING, C.S., WALKER, B., BENGTSSON, J., BERKES, F., COLDING, J., DANELL, K., FALKENMARK, M., GORDON, L., KASPERSON, R., KAUTSKY, N., KINZIG, A., LEVIN, S., MÄLER, K.-G., MOBERG, F., OHLSSON, L., OLSSON, P., OSTROM, E., REID, W., ROCKSTRÖM, J., SAVENIJE, H. and SVEDIN, U. (2002) Resilience and Sustainable Development: Building

Adaptive Capacity in a World of Transformations. Scientific Background Paper on Resilience for the process of The World Summit on Sustainable Development on behalf of The Environmental Advisory Council to the Swedish Government.

FORRESTER, J.W. (1992) System Dynamics and Learner-Centred-Learning in Kindergarten Through 12th Grade Education. *Massachusetts Institute of Technology [Online] URL: ftp://sysdyn.mit.edu/ftp/sdep/Roadmaps/RM1/D–4337.pdf.*

GUNDERSON, L.H., HOLLING, C.S. and LIGHT, S.S. (1995) *Barriers and bridges to the renewal of ecosystems and institutions*, New York: Columbia University Press.

HEYLIGHEN, F., CILLIERS, P. and GERSHENSON, C. (2007) Complexity, Science and Society in BOGG, J. and GEYER, R. (eds) *Philosophy and complexity*, Oxford: Radcliffe Publishing.

HOLLAND, J.H. (1994) *Complexity: the emerging science at the edge of order and chaos*, Harmondsworth: Penguin.

ICMA (2010) First draft of the Inkomati Catchment Management Strategy. Submitted to DWA Minister.

ISON, R.L., RÖLING, N. and WATSON, D. (2007) Challenges to Science and Society in the Sustainable Management and use of Water: Investigating the role of social learning. *Environmental Science and Policy*, 10, 499–511.

LEVIN, S.A. (1999) *Fragile dominion: Complexity and the commons*, Cambridge, MA: Helix Books.

MEADOWS, D.H., MEADOWS, G., RANDERS, J. and BEHRENS, W.W. (1972) *The limits to growth*, New York: Universe Books.

MEADOWS, D.H., RANDERS, J. and MEADOWS, D.L. (1999) *The limits to growth – the 30-year update*, New York, Universe Books.

MURPHREE, M.W. (2004) Communal Approaches to Natural Resource Management in Africa: From whence to where? (Keynote Address). *Breslauer Symposium on Natural Resource Issues in Africa*. Berkeley, CA: Center for African Studies, University of California.

NORMAN, E.S., BAKKER K., COOK, C., DUNN, G. and ALLEN, D. (2010) Water Security: A Primer (Policy Report). Vancouver, BC: UBC Program on Water Governance.

PEARCE, D. (1992) Towards Sustainable Development through Environmental Assessment. Centre for social and economic research on the global environment.

PEJAN, R., ROBERTSON, A., COGGER, J. SEFATSA, D. and EMMERSON, M. (2012) Final report on activities for the legal component of the Shared Rivers Initiative Phase 2, Unpublished report to the Water Research Commision South Africa.

PEJAN, R. (2013). The Application of Assignment and Delegation Within the Context of the National Water Act: The implications for Catchment Management Agencies, *The South African Law Journal*, 130(1).

PEJAN, R., DU TOIT, D. and POLLARD, S. (2013). Using Progressive Realization and Reasonableness to Evaluate Implementation Lags in the South African Water Sector, *Water and the law: Towards sustainability*, IUCN/Edward Elgar

POLLARD, S. and DU TOIT, D. (2011) Towards Adaptive Integrated Water Resources Management in Southern Africa: The role of self-organisation and multi-scale feedbacks for learning and responsiveness in the Letaba and Crocodile catchments, *Water Resources Management*, September.

POLLARD, S., MALLORY, S., RIDDEL, E. and SAWUNYAMA, T. (2011) Towards Improving the Assessment and Implementation of the Reserve: Real-time assessment and implementation of the Ecological Reserve. Final Report to Water Research Commission K8/881/2. March.

POLLARD, S., SHACKLETON, C. and CARRUTHERS, J. (2003) Beyond the Fence: People and the lowveld landscape. In DU TOIT, D., ROGERS, K.H. and BIGGS, H.C. (eds) *The Kruger experience. Ecology and management of Savanna heterogeneity*, Washington, DC: Island Press.

POLLARD, S.R., DU TOIT, D., REDDY, J. and TLOU. T. (2007) Guidelines for the Development of Catchment Management Strategies: Towards equity, efficiency and sustainability in water resources management, Department of Water Affairs and Forestry. February. Pretoria, South Africa.

POLLARD, S.R. and DU TOIT, D. (2005) Achieving Integrated Water Resource Management: The mismatch in boundaries between water resources management and water supply, *African Water Laws: Plural Legislative Frameworks for Rural Water Management in Africa*, Johannesburg, South Africa.

POLLARD, S.R. and DU TOIT, D. (2008) Integrated Water Resources Management in Complex Systems: How the catchment management strategies seek to achieve sustainability and equity in water resources in South Africa, *Water SA Special Edition IWRM*, 34, available at http://www.wrc.org.za.

POLLARD, S.R., MENDIGUREN, J.C.P.D., JOUBERT, A., SHACKLETON, C.M., WALKER, P., POULTER, T. and WHITE, M. (1998) Save the Sand Phase I. Feasibility Study: The development of a proposal for a catchment plan for the Sand River catchment. Pretoria, Department of Water Affairs and Forestry, Department of Agriculture and Land Affairs.

RADZICKI, M.J. and TAYLOR, R.A. (1997) Introduction to System Dynamics A Systems Approach to Understanding Complex Policy Issues, prepared for: U.S. Department of Energy, Office of Policy and International Affairs, Office of Science & Technology Policy and Cooperation.

RAMALINGAM, B. and JONES, H. (2008) Exploring the Science of Complexity: Ideas and implications for development and humanitarian efforts, Overseas Development Institute.

SCHREINER, B. and HASSAN, R. (eds) (2010) *Transforming water management in South Africa: Designing and implementing a new policy framework*, New York: Springer.

SNOWDEN, D.J. (2003) Managing for Serendipity or Why We Should Lay Off "Best Practice" in Knowledge Management, *Journal of Knowledge Management*, ARK.

TYSON, P.D. (1986) *Climatic change and variability in Southern Africa*, Cape Town: Oxford University Press.

WALKER, B. and SALT, D. (2006) *Resilience thinking: Sustaining ecosystems and people in a changing world*, Washington, DC: Island Press.

WALS, A.E.J. (2007) *Social learning towards a sustainable world*, Netherlands: Wageningen Academic Publishers.

WESTER, P., MERREY, D.J. and LANGE, M.D. (2003) Boundaries of Consent: Stakeholder representation in river basin management in Mexico and South Africa, *World Development*, 31, 797–812.

PART V

SOUTH ASIA

13. Managing water in India's federal framework

Rakesh Hooja[1]

13.1 INTRODUCTION

This chapter examines how water is managed in India, with particular focus on the state and sub-state levels and some consideration of the union government's role. It describes the constitutional allocation of responsibilities and the water-related organisations at the different levels, before discussing issues in managing water and reviewing the management of inter-state water disputes. 'Water management' includes the development, operation and maintenance of water delivery systems, as well as issues like sharing between alternate uses and among different regions.

13.1.1 Hydrology of India

Over 70 per cent of annual rainfall in most of India is usually concentrated during the monsoon months of July to September, when most rain falls in 40 days and is often concentrated in a few days of downpour. The national average is 105 cm, but this varies greatly by region. Floods follow the monsoon rains in many areas, while periodic failure of the monsoon results in droughts and famines in significant areas. Thus water is a preoccupation of everyone from the union and state governments down to farmers. Concerns include water storage, dams, embankments to retain waters or prevent floods, water diversion, and groundwater.

India has 16 per cent of the world's population but only 4 per cent of the world's fresh water. Average surface water flows in India are about 1869 billion cubic metres (bcm) of which about 690 bcm may be utilised with appropriate storage; rechargeable groundwater resources are estimated at 432 bcm so that estimated exploitable water resources are 1122 bcm. Table 13.1 shows water availability by river basin. The country's recent utilisation has been a total of 605 bcm (irrigation 501, domestic, including drinking 30, industrial 20, energy 20, and other 34). India's total area is 3,287 263 km^2 of which about 45 per cent is degraded and 25 per cent is

Table 13.1 Water resources potential in the river basins of India

Sl. No.	Name of the River Basin	Average Annual Surface Water Potential Km^3	Estimated Utilisable Flow Excluding Ground-water Km^3	Total replenishable groundwater resources Km^3	Population in 1991 millions	Per Capita Available Surface Water m^3	Per Capita Surface and Groundwater m^3
1	Indus (up to Border)	73.31	46.00	26.49	41.90	1,750	2,382
2	a) Ganga	525.02	250.00	170.99	356.80	1,471	1,951
	b) Brahmaputra, Barak and others	585.60	24.00	53.91	35.24	16,617	18,147
3	Godavari	110.54	76.30	40.65	53.98	2,048	2,801
4	Krishna	78.12	58.00	26.41	60.78	1,285	1,720
5	Cauvery	21.36	19.00	12.30	29.33	728	1,148
6	Pennar	6.32	6.86	4.93	9.70	652	1,160
7	East Flowing Rivers between Mahanadi and Pennar	22.52	13.11	18.22	23.60	954	831
8	East Flowing Rivers between Pennar and Kanyakumari	16.46	16.73		45.20	364	
9	Mahanadi	66.88	49.99	16.46	26.60	2,514	3,133
10	Brahamani and Baitarni	28.48	18.30	4.05	9.77	2,915	3,329

11	Subernarekha	12.37	6.81	1.82	9.46	1,308	1,500
12	Sabarmati	3.81	1.93		10.58	360	1,120
13	Mahi	11.02	3.10	18.42	10.48	1,052	
14	West Flowing Rivers of Kutch, Saurashtra including Luni	15.10	14.98		22.10	683	
15	Narmada	45.64	34.50	10.83	14.70	3,105	3,842
16	Tapi	14.88	14.50	8.27	14.80	1,005	1,564
17	West Flowing Rivers from Tapi to Tadri	87.41	11.94		25.80	3,388	3,744
18	West Flowing Rivers from Tadri to Kanyakumari	113.53	24.27	17.69	32.60	3,483	
19	Area of Inland drainage in Rajasthan Desert	NEG.		7.10			
20	Minor River Basins Draining into Bangladesh and Burma	31.00		2.10		14,762	14,762
	Total	1,869.35	690.31	431.44	842.62	2,218	2731

rain-fed arable land. Net cultivated area is about half of the country, of which the irrigated 40 per cent produces 55 per cent of the country's food, while the rain-fed remainder produces the balance. Within the irrigated area 31 per cent is covered by canal water, 58 per cent by groundwater and 11 per cent by tanks and other sources.

13.1.2 Constitutional Background

Many of India's constitutional arrangements that came into force in 1950 go back to the days of British rule, notably the pre-independence *Government of India Act* of 1935. The Constitution of India simultaneously created the Indian union and the states. The word 'federal' is never mentioned in the Constitution, which was framed just after the painful partition into India and Pakistan. While the Constitution was generally tilted in favour of union power, this was not true for water-related matters.

India has 28 states (which is twice the number at independence) and seven union territories. It is further divided into about 600 districts, which are administrative units of state government. Most rural districts have a *Zila Parishad* (district council) as the highest of three tiers of Panchayati Raj Institutions (PRIs) or rural local bodies; below these are 6094 intermediate level *panchayats* and over 230,000 village panchayats. There are also 3551 urban local bodies (municipal corporations and councils). India's population is over one billion and so states have populations comparable to many middle- to large-scale countries; even the districts within states can be very large.

The Constitution establishes the legislative and functional jurisdiction of the union and state governments. Water is essentially a state subject, and has been since the Government of India Act 1935 transferred control of irrigation to the provinces, which already managed all rights over construction of water projects and over water bodies. Private or community use of water requires a government permit, except extraction from one's own field; groundwater can be sold, but the rights go with the land. Entry 17 of the state list of powers reads: 'Water, that is to say water supplies, irrigation and canals, drainage and embankments, water storage and water power' subject to entry 56 on the union list of powers, which reads: 'Regulation and development of inter-state rivers and river valleys to the extent to which such regulation and development under the control of the Union is declared by Parliament by law to be expedient in the public interest.' Article 262 grants the Union Parliament the right to make laws to 'provide for adjudication of any dispute or complaint with respect to the use, distribution, or control of water of, or in, any inter-State river or river valley'. Only two such laws have been enacted. These are the *River*

Boards Act 1956 and *Inter State Water Disputes Act 1956* (as amended in 2002).

Entry five in the state list, along with the 73rd and 74th Amendments of 1992, empowers state legislatures to determine the responsibilities of the panchayats or municipalities. State legislatures may devolve some responsibilities for minor irrigation, water management and watershed development, and drinking water to various panchayat bodies and some responsibilities for domestic or industrial water supply to municipal bodies. In general such devolution has been quite limited. 'Agriculture', which is closely connected to water management, and 'public health', which includes the provision of drinking water, are also state responsibilities.

The union government is responsible for foreign affairs, treaties and international agreements. While not legally bound to consult affected states when negotiating water-sharing issues with neighbouring countries, politically such consultation may be necessary.

Amendments to legislative powers require a two-thirds approval in the union Parliament plus a majority of the state legislatures. Thus it is not politically feasible to shift 'water' to the concurrent list to enable the union to legislate on water.

For the first 25 years or so after independence India had a political system with the Congress Party in power at both the national Union Government and most states (with a couple of exceptions). Gradually, regional or state-level parties gained in strength as did the state units within national parties, so that states' interests and rights have become more significant while the Constitution remains virtually the same.

13.2 WATER ADMINISTRATION AND ORGANISATION

13.2.1 Water-related Organisations in the Union Government

The Ministry of Water Resources (MOWR) is the national organisation responsible for overall planning and management of the nation's water resources. It was created in 1985 out of the Irrigation Department that was in the Ministry of Agriculture. Its responsibilities include inter-state water disputes to the extent that they engage the union government. The Central Water Commission (CWC), Central Groundwater Board and National Water Development Agency provide overall technical support to MOWR. The CWC also provides technical advice if state governments so require, and examines irrigation projects having a culturable command area over 10,000 ha before granting technical approval. The MOWR operates centrally sponsored schemes (CSSs), in which the union government provides the major funding subject to states providing 'matching funds' in a ratio determined for each CSS. The MOWR establishes guidelines for these, sometimes after consultation with the states. There is also an accelerated CSS irrigation programme where states may propose their projects. Implementation is a state responsibility, while the concerned central ministry monitors progress before releasing further funds.

MOWR services the National Water Resources Council. It is presided over by the Prime Minister and state chief ministers as well as important central ministers; however, it met only four times from 1983 to 2004 and has been largely ineffective, given the need for consensus. (The Inter-State Council of first ministers, which can deal with any issue, suffers the same problems.) A committee of senior union and state officials, the National Water Board, has met somewhat more frequently, but it has also been hamstrung by the need for consensus. The MOWR led the development of the National Water Policies of 1987 and 2002, both of which failed to be implemented because they cannot bind states. A further example of states' independence in managing their waters is that not a single state adopted a model groundwater regulation bill circulated by MOWR.

MOWR coordinates with the Ministry of External Affairs on water issues with neighbouring countries. It is associated with various inter-state management boards through which state governments may coordinate agreed allocations of canal or river waters or sharing costs of maintenance of inter-state infrastructure. These inter-state management boards differ from river basin organisations, which do not exist in India, and they are often politically fractious. Some have advocated river bodies composed of stakeholder members and government representatives, but to no avail.

The Ministry of Rural Development provides funds to states for rural drinking and domestic use projects. In 2006 a National Rain-fed Area Authority was constituted under the Ministry of Agriculture and it developed guidelines for watershed development projects to be implemented by the states with or without central funding under CSS.

The Ministry of Agriculture promotes agricultural extension and technology transfer services for both irrigated and rain-fed areas. Its Indian Council for Agricultural Research has research stations all over the country, though agriculture universities are normally set up by state governments. The Housing and Urban Development Ministry supports state urban drinking and domestic use water projects. The Environment and Forest Ministry regulates water development projects in ecologically sensitive areas and states must refer many water-related works to it for clearance.

The Planning Commission, which is a non-statutory body headed by the

Prime Minister with a full time deputy chairman and members as well as a strong secretariat, feeds into the National Development Council, comprising of the Prime Minister and state chief ministers as well as important central ministers. The NDC meets at least once a year to consider the Five Year Plans and specific policies and projects. These plans propose plan funds for central ministries and states, and include many water-related projects, subject to approval by the central parliament or state legislatures.

13.2.2 Water-related Organisations at State Government Level

States conduct the actual management of the water sector and their irrigation, public works or water resource departments are responsible for the construction, maintenance, and operation of water storage and surface irrigation projects. Most have separate departments for groundwater and for drinking water and sanitation. Water management responsibilities are often held by ministries dealing with public works, public health or internal water transport.

State departments responsible for irrigation (increasingly named 'water resource departments') now normally deal with the allocation of water for alternative uses (irrigation, drinking and domestic use, industrial use), the planning of water development and use, as well as inter-state and centre-state water matters. More and more regions are going from water surplus to scarcity because of growing populations and prosperity, technological changes, and the progressive shift from subsistence to commercial farming.

State departments responsible for agriculture, rural development and local government, animal husbandry and fisheries, water utilisation, and disaster management also deal with water, as do industry, mines, and energy ministries, especially regarding the allocations sought for their sectors. Energy departments, through electricity boards or power generation companies, are responsible for the operation of hydro-electric plants. Environment departments and state pollution boards must provide clearance of water-related projects, as do forestry departments.

Being typically more fiscally stressed than the union government, states look forward to financial assistance through central sector schemes and give priority to them, including in the water sector. (Such special funds are in addition to broader fiscal support through national tax sharing, as well as from grants mandates by the finance and planning commissions.) The CSS for Command Area Development (CAD), which has dedicated departments in states, works to optimise the irrigation potential of a project within a 'command area'. Water courses are constructed to link fields to the outlet of a minor canal, while drains and farm roads are constructed to service farmers' fields. The programme provides agriculture extension services, adaptive agriculture research, and tree plantation. New areas being 'colonised' or converted from rain-fed farming to irrigation may receive addition supports for agriculture, marketing and storage, and even for planning new communities and habitations. Some projects hand over certain irrigation and water management functions to water users associations. The 1987 National Water Policy had advocated that all irrigation projects should adopt the CAD approach as did the 2002 National Water Policy.

The union ministries of agriculture and rural development have CSS programmes that help individual farmers to access bank loans for wells, pump sets, sprinkler systems, and so on along with subsidies to small farmers and those below the poverty line (BPL) to reduce the repayment burden.

In India 'watershed' mostly refers to relatively small catchments in rain-fed areas where there are no rivers or canals, so it does not mean 'river basin'. Watershed development (typically 40 to 300 ha) is largely funded from the centre under CSS. State-level watershed directorates are now attached to state departments of rural development and Panchayati Raj, so that technical aspects regarding appropriate use of common lands and private property now receive less emphasis than participatory and political aspects during watershed development. Because state governments must provide some matching funds to receive CSS funds, state planning departments, which fix departmental budget ceilings, have some influence over water management.

13.2.3 Water-related Organisations at Sub-state Levels

The district is the main administrative unit for state governments in carrying out their numerous regulatory and developmental tasks. Various departments have district level officers (DLOs), who play important roles in all water-related activities. The senior official is a district collector (or deputy commissioner or district magistrate in some states), who is an officer of the Indian Administrative Service posted by the state government to head the general administration of the state government in the district. The collector coordinates departmental activities through some 70 or more committees and thus significantly influences developments (including loans) for agriculture, industry, irrigation, drinking water, drought proofing, and flood protection as well as preparation of the district plan. The collector heads famine relief works and flood control and relief operations. He has certain supervisory powers over local bodies, and heads the lands administration, which acquires lands for water and other projects. Before each crop season the collector chairs a meeting to assess the likely water available to determine when water is to be let into the canals, the number of waterings, and the canal running programme for the crop season. He coordinates the water awareness marches to promote better use of water and suggests to the state government priorities for drinking water projects. The collector is responsible for the maintenance of law and order, including potential or existing conflicts over water use (with water rights being determined by courts or by the water resources department) and he can be drawn into grievances against departments and loans officials.

Integrated CAD projects, including irrigation, can be an important organisation at the sub-state level. A single officer heads such a project with officers from irrigation and other departments seconded to it. Project budgets and activities are implemented in an integrated manner within the command area. The boundaries of a command unit are determined topographically and often cut across political and administrative boundaries and cover more than one district. Command area development authorities normally comprise district collectors, members of the state legislature, heads of the Zila Parishads concerned, and senior state government officials. Larger integrated CAD projects are headed by a member of the Indian Administrative Service appointed by the state as an area development commissioner, while smaller projects may be headed by an irrigation engineer, agriculture officer, or state administrative service officer.

13.2.4 Third Tier of Indian Federalism

This tier is comprised of rural local bodies, including the various panchayats and urban bodies, with both elected representatives and government functionaries. Since the 73rd and 74th Constitutional Amendments of 1992, these bodies have constitutional status though they are subject to state legislation under the Constitution. Each local body is independent and there is no hierarchy of local bodies. Despite the constitutional changes of 1992–1993, state governments have shown little enthusiasm for devolving functions to the local bodies and local bodies are reluctant to raise tax revenues or fees, which they could, so they rely heavily on state funds and functionaries. Understandably, many state departments treat them as implementing agencies subject to detailed guidelines for funds they receive, including for watershed development schemes and employment programmes for water-related works. Rural drinking schemes are normally prepared and implemented by the health department. There have been efforts to get the lowest level of local government to take on more responsibilities, such as for community wells, but they often prefer that the public health engineering department does this work.

Typically the three tiers of panchayats avoid water management responsibilities and prefer instead to influence the field staff of state departments or to obtain state funds for small works, though some enlightened panchayat bodies are more active. Even in cities and towns, state public health engineering departments are more likely to operate the drinking water and sanitation systems and only they have the capacity to manage long distance water schemes.

13.2.5 Fourth Tier of Indian Federalism

Certain community-based user or beneficiary organisations, such as water users, watershed committees, forestry groups, or residents' groups can be viewed as a sort of fourth tier in India's federal structure. Some water-users associations on canal irrigation systems enjoy statutory status, while others may be registered or cooperative societies, administrative creations or informal groups. They take on small water management tasks or lobby government agencies for improved service delivery. Their boundaries are often hydrologically or topographically defined, and normally much smaller than political or administrative units. Some advocate placing them under the control of panchayat bodies, but this writer believes they are better being autonomous.

13.3 SOME ISSUES OF MANAGING WATER IN INDIA

The states' paramountcy over water resources dates from their assumption of the main role in irrigation development in 1866. The Constitution, as well as central and state laws, has helped secure this right. Neither riparian rights, nor prior appropriation rights of private parties including farmers are secured. Because of this, private initiatives in water resource development are limited to private lands through wells and tubewells.

British policy in India favoured canal systems for extensive irrigation, with water spread over larger areas with longer canals and lower water allowances. The regime suited subsistence more than commercial farming and it carried over into independent India. The result is that all farmers feel short of water, particularly since conditions now favour more water-intensive commercial farming. This leads to head and tail disputes which can turn violent and may be complicated if multiple districts and even states are involved or if farmers in the head reaches get accustomed, while major canals are being built, to using more water than their allocated water allowance.

State governments, collectors, district administrations, and water resources departments often have to resolve such head–tail disputes. Under the British, water resource departments used military and civil engineers for such works in contrast with the Spanish and Dutch colonies where the irrigation departments were more multidisciplinary. Civil engineers tended to focus on construction and not interact with the irrigator-farmers, and they still feel uneasy communicating with agriculture department staff about projects to irrigate agriculture. So water resources departments seek technical solutions even when many issues require socio-political interventions. Such engineers tend to overlook long-established community arrangements for water management (such as old tank irrigation systems in south India or traditional water harvesting systems in drought-prone and desert areas) and impose departmental norms instead. Thus traditional community managed schemes, including irrigation management by users' associations, have often languished. These engineers are resistant to multidisciplinary management of integrated command area projects, despite the need for large projects to balance competing uses of water, as well as to be coordinated with the development of drainage, roads, schools, hospitals, storage and marketing facilities.

State charges for irrigation water and for domestic and drinking water have been kept very low and no state government has shown the political courage to raise them significantly – not even to meet maintenance costs. This naturally limits governments' capacity for wells development and management.

State water resources and groundwater departments have had little coordination so that surface and groundwater use has remained largely delinked, with inadequate regional assessments of overall water availability. Harnessing of groundwater is mostly done by individual farmers, though the PHED uses significant groundwater for drinking water schemes. Groundwater departments largely undertake exploration surveys and projections of potential availability. Farmers, industrialists and communities may all dig wells in the same area, which leads to over-exploitation of groundwater that no government department can control.

State water resource departments seem to have an unintentional (and historic) bias in favour of surface irrigation, while neglecting industrial water supply, domestic and drinking water supply, environmental protection and water quality in designing surface irrigation projects (though small drinking water projects were sometimes constructed separately). In earlier periods, groundwater was a limited option as electricity was mostly confined to urban areas and financing arrangements for wells and pumps were lacking. Nowadays most state governments give top priority to drinking water. Thus many irrigation projects have to reallocate water from irrigation to drinking and domestic-use water – water from some entire projects has been reallocated in this way.

State departments are expected to deal simultaneously with development, management and regulation, but they tend to favour development over management, and are largely ineffective in regulation. Unfortunately they too often do not give equal attention to all three, and they do not see irrigated areas (surface or groundwater) and rain-fed or dry farming areas as forming an interrelated system.

When panchayats take up small local works, they tend to try to tap and store all water, rain-fed or from surface flows, so that 'not a single drop of water should be permitted to flow out of the village'. They can construct small dams to store water on the spot, causing local groundwater levels to increase, and adversely affecting downstream dams, reservoirs and rivers. State governments often must destroy such small dams and structures.

Regrettably, the philosophy of state level leaders regarding neighbouring states resembles that of village leaders – not wanting any of their water flowing to a downstream state, which they view as a waste. Downstream states accustomed to receiving such waters for centuries necessarily protest attempts to divert or block such waters. Many inter-state water disputes result from such thinking. Some state governments even try to evade existing inter-state agreements that specify the rights of downstream states.

State agriculture departments provide information services on subjects such as on-field water application techniques, optimum water quantities and appropriate times for water application. But no agencies provide extension services for water resource development and management, technology transfer, recommended methodologies of operating canals, optimal use and conservation of drinking and domestic water, or efficient use of water for industrial purposes including control of pollution. Such extension services should be provided by appropriate departments.

While state departments benefit from central and multilateral funding for a wide range of water projects, they cannot deal directly with foreign donor agencies. The Department of Economic Affairs in the union Ministry of Finance channels all foreign aid to the states. This provides the union government influence over allocative decisions, and the DEA works with other central ministries in steering foreign loans or aids, but it lacks the feel for ground realities that state departments have.

Since the earliest days of independence there has been the view that the union government should have a more effective role in water management. Thus there are periodic proposals for a constitutional amendment to make water a concurrent jurisdiction. Such suggestions were made in 1987 and 2002 when the National Water Policies were drafted but failed because they were not binding upon the states. While it can be debated whether such an amendment would be desirable, it would be virtually impossible today when different parties or coalitions are in power at the centre and in different states and even the political parties are often divided on inter-state water issues. It might have been feasible up to the early 1960s when the Congress Party was in power at the centre and all the states, but that window has long since closed.

As it is, the union's influence over the states is largely through the power of the purse – both through CSSs and the routeing of external aid through the central government – as well as the requirement for larger projects to be technically cleared by CWC. In practice, the union's influence over the states has increased as regards water (and other sectors), though there is now a political trend of state governments increasingly asserting themselves vis-à-vis the union.

With water scarcity gradually assuming significance, there is need for massive investment in development of water sources. With the easy-to-develop sources already largely exploited, the remaining significant projects tend to be technically more difficult and costly and often have inter-state ramifications. Massive investment is also required for deferred maintenance, repair or rehabilitation, and modernisation of projects originally developed in the nineteenth and twentieth centuries that have not been maintained regularly. Many of these works have inter-state implications of cost sharing, canal closures for repair works, and ensuring that water shares are not affected by the modernisation and deferred maintenance works, but, difficult as it can be, such investment in development of water sources is needed to reduce the likelihood of future water conflicts.

13.4 INTER-STATE WATER SHARING AND DISPUTES

This chapter has left water-sharing disputes between states last to highlight the importance of the institutional arrangements discussed above. Sixteen of India's 18 major basins are inter-state and they drain more than 90 per cent of the country's area. Except for the Brahmaputra, all major river basins are water stressed or water scarce. The union government has shown interest in having a greater say over inter-state river development, which the state governments consistently resist as attempts to subvert their rights.

As of 2000, there were 58 inter-state water agreements, bilateral or multilateral: 39 about costs and benefits of shared projects, and 19 about water sharing. Such agreements can have problems, such as in not factoring in groundwater, leaving out other states further downstream, or because of the division of previous states. States may resist honouring agreements or change their position on their water share.

The River Boards Act 1956 provided for the creation of advisory boards on integrated development of inter-state basins, but it has remained a dead letter because states would not agree. Similarly, attempts in the 1990s and later to agree on national water-sharing principles have failed. States take a narrow view of their interests with no give-and-take. There is a competitive unsustainable demand for water (Iyer 2007). Availability of water favours more water-intensive crops and cropping patterns that require more water and more irrigated areas. Increased industrialisation, urbanisation and population also drive demand. Environmental requirements for water must also now be factored in. So far states have responded through increased exploitation of both surface and groundwater, but more areas are becoming water stressed.

In response, the BJP-led coalition government between 2000 and 2004 advocated a country-wide inter-linking of rivers project to transfer water from 'surplus' to 'deficit' areas. But there are no established criteria for these concepts and no state accepts that it has a surplus. States have invested in major water infrastructure to ensure they utilise their water shares and avoid charges of having a surplus. Perceptions are more important than reality in that even states experiencing floods deny having surpluses. No state is likely to agree to losing water through the inter-linking of rivers.

The Inter State Water Disputes Act 1956 allows a state to ask the union government to refer a dispute to a tribunal comprised of three senior judges. Tribunals can retain experts and are subject to time limits. Their awards are final, with the same force as a Supreme Court decision. There have been only five tribunals since 1956. The Godavari and Narmada river awards were successful (though in both cases facilitated by parallel political agreements). Neither the interim nor final awards in the Cavery dispute were well received. The tribunal on the Krishna gave its award in 1976 but the dispute festered so a new tribunal was named in 2004. The Ravi-Beas award has been contested by Punjab. Increasingly states resist compliance with awards, despite express provisions in the Constitution, and there is no authority to enforce implementation.

There are serious criticisms of the cumbersome procedures adopted by tribunals. They require adjudication when negotiation may be better, and their adversarial nature is costly and divisive. There are no agreed water-sharing principles to guide awards. Parties left with a grievance with the award have no legal remedy (Iyer 2007). However the act does not require adversarial proceedings so tribunals could adopt solution-exploring procedures. Unfortunately, such alternate procedures are not being actively considered.

The prospects are for states to become increasingly aggressive over water shares, with little likelihood of any amendments to the Constitution. States are likely to resist integrated river basin management that implies any transfer of authority to a river basin organisation. Even within states, integrated water resource management is proving difficult.

13.5 CONCLUSION

The pressure on water resources in India is growing as the population increases, prosperity rises, and society become less agrarian (and agriculture more commercial). The surface or groundwater sources that were easier to tap have been developed, and new sources are more costly and may involve complex technical, environmental and political issues. Much of the existing water infrastructure requires major renovation. Competition for water, and even conflicts over sharing water, have been gradually emerging. There are significant challenges for storage and distribution of water through canals, pumping of groundwater, and rainwater harvesting and watershed development. The entire governance and management of the water sector at state, district, water development project, and the local level needs to be looked at afresh along with the role of the union government, so that India has systems appropriate to address the scale of the challenge.

13.6 NOTE

1. This chapter was edited by George Anderson to substantially reduce its length after the death of Dr Hooja.

13.7 REFERENCES

Briscoe, John and R.P.S Malik (eds) 2006. *India's Water Economy – Bracing for a Turbulent Future*, The World Bank and Oxford University Press

Briscoe, John and R.P.S. Malik (eds) 2007. *Handbook of Water Resources in India: Development, Management and Strategies*, New Delhi: Oxford University Press New Delhi; especially articles by A.D. Mohile, A Sekher and Tushar Shah.

Dinesh Kumar, M. 2007. *Groundwater Management in India*, New Delhi: Sage Publications.

Hooja, R. 2010. Managing Water Within India's Federal Framework, Occasional

Paper No. 10, HCM Rajasthan State Institute of Public Administration, Jaipur.

Indian Journal of Public Administration 2003. Special Number on Water Resource Management, 49(3), July–December.

Iyer, R.R. 2007. 'Water Sharing in a Federal Context' (mimeo), Paper presented at Regional Conference organised by Inter State Council at Kochi, 26–27 July 2007 in preparation for 4th International Conference on Federalism.

Second Administrative Reforms Commission 2008. Seventh Report on *Capacity Building for Conflict Resolution – Friction to Fusion*, New Delhi; Chapter 5 'Water Related Issues'.

14. The hydro-institutional challenge of managing water economies of federal rivers: a case study of Narmada River Basin, India

M. Dinesh Kumar[1]

14.1 INTRODUCTION

Many Indian river basins falling in the semi-arid and arid parts of India are physically water scarce with demands for water far exceeding the utilizable renewable water resources. The most important of them are Narmada, Cauvery, Krishna, Sabarmati and Pennar basins (Kumar et al., 2012). The demand for water in India continues to grow as a result of rising population, rapid urbanization and industrialization, more advanced environmental services and the changing livestock and farming economies (Kumar, 2010; Kumar et al., 2012), while utilizable fresh water resources are dwindling due to groundwater depletion and quality deterioration and pollution of water bodies from point and non-point sources (Mukherjee and Chakraborty, 2012). With excessive diversion of surface water and uncontrolled mining of groundwater, these basins are facing severe environmental water stress (Smakhtin et al., 2004).

These are all inter-state basins, with their drainage areas falling in two to four states, so they fall under the concurrent list of the constitution (GOI, 1999), and competition and conflicts over the use of their water have been increasing (Kumar, 2010). Over the past four decades or more, 'water disputes tribunals' were set up as constitutional bodies to resolve disputes over sharing of water from these inter-state river basins, and methodologies and institutions were created to enforce the 'awards' by these tribunals.

While equity, efficiency and sustainability are the key principles in basin-wide water allocation in water-scarce regions, the overwhelming challenge is in assessing the legitimate water demands of the contesting states, and evolving methodologies for determining their share of water

(Wang et al., 2007) from the estimated basin yields. Therefore, arriving at water allocation decisions involves long-drawn legal and institutional processes.

The Helsinki Rules drafted in 1967, by the International Law Association (ILA), represent an early attempt at codifying customary international law pertaining to trans-boundary water resources (Eckstein, 1998, p. 92). These state that 'groundwater is connected to surface water' (Krishna and Salman, 1999, p. 170). Their interdependency heightens with increasing pressure on water resources and demands an integrated approach and institutional arrangement to governance of river basins (Molle et al., 2007).

In contrast, such inter-dependencies are not considered in basin water allocation decisions by the tribunals in India, which are pre-occupied with river flows only. The tribunals also miss out the important impact of changing land use in upper riparian states.

The Narmada River flows west, dividing north and south India, with the states of Madhya Pradesh, Gujarat and Maharashtra in its basin. Discord between these states surfaced in the 1960s over Gujarat's proposal to increase the height of the terminal dam in a way that would submerge potential hydro sites in the two upstream states (Paranjape, 1990, p. 31). In response, the Government of India established the Narmada Water Disputes Tribunal in 1969; in 1979, the tribunal assessed the dependable yield of the basin as 28 million acre feet (MAF) for arriving at water allocation decisions for the four party states. Since then, the Narmada Basin had undergone several hydrological, ecological and socio-economic changes, notably increased groundwater draft in the upper riparian areas in Madhya Pradesh and increased net and gross cultivated areas. This chapter shows how the neglect of these changes undermine the viability of water sharing agreements between Gujarat and Madhya Pradesh and put water and energy security in the lower riparian state of Gujarat at risk. It also provides some suggestions for increasing the effectiveness of existing institutional arrangement for trans-boundary water management in the basin context.

14.2 DISPUTE RESOLUTION FOR WATER ALLOCATION IN INDIA'S FEDERAL RIVERS

The need to cooperate on issues for freshwater systems crossing state borders has been broadly accepted for many years. Since 1814, more than 300 bilateral and multilateral agreements on the cooperative use of trans-boundary waters have been concluded (Gerlak, 2007, p. 2), with approximately 200 agreements on the non-navigational uses of transboundary waters made within 50 years until 2003 (WWAP, 2003, p. 316). Collaborative, adaptive and multi-level institutional arrangements are needed for governance and management of water in river basins to cope with competing values and interests in a context of increasing water scarcity and climate change (Molle et al., 2007).

The Cauvery Water Disputes Tribunal was constituted by the Government of India in 1990 to adjudicate the water-sharing dispute of the states of Karnataka and Tamil Nadu. The tribunal had passed an interim order in 1991 and further orders in 1992 and 1995. It submitted its reports and decision to the union government in 2007. The party states and the central government have sought clarification and guidelines under Section 5(3) of the Interstate Water Disputes Act and so the terms of the tribunal were extended up to 2 November 2008. The party states took the case to the Supreme Court, which ruled in November 2008 that the matter may be listed before a three-judge bench. The tribunal's final award was 30 TMC for Kerala, 270 TMC for Karnataka, 419 TMC for Tamil Nadu and 7 TMC for Pondicherry. The Inter-State Waters Dispute Act, 1956 bars any court, including the Supreme Court, from having or exercising jurisdiction, after the publication of the award in the *Official Gazette* by the central government. In the three years since the final award, it has not entered into force since it has not been gazetted because the matter is sub-judice.

The Krishna is the second largest river in peninsular India, and is shared by the states of Maharashtra, Karnataka and Andhra Pradesh. When disputes arose among them, the government of India constituted the Krishna Waters Disputes Tribunal in 1969, which made its award on the basis of 75 per cent dependable yield of 2160 TCM estimated for the basin, with allocations of 560 TMC, 700 TMC and 800 TMC, respectively for Maharashtra, Karnataka and Andhra Pradesh. The tribunal award factored in return flow and the states were allowed to use it to the extent of 25, 34 and 11 TMC, respectively, unlike in the case of Cauvery. Further, the tribunal allowed the states to fully utilize their water shares for planned projects in any water year (including a deficit year) by utilizing carry-over storage. A state can create carry-over storage during years when the river's water yield exceeds 2130 TMC to use in a water year when the yield is less than that. Thus the tribunal allocated water use up to 2130 TMC based on an average yield in the river and not subject to water availability in a 75 per cent dependable year. Another part of the award provided for the creation of the Krishna Valley Authority to ensure implementation of a water allocation scheme.

On a request from all the three states, the central government constituted a second Krishna Water Disputes Tribunal in September 2003, which gave its verdict on 31 December 2010. Its allocation of available water was done according to 65 per cent dependability, considering the records of flow of water for the previous 47 years. Under this second tribunal's award, Andhra Pradesh got 1001 TMC of water, Karnataka 911 and Maharashtra 666. Thus each state's allocation increased given the revised estimates of stream flows and degrees of dependability.

14.3 EMERGING ISSUES IN TRANS-BOUNDARY WATER ALLOCATIONS

The upper catchments of the Krishna, Godavari, Cauvery, Narmada, Sabarmati, Banas, Pennar, Mahanadi, Ganges and Indus river basins have hilly or mountainous terrain underlain by hard rock (crystalline and basalt). These regions are not ideal for groundwater exploitation and so receive low priority for institutional financing for well development. These upper catchments have higher rainfall and lower aridity than downstream areas and have larger areas under forests and trees, with less arable land (Kumar et al., 2006 and 2008). Importantly, they are inhabited by socio-economically backward tribal populations, which largely depend on subsistence farming, with low adoption of green revolution technologies (Phansalkar and Verma, 2005). Irrigation needs for their traditional crops could be met from manually operated open wells.

These factors traditionally militated against intensive groundwater exploitation, and the local communities largely depended on rain-fed crops. But this started changing in the early 1980s, with growing pressure on land, and declining income from forest resources as farmers started adopting energized wells for intensifying cropping. Rural electrification and subsidies for well digging and pump sets accelerated this. By this time, the Krishna and Narmada dispute tribunals had already declared their 'awards', which could not anticipate the impact of rising groundwater use in upper catchment areas on basin stream flows.

Surface water and groundwater resources interact in many basins (Hernández-Mora et al., 1999; Kumar et al., 2006; Kumar, 2010). Many in India experienced phenomenal rises in groundwater draft in recent decades. New land and water use patterns also change the basin hydrology. Over long periods, green water use in a basin can go up with increasing net cropped area and adoption of longer duration crops. Even so, with an increasing development, the blue water contribution to crop production in a basin can increase compared to green water (Kumar, 2010). These changes can affect surface water flows available for downstream uses and local groundwater recharge, with serious implications for the viability of trans-boundary water allocation decisions in large river basins (Falkenmark, 2004).

The 1990s saw major initiatives from the government and non-governmental organizations to promote decentralized water management (Talati et al., 2005). The upstream catchments of river basins provided optimal conditions for intensive water harvesting and watershed development, because of the hilly topography, high rates of soil erosion and high drainage density. The regions being largely drought-prone and backward ensured government funds under various schemes (such as the tribal sub-plans in the national watershed development and drought prone area program). Intensive water harvesting and watershed development activities changed the landscape of the upper basin areas, with water captured in tiny reservoirs. Many areas in Gujarat, Rajasthan, Madhya Pradesh, Maharashtra, Karnataka and Andhra Pradesh saw intensive water harvesting. Impoundment in the upper basin areas led to reduced flows in the downstream, sometimes with major declines in the storage of reservoirs downstream (Talati et al., 2005; Ray and Bijarnia, 2006; Kumar et al., 2006 and 2008). The increase in water storage area increased evaporation, adversely affecting the basin water balance (Kumar et al., 2006).

14.4 THE NARMADA CASE STUDY

The Narmada River Basin is located in central India. The State of Madhya Pradesh can be divided into 11 agro-climatic zones (GOMP, 2001), nine of which fall partly or fully in the Narmada Basin. The Central Narmada Valley receives about 1200–1600 mm precipitation. The northern hill region of Chhattisgarh receives 1200–1600 mm of rainfall, and the Chhattisgarh plain also receives 1200–1600 mm precipitation (GOMP, 2001). The Kymore plateau and Satpura hills receive 1000–1400 mm of average annual rainfall. Satpura plateau has two districts (Betul and Chhindwara) and gets 1000–1200 mm of annual average rainfall. Vindhya plateau has four districts (Sehor, Raisen, Sagar, and Damoh) with normal precipitation of 1200–1400 mm. Rainfall in Malwa plateau ranges between 800–1200 mm. Nimar plain receives 800–1000 mm rainfall. Jhabua district receives 800–1000 mm rainfall (GOMP, 2001).

The northern hills of Chhattisgarh have a moist sub-humid to dry-sub-humid climate, while the Kymore plateau and Satpura hills, Vindhya plateau, Satpura plateau and Central Narmada Valley have a dry sub-humid climate. Malwal plateau and Nimar plain have semi-arid climatic conditions. The district of Jhabua, which falls in the region called 'Jhabua hills', is semi-arid (GOMP, 2001).

The Narmada Basin has seen a decade-long legal battle over water allocation among Gujarat, Madhya Pradesh, Maharashtra and Rajasthan (Kumar, 2010). The Narmada Water Disputes Tribunal came into being in 1969 to adjudicate the dispute over sharing of irrigation and hydropower benefits between Gujarat and Madhya Pradesh. The tribunal assessed the basin's utilizable yield of 75 per cent dependability as 28 MAF in 1979 in deciding on water allocations for the four party states. Groundwater was not within the ambit of water sharing. The tribunal determined that 9 MAF be provided for Gujarat and 0.5 MAF for Rajasthan at Sardar Sarovar. The requirements at Sardar Sarovar have to be met by releases by Madhya Pradesh and by inflows from the intermediate catchment, surplus to the requirements of Madhya Pradesh below Narmada Sagar and Maheshwar. The releases from Maheshwar work out to 8.12 MAF, with uniform monthly releases.

The tribunal award provides that utilizable flows in a water year be shared as follows: Madhya Pradesh, 18.25 MAF; Gujarat, 9.00 MAF; Rajasthan, 0.50 MAF; Maharashtra, 0.25 MAF. Utilizable surplus or deficit supplies in a water year are to be shared to the extent feasible by the party States in the same proportions. Surplus water shall first be utilized for filling reservoirs to capacity and only then shall excess water be utilized for irrigation and other purposes (GOI, 1979).

As required by the tribunal, the Narmada Control Authority (NCA) was set up in 1980 as an inter-state mechanism for the implementation of the tribunal's decisions. It is a body corporate with representatives from the four party states and the government of India. It is funded in equal proportions by all four states. The Secretary of the Water Resources department in Delhi is the ex-officio chairman of the authority, with routine administration being the responsibility of the authority's staff. It coordinates and directs all water infrastructure projects, including engineering, environmental protection and rehabilitation measures in compliance with terms stipulated by the central government. The NCA can implement orders of the tribunal with respect to: the storage, apportionment regulation and control of the Narmada waters; sharing of power benefits from Sardar Sarovar Project (SSP); regulated releases by Madhya Pradesh; acquisition by the concerned states for SSP of lands and properties likely to be submerged under Sardar Sarovar; compensation and rehabilitation and settlement of oustees; and sharing of costs.[2]

The NCA has created seven discipline-based sub-groups/committees dealing with resettlement and rehabilitation, the environment, rehabilitation, the main Narmada canal, hydro-meteorology, power and the regulation of the Sardar Sarovar reservoir. It has from time to time granted permission to raise the height of the Sardar Sarovar dam up to 110.64 m, following the Supreme Court's decision.

The Narmada Basin's population increased from 7.8 million to 15.38 million between 1971 and 2001, driving increased water demand for urban areas. In rural areas, the number of cattle rose from 4.11 to 4.75 million from 1975 to 2000 and the areas under crops and irrigation both increased.[3] Short duration rain-fed crops were replaced by longer duration crops, which required supplementary irrigation (Kumar, 2010). There has been an explosion in groundwater irrigation in the basin, with a huge increase in open wells and bore wells for irrigation, as well as the energization of wells (Kumar, 2007; Kumar and Ranade, 2004; Ranade, 2005). Several new surface irrigation schemes came up in Madhya Pradesh part of the basin during the past few years (Kumar and Ranade, 2004).

Analysis of groundwater hydrographs shows a significant long-term trend in the annual rate of change in groundwater levels. Many observation wells, which previously showed an annual rise in water levels, now show a higher rate of annual rise, while many other observation wells, which previously showed an annual drop, now show a reduced rate of drop. While groundwater abstraction in Narmada Basin substantially increased, from 1008 MCM in 1985 to 2582 MCM in 1999, the long-term trends in water levels indicate that net annual groundwater storage in the basin, which takes into account all inflows and outflows, has become positive (Kumar, 2010). This is because the groundwater outflow into the surface streams during the lean season would have significantly reduced from the time when the exploitation of aquifers was low.

Analysis of estimated flow series for 1965–2000, 1965–1990 and actual flows for 1980–2000 in relation to rainfall, using the rainfall–runoff relationship for the basin, shows that the stream flows had reduced for reasons other than rainfall.[4] Mean annual flows in the period 1980–2001 are estimated to be only 34.718 BCM, compared with estimated mean annual flows from 1891 to 1990. The dependable yield (75 per cent) had also reduced (Kumar, 2010). Thus, with the increasing groundwater pumping and lowering of water levels (annual draw-downs), groundwater outflows into the surface streams had reduced, leading to a reduction in surface flows in the basin (Kumar, 2010).

Out of total water consumption of 23,423 MCM in the basin, 94 per cent is for crop production, with the balance being for domestic needs and industry. Of the total diverted water used from the MP part of the basin, 76.7 per cent is used in agriculture, whereas 30 years ago, only a small percentage of diverted water was used for agriculture (Kumar, 2010).

The use of green water or rain water directly consumed by the crop from the soil profile, accounts for a major chunk of the total consumptive water use in agriculture in the MP part of the basin, but it has increased slowly compared to blue or diverted water, in spite of the significant increase in cropping intensity. The increase in the use of effective rainfall rose only from 16.58 BCM (billion cubic meters) to 17.52 BCM, a mere 940 MCM increase. This is because the increasing cropping intensity has been mainly from expansion in irrigated area, and not rain-fed area. Nevertheless, the rise in green water use is quite significant when compared to the current blue water use, which is only 4530 MCM. Increased *in situ* use of rainfall affects the availability of surface flows and groundwater first by increasing evapo-transpiration from crop land which reduces surface runoff and second by increasing the infiltration of rainwater into the soil. But the extent to which this results in increased groundwater recharge depends on how irrigation water is applied and how much rainwater is available against the crop water demand. Optimum use of rainwater and irrigation water through proper irrigation scheduling can result in reduced percolation of applied water and rainwater from the cropland (Kumar, 2010).

The change in use of blue water for agriculture production is rather dramatic, from 600 MCM to 4530 MCM over 30 years. This is facilitated by increased access to well irrigation and large-scale investment in surface water development. The net increase in groundwater draft is 1541 MCM (Kumar, 2010).

The effective surface-water diversion from Narmada River and its tributaries to Garudeshwar was 3.59 BCM in 1999–2000, of which nearly 57 per cent went to agriculture. The remaining 43 per cent goes in percolation, which is largely available for reuse. But estimates of the re-usable component of water available from surface lift irrigation, including private lifting, suggest under-reporting of water lifting from the main river and its tributaries by farmers through diesel engines. The increase in cultivated land and the increased diversion from rivers and streams can reduce the stream flows (Kumar, 2010).

Scholars have provided empirical evidence to illustrate the scale effects in small water harvesting (Batchelor et al., 2002; Kumar et al., 2006 and 2008; Ray and Bijarnia, 2006) and watershed development (Syme et al., 2012). Madhya Pradesh is known for decentralized water harvesting and watershed management activities implemented by the government agencies and NGOs. The Narmada Basin also witnessed large-scale implementation of decentralized water harvesting. Unfortunately, the planning of these schemes did not involve any hydrological or geo-hydrological considerations (Talati et al., 2005). Essentially, the total amount of water in the basins and the amount of un-committed flows were not studied.

A study (Talati et al., 2005) using primary data from the Hathni and Kundi sub-basins of Narmada, and two micro-watersheds in one of the basins, showed that after watershed treatment activities, the stream flows in the two sub-basins reduced significantly, while the groundwater recharge fraction[5] increased. Further, it was observed that the recharge fraction for the treated watershed was higher than that of untreated watershed in the same basin for higher magnitudes of rainfall (Talati et al., 2005). As regards the reduction in stream flows, this was found to be due to reduction in runoff coefficient after the watershed treatments (Talati et al., 2005).

14.5 KEY CHALLENGES IN TRANS-BOUNDARY WATER MANAGEMENT IN NARMADA BASIN

From a hydrological perspective, trans-boundary water management faces three major challenges in understanding:

- the changing interaction between groundwater and surface water due to changes in hydrological stresses, and its effects on stream flows in a basin;
- a basin's green water economy and how changes in green water use resulting from changes in land use, affect the basin hydrology; and
- the scale of water harvesting, watershed management and river water lifting by farmers and other water users and their effects on basin hydrology.

In addition, there is an institutional challenge of monitoring the use of groundwater by private well owners and public utilities as well as the direct use of rain or green water for crop production, and enforcing controls on such uses.

Basin-wide hydrological studies would help address the scientific challenges, though they would require enormous amount of real-time data on the physical and socio-economic features of the basin. As for monitoring and controls, some checks on water harvesting and watershed development can be exercised by monitoring agencies through remote sensing and so on, but enforcing restrictions on the volumetric use of groundwater and the use of green or rain water for crop production are particularly challenging.

As regards control on the use of groundwater, the problem is that wells are in the private domain (Kumar, 2007).[6] Further, groundwater is a state subject. So, enforcing restrictions on groundwater pumping by well owners would fall to a state government, which would be politically challenging if done to accommodate a neighbouring state, even if it were required under an interstate agreement or a tribunal award, given the importance of access to groundwater for the survival and livelihoods of rural communities. Further, the groundwater agencies lack the institutional capability to monitor and regulate hundreds of thousands of wells spread across rural areas.

As regards green – or rain – water, it is difficult to exercise any control on land use with millions of tiny operational holdings in each river basin in India and the complications over the legal ownership of precipitation falling on land.

Another important challenge is in predicting the impact of climate change on basin hydrology and water resources. There are great uncertainties involved in the prediction of monsoon rainfall trends in India (Jain, 2012) and these are even greater under climate change. Past trends do not show any consistency across physiographic units and time (Sontakke et al., 2007). Temperature rise would have differential impacts on rainfall in different climates. For example, in arid regions, a rise in temperature could lower rainfall, given the aridity and lack of soil moisture during the summer months (Trenberth, 1999). Hydrological predictions are very challenging in that many of the federal rivers (like the Ganges, Indus, Krishna, Narmada, Godavari, Cauvery, Mahanadi) have widely varying rainfall and climate within their basins.

14.6 SUGGESTIONS FOR IMPROVING INSTITUTIONAL CAPACITIES FOR TRANS-BOUNDARY WATER MANAGEMENT

The difficulties in exercising direct control over groundwater withdrawal and land use mean that the best strategy for protecting the viability of interstate water sharing agreements in federal river basins would be to have an independent regulatory authority, appointed by the Ministry of Water Resources, to periodically examine the need for reviewing such agreements. This federal agency could map temporal changes in land use in a basin through techniques such as remote sensing, or even make it mandatory for the state agencies to provide realistic data on areas under well irrigation and rain-fed production, the number of tiny reservoirs, and so on. The regulatory agency could then provide estimates of the effects on surface water flows to the Ministry of Water Resources and the relevant states. Robust hydrological models would be required to analyse the effect on basin stream flows of various hydrological stresses, such as changes in areas under cropping, in groundwater draft, in the scale of water harvesting, in watershed management as well as stresses induced by climate change.

The Ministry of Water Resources could then consider whether land use changes bearing on stream flows merited a revision of the water-sharing arrangements in an inter-state agreement. If so, it could suggest setting up a new tribunal for working out modified agreements. A new tribunal could consider additional developments with regard to groundwater use, small water harvesting and direct use of rainwater in the basin in preparing a revised allocation, based on the principle of inter-regional equity. Thus upstream states would have to eventually forgo a share of their previous 'share' for the downstream states, or to compensate them for their losses should they retain their previous allocation.

In basins that are not covered by any earlier agreement, new agreements on water allocation could initially be done for stream flows on the principle of 'prior appropriation rights', and therefore ignore the use of base-flows and runoff by the upper riparian states, but with explicit provision about the stage of development of groundwater, and extent of rain-fed crop area and small water harvesting so that the agreement would remain unchallenged, to avoid future conflicts.

Obtaining the relevant data from the respective states will not be easy. Experience with Cauvery basin water sharing shows that state governments often refuse to share reliable data with tribunals required for them to make just and equitable water allocation decisions because they fear a reduced share of water and a political backlash. Therefore, the existing inter-state water disputes Act (1956) may need amendments requiring state governments to share such data with federal agencies, and to enable the agency to carry out field inspections if required. But, as noted by Iyer (2007), a greater challenge facing trans-boundary water management is in making the basin states agree to implement the allocation decisions of tribunals.

14.7 CONCLUSIONS

In India, inter-state water sharing agreements have so far considered only surface water resources. No water dispute tribunal has been able to consider the effect of groundwater-surface water interactions, and the changing land use and water use patterns on surface water availability, which undermine the viability of inter-state water sharing agreements.[7] The Narmada River Basin case shows that inter-state water sharing agreements cannot be static over long time horizons, but need to change to address evolving physical, climatic and socio-economic conditions. If federal agencies are to help enforce water dispute tribunal 'awards', their scientific capabilities to review awards periodically need to be strengthened.

If a review of an existing agreement is necessary, a new tribunal could also consider additional developments in the basin in working out a revised allocation, based on the principle of equity. In practice, upstream basin states would have to forgo part of their previous allocation for the downstream states or to compensate them for their losses if they retain this previous allocation. But, obtaining relevant information from respective states will not be easy, so the existing Inter-state Water Disputes Act (1956) may require amendments to make it mandatory for state governments to share required data with the federal agencies and to enable federal agencies to carry out field inspections if required. Such amendments may prove difficult, but the greatest challenge facing trans-boundary water management is in getting the basin states to accept the inter-state water allocation decisions arrived at by tribunals, including revision of 'awards' made by earlier tribunals.

14.8 NOTES

1. Executive Director, Institute for Resource Analysis and Policy, Hyderabad, India. Email: dinesh@irapindia.org/dineshcgiar@gmail.com.
2. Construction costs of storage, power installations, diversion works, head works and canal networks are wholly borne by the state government in whose territory the work is located except for works whose cost has been ordered by the tribunal to be shared between two or more party states. Where the capital cost is thus shared, the operation and maintenance cost is also shared in the same proportion.
3. While the percentage area under irrigation of food crops increased from 4.6 per cent to 35.1 per cent during 1970–1971 to 1999–2000, the percentage area under irrigation of cotton increased from 1.95 to 41.2 per cent and that under non-food crops increased from 0.8 per cent to 9.4 per cent (Kumar, 2010).
4. Analysis of historical data of point rainfall in different locations across Narmada Basin shows that there has not been any statistically significant trend in the rainfall in the region. At the same time, spatial variation is rainfall is quite significant.
5. Recharge fraction is estimated as the ratio of the average water level fluctuation (during monsoon) and the annual rainfall.
6. In India, *de jure* rights in groundwater are not clear, but de facto, rights to groundwater are attached to land ownership rights.
7. It is important to mention here that return flow from irrigation was considered in the KWDT II award.

14.9 REFERENCES

Batchelor, Charles, Ashok Singh, MS Rama Mohan Rao and Johan Butterworth (2002) Mitigating the Potential Unintended Impacts of Water Harvesting, paper presented at the IWRA International Regional Symposium 'Water for Human Survival', 26–29 November 2002, Hotel Taj Palace, New Delhi.
Eckstein, Gabriel (1998) Hydrologic Reality: International Water Law and

Transboundary Ground-Water Resources, in: *Water: Dispute Prevention and Development*, Washington, DC: American University Center for the Global South.

Falkenmark, Malin (2004) Towards Integrated Catchment Management: Opening the Paradigm Locks between Hydrology, Ecology and Policy Making, *International Journal of Water Resources Development*, 20 (3), 275–282.

Gerlak, A.K. (2007) Regional Water Governance and Institutional Arrangements Around Transboundary Waters, paper presented at the annual meeting of the International Studies Association 48th Annual Convention. http://www.allacademic.com/meta/p179029_index.html (accessed 1 December 2008).

Government of India (GOI) (1979) Further Report of the Narmada Water Disputes Tribunal, Vol. II, Narmada Water Disputes Tribunal, Government of India, Narmada Water Disputes Tribunal, New Delhi.

Government of India (GOI) (1999) Integrated Water Resource Development: A Plan for Action, Report of the National Commission on Integrated Water Resources Development, Vol. I, Ministry of Water Resources.

Government of Madhya Pradesh (GOMP) (2001) Government of Madhya Pradesh Compendium of Agricultural Statistics, M.P. State Agricultural Marketing Board, M.P.

Hernández-Mora, N., M.R. Llamas and L. Martínez (1999) Misconceptions in Aquifer Overexploitation; Implications for Water Policy in Southern Europe, 3rd Workshop SAGA. Milan, December (1999) Preprint p. 22.

Iyer, Ramaswamy, R. (2007) Water Sharing in a Federal Context (mimeo), paper presented at Regional Conference organized by Inter State Council at Kochi on 26–27 July 2007 in preparation for 4th International Conference on Federalism.

Jain, Sharad K. (2012) Sustainable Water Management in India Considering Likely Climate and Other Changes, *Current Science*, 102 (2): 177–188.

Krishna, R. and S. Salman (1999) International Groundwater Law and the World Bank Policy for Projects on Transboundary Groundwater, in: Groundwater: Legal and Policy Perspectives: Proceedings of a World Bank Seminar, World Bank Technical Paper No. 456, pp. 163–188.

Kumar, M. Dinesh (2007) *Groundwater Management in India: Physical, Institutional and Policy Alternatives*, New Delhi: Sage Publications.

Kumar, M. Dinesh (2010) *Managing Water in River Basins: Hydrology, Economics and Institutions*, New Delhi: Oxford University Press.

Kumar, M. Dinesh and Rahul Ranade (2004) Large Water Projects in The Face Of Hydro Ecological and Socioeconomic Changes in Narmada Valley: Future Prospects and Challenges, *Journal of Indian Water Resources Society*, 24 (3).

Kumar, M. Dinesh, Shatanu Ghosh, Ankit Patel, O.P. Singh and R. Ravindranath (2006) Rainwater Harvesting in India: Some Critical Issues for Basin Planning and Research, *Land Use and Water Resource Research*, 6 (2006): 1–17.

Kumar, M. Dinesh, Ankit Patel, R. Ravindranath and O.P. Singh (2008) Chasing a Mirage: Water Harvesting and Artificial Recharge in Naturally Water-Scarce Regions, *Economic and Political Weekly*, 43 (35) 61–71.

Kumar, M. Dinesh, M.V.K. Sivamohan and A. Narayanamoorthy (2012) The Food Security Challenge of the Food-Water-Land Nexus in India, *Journal of Food Security*, 4 (4) 539–556.

Llamas, M.R., W. Back and J. Margat (1992) Groundwater Use: Equilibrium Between Social Benefits and Potential Environmental Costs, *Applied Hydrogeology, Heise Verlag*, 1 (2): 3–14.

Molle, Francois, P. Wester, P. Hirsch, J.R. Jensen, H. Murray-Rust, V. Paranjape, V.P. Sharon and P. van der Zaag (2007) River Basin Development and Management, in Molden, David (ed.) *Water for Food, Water for Life: A Comprehensive Assessment of Water Management in Agriculture*. London, UK: Earthscan; Colombo, Sri Lanka: IWMI, pp. 585–625.

Mukherjee, S. and D. Chakraborty (eds) (2012) *Environmental Scenario in India: Successes and Predicaments*, London: Routledge Studies in Ecological Economics.

Paranjape, Vijay (1990) in N.D. Jayal (ed.) *High Dams on the Narmada – A Holistic Analysis of the River Valley Projects*, New Delhi: Indian National Trust for Art and Cultural Heritage, New Delhi.

Phansalkar, Sanjiv and Shilp Verma (2005) *Mainstreaming the Margins: Water-centric Livelihood Strategies for Revitalizing Tribal Agriculture in Central India*, New Delhi: Books for Change.

Ranade, Rahul (2005) Out of Sight, Out of Mind, special article, *Economic and Political Weekly*, 40 (21): 2172–2175.

Ray, Sunil and Mahesh Bijarnia (2006) Upstream vs Downstream: Groundwater Management and Rainwater Harvesting, *Economic & Political Weekly*, 10 July.

Sahuquillo, Andres (1985): Groundwater in Water Resource Planning: Conjunctive Use, *Water International*, 10 (1985): 57–63.

Smakhtin, Vladimir, Carmen Revenga and Petra Döll (2004) *Taking into Account Environmental Water Requirements in Global-scale Water Resources Assessments*, Comprehensive Assessment Research Report 2, Comprehensive Assessment Secretariat, Colombo.

Smith, Z.A. (2002) Inter-state and International Competition for Groundwater Resources: Revisiting a Groundwater Management Issue 15 Years Later, *Water International*, 27 (2002): 485–491.

Sontakke, N.A., H.N. Singh and Nityanand Singh (2007) Chief Features of Physiographic Rainfall Variations across India during Instrumental Period (1813–2006), Research Report 121, Indian Institute of Tropical Meteorology, New Delhi.

Syme, G.J., V.R. Reddy, P. Pavelik, B. Croke and R. Ranjan (2012) Confronting Scale in Watershed Development in India, *Hydrogeology Journal*, 20 (5) 985–993.

Talati, Jayesh, M. Dinesh Kumar and R. Ravindranath (2005) *Local and Sub-Basin Level Impacts of Local Watershed Development Projects: Hydrological and Socio-economic Analysis of Two Sub-Basins of Narmada*, Water Policy Research Highlight 15, IWMI-Tata Water Policy Research Program, Anand, Gujarat, India.

Trenberth, K.E. (1999) The Extreme Weather Events of 1997 and 1998, *Consequences*, 5 (1): 2–15.

Wang, Lizhong, Liping Fang and Keith W. Hipel (2007) Basin-wide Cooperative Water Resource Allocation, *European Journal of Operational Research*, 190 (3): 798–817.

WWAP (2003) *United Nations World Water Development Report 1: Water for People, Water for Life*, Paris/London: UNESCO/Berghahn Books.

15. Inter-jurisdictional water management in Pakistan's Indus Basin

Shahid Ahmad, Khalid Aziz, and Mujib Khan

15.1 OVERVIEW OF THE INDUS BASIN

The Indus Basin Irrigation system (IBIS) is the world's largest contiguous irrigation network managed under a federal arrangement (Figure 15.1). While constitutionally water is a federal subject and provinces are responsible for irrigation and drainage, domestic water supply and sanitation, major decisions have always involved both jurisdictions politically and other constitutional provisions ensure they are administered jointly through the Council of Common Interests (CCI), which brings together all the country's heads of government.

IBIS is a continuous-flow and fixed-rotation irrigation system comprising hydro-power dams, head works and canals. Its management becomes difficult because of physical and hydrological limitations. It plays a vital role in the livelihood of Pakistanis and is the country's largest infrastructural enterprise worth US$300 billion and contributing US$18 billion, or over 21 per cent to Pakistan's GDP. Most of country's population is concentrated in irrigation districts of IBIS. Pakistan lies mainly in an arid region. Mean annual rainfall varies from less than 100 mm in parts of Balochistan and Sindh to more than 1500 mm in wet mountains. In Gilgit-Baltistan, at altitudes greater than 5000 m, snowfall exceeds 5000 mm. Gilgit-Baltistan also contains the largest area of perennial glaciers outside the polar regions – 28 per cent of its area is glaciated, with over 100 glaciers more than 10 km in length (Ahmad 2008a). Winter snow covers 30–40 per cent of land area. Glacier- and snow-melts are major contributors for river flows during the dry period.

River flows are highly variable and storage is essential to regulate them (Ahmad 1999). Western Rivers[1] provide 136.1 MAF (million acre feet) in an average year at the rim stations.[2] Flows in wettest years are twice those of driest years. Summer flows are five times those of winter. Variability

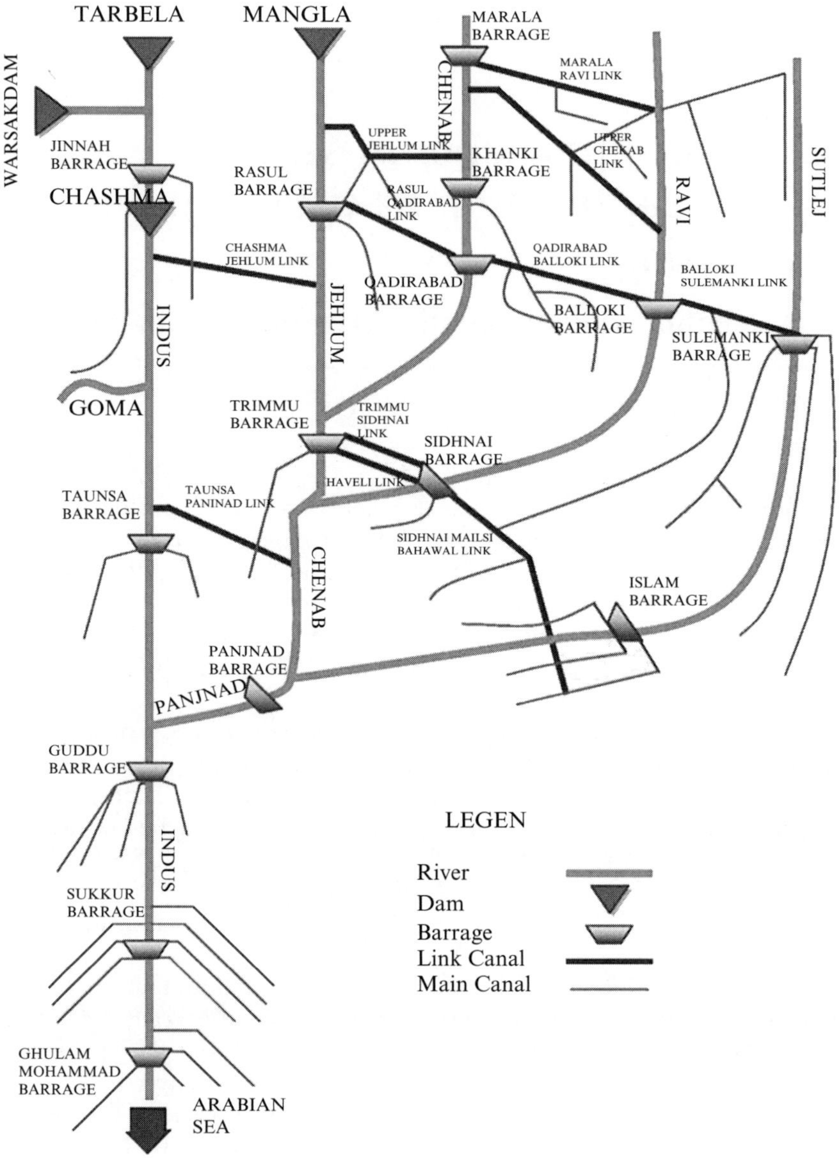

Figure 15.1 IBIS contiguous network

Table 15.1 Per capita water availability in the IBIS

Indus Basin	Total Renewable Water Resources[1] MAF (km^3)	Per Capita Water Availability (m^3/person)			
		1990	2000	2025	2050
Indus-India	78.6 (97.0)	2,487	2,109	1,590	1,132
Indus-Pakistan	154 (190.0)	1,713	1,332	761	545

Note: [1]. Renewable water resources represent the long-term average annual flow of rivers and groundwater. External renewable water resources are those which enter from upstream countries through rivers or aquifers.

in flows of the Eastern Rivers was even higher but these are largely in the summer season. Mean total annual actual flow of IBIS is 143.1 MAF at the rim stations (Ahmad 2008a).

IBIS extends over an area of 1.166 million km^2 and is divided among various countries: Pakistan 0.693 million km^2; Afghanistan and China 0.015 million km^2; and India 0.321 million km^2. IBIS-Pakistan averages annual renewable water resources of 151.54 MAF as reported by Sharma et al. (2008), while comparing renewable water resources in Pakistan and India, using data from Food and Agriculture Organization (FAO) AQUASTAT. Pakistan's renewable water resources will always be higher than actual flows at rim stations located at the Mangla and Tarbela dams and there are some losses in the upstream.

Pakistan's population was 132 million in 1998 and will grow to about 209 million by 2025, which will increase needed water availability by 23 per cent (Ahmad 2008b). Projected per capita water availability in Indus-Pakistan will reduce to 761 m^3 by 2025, a water scarce basin. Indus-India will have per capita water availability of greater than 1000 m^3 even beyond 2050 (Table 15.1).

Per capita water storage capacity available in Pakistan is 144 m^3 (equivalent to only 30 days of flows compared to 900 days of storage for the Colorado River).

There are one million tubewells: electric (13 per cent) and diesel (87 per cent). Electric tubewells decreased due to increased tariffs, repeated power failures, load shedding and voltage fluctuations. Large subsidies (around 80 million US$ per annum) have been provided to over 15,626 deep tube-wells in Balochistan representing only 1.5 per cent of total tubewells' population; the Balochistan government pays 90 per cent of electricity costs and overuse has resulted in lower water tables. Farmers have no incentive to use tubewell water efficiently. Diesel prices were at least one-third less than petrol, but in the last few years have become more expensive. From

1976 to 2006, groundwater's contribution doubled from 25.6 to 50.3 MAF due to rapid growth of tubewells. Currently, it contributes 48 per cent of available surface water at the canal head (GoP 2011).

Overall irrigation efficiency in IBIS is 41.5 per cent against a potential of 55 per cent within the existing system. The net crop water requirement is 101.7 MAF. Canal diversion provides 99 MAF but 44.25 MAF is lost in conveyance but this recharges the groundwater. As a result, only 54.75 MAF is available at the farm head with an additional 50.3 MAF from groundwater, totalling a net of 105.05 MAF. Due to field application losses only 78.79 MAF is used for crop consumption. Assuming rainfall contribution of 13.4 MAF, net water available is 92.3 MAF. The shortfall in water for agriculture ranges between 9.51 and 22.91 MAF (Ahmad 2008c).

15.2 FEDERAL SYSTEM

In 1955, West Pakistan's four provinces were consolidated into one in the then two-province federation of Pakistan (including East Pakistan). The unitary system of water development for West Pakistan was thus strengthened with the promulgation of the Water and Power Development Authority Act (WAPDA) in 1958 as an apex institution. After signing the Indus Water Treaty (IWT) in 1960 with India, Pakistan initiated the Indus Basin Project (IBP), the largest engineering project of the time, when three hydro-power dams, nine link canals and six barrages were built as replacement works of the commands of Eastern rivers and additional infrastructure for other commands.

In 1970 the four provinces of West Pakistan were resurrected in a new federal arrangement, and in 1971, with the loss of East Pakistan, the federation assumed its current status with four provinces. Following the recreation of the provinces, the West Pakistan Irrigation and Power Department was decentralized, with the provinces becoming responsible for the operation and maintenance (O&M) of the irrigation system. However, by then the capacity of provincial Irrigation and Power Departments had eroded, because most of the top irrigation experts had joined WAPDA. Tensions on water apportionment continued between Punjab and Sindh, and Sindh and Balochistan until 1991 when the Water Apportionment Accord was signed by the four provincial chief ministers.

The federal Ministry of Water and Power controls key institutions including WAPDA, Federal Flood Commission, Indus River System Authority (IRSA), and Office of the Chief Engineering Advisor. Development of large infrastructure is the responsibility of WAPDA. The IRSA was created in 1992 to administer the Water Apportionment Accord as the body responsible for allocating available supplies and resolving disputes. At the provincial level, the Departments of Irrigation and Power, and the Irrigation and Drainage Authorities manage the irrigation and drainage network. Departments of Agriculture provide services to farmers in On-Farm Water Management.

While concerns on water apportionment between Punjab and Sindh continued even after the accord, the two provinces with the support of IRSA were successfully sharing waters for cotton plantation, where Punjab provided its share to Sindh during April and part of May and then Sindh compensated Punjab in May. This is the achievement side of the accord.

Provinces may take conflicts on water sharing to the Council of Common Interests. The CCI is the highest forum for federal–provincial relations. It has existed for many years and its role, composition and functioning were strengthened in the 18th Amendment of the Constitution in 2010. It is chaired by the Prime Minister and has three federal ministers and the four provincial chief ministers as members; it must meet at least once every 90 days. Its decisions 'shall be expressed in terms of the opinion of the majority'[3] though politically consensus is more often sought. The CCI played a major role in negotiating the accord of 1991. Under the 18th Amendment it has responsibility for policies and institutions under Part II of the federal legislative list, which includes WAPDA, and disputes relating to interference in water supplies or construction may be submitted to it. As well, the IRSA is a subject under the Ministry of Inter-Provincial Coordination, so that it falls within the purview of the CCI.

Technically, the provinces rely heavily on the IRSA; however, it is far from becoming a full custodian of the irrigation system despite the provinces' willingness to make it more effective so that they could benefit from more decentralized management. The federal and provincial laws that impact water are shown in Table 15.2.

15.3 EVOLUTION OF WATER MANAGEMENT WITHIN FEDERAL PAKISTAN

A more institutionalized regime of water entitlements, development, and management started in the twentieth century, when weir-controlled structures were installed to regulate canals in united India. The regime has evolved through four phases: pre-Indus Water Treaty (IWT); IWT and the era of water development; inter-provincial water disputes and the Water Apportionment Accord; and the post-accord era.

Table 15.2 Federal and provincial water legislations in Pakistan

Water Legislations	Year of Promulgation
Federal Legislation	
Pakistan Penal Code	1860
Canal and Drainage Act	1873
Factories Act	1934
Water and Power Development Authority Act	1958
Indus River System Authority Act	1992
National Environmental Quality Standards	1993
Pakistan Environmental Protection Act	1997
Provincial Legislation	
Sindh Irrigation Act	1879
Punjab Minor Canal Act	1905
Canal and Drainage Act NWFP Amendment	1948
Punjab Soil reclamation Act	1952
Balochistan Groundwater Rights Administration Ordinance	1978
Water Users' Association Ordinance	1982
Provincial Irrigation and Drainage Authority Act	1997
Community Irrigation and Farmers' Organizations Regulations	2000
Amendment to Balochistan Canal and Drainage Act	2006

15.3.1 Pre-Indus Water Treaty

Conflicts between the upper and lower riparian were not uncommon in the undivided India, for example Sindh and Punjab raised objections to Sutlej Valley and Sukkar projects respectively as having adverse impact on their water availability, and the central government in India had to appoint commissions to address these issues. The resultant agreement between the chief engineers of Sindh and Punjab in 1945 resolved most of these conflicts.

At the time of partition in 1947, water apportionment between Indian and Pakistani Punjab was maintained as before; however, as a result of the Radcliffe Award, the control of canal headworks that was vital for Pakistan remained in India. Thus, in April 1948, the East Punjab government in India cut off canal supplies serving Pakistan-Punjab; this was quickly resolved that May, when India and Pakistan signed an agreement on the progressive curtailment of water supply to canals in Pakistan-Punjab, while providing reasonable time for Pakistan to develop alternative arrangements.

Internally, sharing of Indus basin water among Pakistan's provinces until 1955 was largely based on the Sindh-Punjab Agreement of 1945, and this effectively continued to resolve the internal issue when the provinces of the then West Pakistan were merged in 1955. However, the serious economic consequences of water issues with India were such that the World Bank brokered negotiations between the two countries, resulting in the IWT of 1960.

15.3.2 IWT and Era of Water Development

Since trans-boundary conflicts occurred on all tributaries of the Indus River system, the IWT allocated Western Rivers (Indus, Jhelum and Chenab rivers) to Pakistan with some water entitlements to India, and exclusive rights on the Eastern Rivers (Ravi, Sutlej, and Beas) to India. Allocation of entire rivers, rather than partial flows, led to a situation that required independent, rather than cooperative, river management. India's right to the use of Western Rivers resulted in differing interpretations and the subsequent construction of Indian hydroelectric projects has affected Pakistan's water availability.

The Indus Basin Project was initiated during the 1960s to construct replacement works to provide water previously under the command areas of the Eastern rivers and additional supplies to other commands for increasing cropping intensity and irrigation of new areas.

While the IWT was successful in resolving trans-boundary issues, since the 1990s the issues of bi-national apportionment of rivers and water are becoming more complex due to climatic variability and change, rising water demand and environmental issues. India has developed extensive plans for new hydropower projects and completed the Salal and Baglihar projects, which would have negative cumulative impacts on the flows of the Chenab River into Pakistan, especially during dry periods. The treaty does not specify how India is to share shortages of water during dry periods, because it fixes India's entitlements on the Western Rivers. The problem is that flows are variable so it is hard to fix water entitlements equitably; the dispute over this is emerging as a threat to peace in the sub-continent. A further unresolved issue has to do with a problem on which the IWT is silent: India got fresh water from the three rivers, while Pakistan was compensated by the construction of reservoirs; unfortunately, these have now silted up and the capacity needs to be replenished but there is no agreement as to how.

15.3.3 Inter-provincial Disputes over Water and Water Apportionment Accord

While provinces didn't exist in West Pakistan from 1955 to 1970, issues relating to water apportionment continued to arise around water releases to barrages and canals. The government appointed various commissions to look into these issues, which became more politically heated with the re-creation of the four provinces; however, reasonably sufficient water flows kept these issues within manageable limits.

Lack of consensus among the provinces on equitable use of water meant that no significant surface water development or hydropower projects were taken up in the Indus Basin after Tarbela Dam was commissioned in late 1975. The country witnessed serious and persistent crises of diminished supply of irrigation water and a stagnant power generation capacity being vastly overtaken by burgeoning demand. Prolonged power outages and irregular canal water supplies due to increased variability in river flows accentuated the crisis. An inter-provincial resolution was needed to the disputes over canal water entitlements, allocation of future river supplies and the distribution of surplus floodwater or additional supplies from the construction of new storages.

Finally the four provincial chief ministers, with the support of the federal government, working within the CCI, were able to come to a conclusion over water apportionment on 16 March 1991. The Water Apportionment Accord protected existing uses of canal water in each province, with recognition of the need for escapage below Kotri for environmental purposes, and apportioned the balance of river supplies, including flood surpluses and additional supplies from future storages. It clearly defined provincial water entitlements based on water availability of 114.35 MAF, assuming that new water storages will provide additional water of around 10 MAF.

While the distribution of existing water proceeded somewhat cooperatively during the water sufficient years, bitter disputes between Punjab on the one hand and Sindh and Khyber Pakhtunkhwa on the other continued over construction of new storage dams, particularly over the proposed Kalabagh Dam. This eroded Pakistan's capacity to develop its hydel-power and water development projects. The stagnant production of hydel-power and reduced water supply due to the loss of one-third of the live storage capacity of the Tarbela and Mangla reservoirs has adversely affected the productivity of irrigated agriculture, particularly in Punjab and Balochistan due to their extensive use of groundwater. In Punjab almost half of water available for crop consumptive use is provided from groundwater.

Thus the accord resolved much of the dispute over apportionment of existing water among provinces, though some concerns continued related to new water storages and hydel-power developments.

15.3.4 Post Accord Era

Since the accord, key issues are being addressed in an ad hoc manner, largely because of a lack of an integrated water policy at the national level. In 2006, Balochistan became the only province having integrated water resources management policy relating to its responsibilities. The best way forward for the other provinces might be to formulate their own integrated water resources management policies so that the federation could address its constitutional roles while formulating an integrated national water policy. The most remarkable contributions on water policy in this period were the Water Vision 2025 by WAPDA and Pakistan Vision 2030 by the Planning Commission. Federal budgets have also included water development and management in the ten year plan, resulting in the construction of Kacchi, Rennie and Thal canals and recently the Basha Dam, which is the only major water development project approved since the 1980s.

15.4 INTER-PROVINCIAL WATER MANAGEMENT – IS FEDERALISM WORKING

15.4.1 Limitations of the Water Apportionment Accord

15.4.1.1 Overoptimistic assumptions

The accord assumed, for making water entitlements to the provinces for canal water diversions, an annual water availability of 114.35 MAF against the annual flows of Western rivers between 91.2 to 187.7 MAF during 1937–2007. Mean annual river flows at 50 per cent probability is 136.1 MAF. When usable rivers flows are less than entitlements, the shortages must be shared proportionately and these periods spark severe conflicts among the provinces. The intent when the Accord was negotiated was that the possibility of higher flow levels would spur the development of additional reservoirs so as to transfer water from wet to dry years; however, the current storage is used only for the transfer of summer water to the winter season. Additional storage would help limit flood risks; there has been huge flood damage in recent years and it risks getting worse with climate change. All river flows cannot be diverted to canals because of minimum allocated flows of 10 MAF below the Kotri Barrage, though this could change if the federal government decides otherwise following a current study. In reality, average canal diversions in post-Tarbela periods have only been 103 MAF, much less than the accord's entitlements. This has created problems among lower riparian when they have to share shortages. Another inadequacy in the accord is the non-provision of a clause authorizing provinces to market their unutilized share of water; this introduces perverse incentives, leading to the extravagant use of available water and disagreement on the development of infrastructure.

15.4.1.2 Issues in implementation

The implementation of the accord was fraught with different interpretations. Differences arose within six months of the accord being signed, when the CCI met to determine ten-day average water use. Punjab wanted allocations to be based on the 'historic use' between 1977 and 1982. This was resisted by Sindh and the other two provinces, which alleged that the ad hoc allocations by the federal government during that period had favoured Punjab, and it was rejected by the CCI. Another attempt by Punjab in 1994, to have recourse to historic use was resisted by the smaller provinces on the ground that Punjab would use its upstream position and control of water infrastructure to implement a sharing formula on its terms. Sindh particularly fears reduced overall water availability and reduced flows to meet minimum sea escapage to counter sea water intrusion. Seawater has now moved up to 100 km inland, which has resulted in increased salinization of agricultural lands in lower Sindh, with adverse consequences for the ecosystems and deterioration in the quality of water supply to Karachi. Karachi's mangrove forest, once sixth largest in the world, shrunk by 38 per cent between 1977 and 1990 due to excessive cutting and lack of adequate fresh water inflows; the livelihood of thousands of people is threatened. Transport of effluents to downstream areas and impacts of pollutants on bio-diversity and ecology of the delta ecosystem is now a serious concern.

Sindh and Balochistan have disputes over Khirther and Pat Feeder Canals. Balochistan the lower riparian, cannot use its allocated share due to inadequate irrigation infrastructure even after the construction of the Kacchi canal. Who uses Balochistan's share must be addressed.

15.4.1.3 No water entitlements for special areas

No water allocations are made in the accord for special areas – Gilgit-Baltistan, FATA,[4] Azad Jammu and Kashmir and Islamabad Capital Territory (ICT). They too will claim water entitlements given the rising water demand because of population growth and economic development. For example, the Munda multi-purpose dam is being constructed in Malakand Agency of FATA and there is an issue how FATA will ensure their rights for water without having any formal entitlement. Similarly, the Capital Development Authority has been proposing options to divert water from the Jhelum or Indus rivers to meet Islamabad's growing demand. The ICT can implement such a proposal only after approval of provinces. Furthermore, the accord has no provision for provinces to market their water entitlements to other regions.

15.4.1.4 Need for storages and opposition from provinces

Storage is inadequate on Indus Main to regulate water supply, minimize water wastage, and ensure sufficient water in times of shortage. Current storage availability is inadequate to meet future water demand and to ensure availability of water to the provinces as per Accord.

While the proposed Kalabagh Dam is intended to address these issues, its construction has been hotly debated for 25 years, with Punjab in favour and the other provinces opposed. Khyber Pakhtunkhwa (KPK) fears that the Kalabagh Dam would dam up the flow in Kabul River and risk heavy flooding in the Peshawar valley and that it would threaten to raise water logging in Mardan and Charsada, thus compromising agriculture in these heavily populated districts (this contrasts with KPK's support for the Basha dam, discussed above, which did not give rise to environmental concerns; moreover, it could well profit from Basha's power generation). Sindh opposes the Kalabagh Dam on the basis that any more extraction from the main Indus stem will compromise water flow to the delta and thus invite salt water intrusion from the Indian Ocean. Achieving a fair distribution of power royalties from hydro projects among provinces is a serious issue for the federal government and CCI. An alternative to the Kalabagh Dam that has been suggested would be a high level Kalabagh Barrage that will provide replacement water to Punjab and KPK as well as provide about 800 MW of power annually. IRSA has not yet generated the required analysis on how to make the accord more effective and acceptable to all stakeholders so that water and power development can be initiated. The real issue is to address the fears of the provinces highlighted above. A possible consensus option might be to utilize the huge variability in annual river flows (from 91 to 188 MAF) to store water in flood years of over 140 MAF and release it in dry years in future storage dams, which can be built above Tarbela thus avoiding the problems of below-Tarbela dams like Kalabagh.

15.4.1.5 Debate of small versus large dams

The last decade has seen a strong anti-dam lobby argue for small dams instead of large ones, even though there is no comparison between the two. On the Indus Main only large hydro-power dams having storage of over 6 MAF can be constructed at the potential sites. Small dams are constructed outside the Indus Basin, but they have very little storage of 0.04 to 0.08 MAF and an effective life of around 10–20 years; they are still justified for areas outside the basin where the poorest-of-the-poor live. The country's inability to make decisions on large dams is a significant part of the current major crisis in electricity supply.

15.4.1.6 Water politics

IRSA in 2004 installed a satellite-based telemetric system to provide real-time transparent data of river flows to build trust between the provinces. The system is still not functional largely due to: incompetence of IRSA, WAPDA's continued involvement in measuring river flows (despite its having no rapport in Sindh), and the vested interest of provincial bureaucratic and political leadership. There are perpetual tensions between Sindh and Punjab on the operation of link canals in Punjab and on the whole issue of Sindh's allocated water share not getting to the province. Again, there could have been a substantive basis for arbitrating those conflicts if transparent data from the telemetric system were available (Mustafa 2010).

The Indus waters distribution controversy, at the moment, is limited to sloganeering and street protests by part of the populace of Sindh and to a lesser extent of KPK, as well as heated debates among water managers and provincial governments. Incidentally, in the Punjab, which stands to benefit the most from the construction of Kalabagh Dam and other water development projects, grassroots opinion is uninterested at best, in contrast to Sindh, where there may have been a fusion of hydropolitics with identity politics of Sindhi nationalists. Consequently, given the shallowness of popular support for additional water development on the Indus River, there is a need for a more enlightened and multidimensional dialogue with the political leadership to resolve the controversy. The conflict could possibly be narrowed if the discourse were widened from purely engineering concerns to wider social, cultural, environmental, equity- and justice-related water concerns (Mustafa 2010).

In summary, water problems result largely from poor management as well as water shortage. These problems are accentuated, both materially and politically, by the narrow approaches of the provinces. Furthermore, the federal water agencies failed badly in managing the flood peaks during the very severe floods of 2010 that adversely affected the rural population and its livelihood. This reflects the fact that flood peak attenuation has not been an objective of managing the large reservoirs. The issue of floods is now being used by the pro-dams lobby to call for construction of more storage on the Indus, but this is still different from an integrated strategy that would balance between water development and management in the interest of all.

15.4.2 Federal Institutions and Expected Role

15.4.2.1 Council of Common Interest (CCI)

Since the 18th Amendment, the CCI is now responsible for formulation and regulation of policies as enumerated in Part II of the Federal Legislative List, and to exercise supervision and control over related institutions. Part of the constitutional reform involved the transfer of important subjects from the list of concurrent subjects (which has been deleted) to Part II of the Federal Legislative List for the consideration of and decision by the CCI. The CCI will also have policy control over reservoirs in addition to natural sources of water supply. The Federation shall not build new hydro-power stations in any province except after consultation with that province. The fact that the CCI must meet frequently should help resolve the conflicts between the provinces.

15.4.2.2 Senate of Pakistan

The Senate's role has been substantially enhanced since the 18th Amendment. The annual report on implementation of Principles of Policy is to be placed before the Senate, and the Prime Minister and Cabinet will henceforth be collectively responsible both to the Senate and the National Assembly, not just the lower house. The existing Standing Committee of Senate on Water and Power will now be joined by a new Standing Committee for Inter-provincial Coordination. There are merits of shifting the issues related to IRSA to the Standing Committee on Inter-provincial Coordination, as this ministry also serves as the secretariat for the CCI.

15.4.2.3 Indus River System Authority

Under Clause (4) of the IRSA Act, the authority consists of five members, one each to be nominated by each province and the federal government from among high-ranking irrigation engineers. The first chairman was nominated by the government of Balochistan to be followed by the nominees of the governments of KPK, Punjab, Sindh and the federal government and thereafter in that order. The term of office of the chairman shall be one year and that of a member three years. The chairman of WAPDA and chief engineering adviser are the ex-officio members, but they shall have no right to vote.

Clause 3 of IRSA Regulations provides for meetings to be convened by the chairman, in consultation with the members, for the purpose of:

- making regulations and distribution of the flow-cum-storage waters;
- settling any question in respect of distribution of river and reservoir waters, which may arise between two or more provinces;
- considering and making recommendations on the availability of

water, against the allocated share of the province concerned, whenever a new water project is received;
- settling any question that may arise in respect of implementation of the accord; and
- considering such matters as are to be referred to the Advisory Committee, at the start of Kharif and Rabi seasons.

Decisions of the authority are by the majority of members, with the chairman voting to break a tie. The IRSA has been too narrowly technical, which reflects its staffing by retired irrigation engineers as representatives of the provinces. Its staffing and orientation need to be rethought if it is to make the necessary contribution to resolving the country's challenges.

15.5 CONCLUSIONS

Pakistan faces both a water crisis and a power crisis. While there are gaps in policy and the roles of institutions, the basic institutional structures needed are present. The weaknesses have been in their functioning. The technical agencies have too often failed to provide the information needed to help resolve issues and they have shown a lack of political sensitivity because they are dominated by engineers with a narrow focus. The main problems are political. The provinces are very vocal and mistrustful of one another (and often of the federal government and of the technical agencies). They tend to adopt narrow perspectives and engage in positional, rather than interest based, bargaining. The federal government has proven unable to develop an integrated water policy or to lead in the development of agreements on strategic projects. The new and stronger role for the CCI may prove helpful in this regard, but to date there has been no breakthrough.

Balochistan's success in formulating an integrated water management policy and supporting its implementation may provide an example of one option for initiating a more positive process of inter-jurisdictional water management and development in the IBIS. A start may be made by formulating water management policies for KPK, Punjab and Sindh, which could then serve as a basis for a federal integrated water policy that would reflect the respective constitutional roles amongst the provinces and federal government. The focus of federal water policy would be on legislation, water entitlements, management and development, effective distribution and use of water, and improving performance of water institutions at all levels. New policy, legislation and planned infrastructure developments will help only if the country has an effective system of water measurement accepted by all the stakeholders for resolving water conflicts and building trust.

A consensus on future water development projects including large hydro-power dams on Indus Main could be made by making water as business-for-everyone including political leadership, civil society, water managers, legislatures, and so on. Substantive measures would include a fair distribution of water as per the accord, the CCI developing a revised policy for distributing power royalties amongst the provinces, and permitting provinces having excess water to auction it to other provinces.

The issue is of political, cultural and historic nature related to long-standing conflicts, especially between Punjab and Sindh. It cannot be addressed until the federal and provincial top-leadership is fully committed to resolving the conflicts and finding a way ahead to develop and manage water resources. The energy crisis, with it huge costs to industry, requires that hydro-power be developed utilizing the available potential; doing so would help address the larger issues of unemployment, insecurity and terrorism. Without this Pakistan will face difficulties in resolving trans-boundary issues with India and expected issues with Afghanistan.

The Senate of Pakistan could constitute a 'Sub-Committee on Consensus Building for Water Development and Management' and then submit its recommendations to the CCI. As there is equal representation of provinces in the Senate, such a sub-committee might be able to formulate implementable recommendations through consensus building.

The IRSA that manages the Indus water system may be made more effective and autonomous. It could be managed directly by the CCI and placed under the Ministry of Inter-Provincial Coordination rather than the technically oriented Ministry of Water and Power. Adding additional members in a personal capacity with expertize in the areas of water diplomacy, water laws, water entitlements, conflict resolution and consensus building would help transform its orientation. Such an enlarged Board could be complemented by an executive committee of experts mandated to implement decisions of the board and to assist the board by providing analysis of conflicts and policy options for future water development and management.

Thus a combination of innovations both in terms of the technical and political arrangements and the approaches to water management are urgently required to help Pakistan emerge from its water and power crisis.

15.6 NOTES

1. Western Rivers include the Indus, Jhelum and Chenab, which under the Indus Basin Treaty of 1960 were reserved for exclusive use by Pakistan, barring some non-consumptive use and limited consumptive use by India.

2. Rim stations are established at points where real-time actual river flows are measured for the assessment of water availability and diversion, that is, Tarbela for the Indus Main, Mangla for Jhelum, Marala for Chenab, Baloki for Ravi and Sulaimanki for Sutlej.
3. Article 153, Constitution of Pakistan.
4. FATA refers to Federally Administered Tribal Areas covering seven tribal agencies, namely Bajour, Mohmand, Khyber, Orakzai, Kurrum, North Waziristan and South Waziristan. The FATA Secretariat is located in Peshawar.

15.8 REFERENCES

Ahmad, S. 1999. Achievements and issues of irrigation in the 20th century. Proceedings of the National Workshop on 'Water Resources Achievements and Issues in the 20th Century and Challenge of the Next Millennium', Islamabad. PCRWR/UNESCO, June 28–30, pp. 188–201.

Ahmed, S. 2008a. Northern Areas Economic Report. Background Paper on Water. World Bank Resident Mission, Islamabad.

Ahmad, S. 2008b. Keynote Address: Water Shortage and Future Agriculture in Pakistan: Challenges and Opportunities. Proceedings of the National Conference on 'Water Shortage and Future Agriculture – Challenges and Opportunities', Agriculture Foundation of Pakistan. Islamabad.

Ahmad, S. 2008c. Land and Water Resources in Pakistan – A Critical Appraisal. Paper presented in the 23rd Annual General Meeting & Conference of Pakistan Society of Development Economists (PSDE). Broad theme of 'Natural Resource Management: Issues and Challenges'. March 2008, Islamabad, Pakistan.

GOP 1992. IRSA Act No. XXII of 1992. Rules and Regulations. Government of Pakistan Islamabad http://www.pakirsa.gov.pk/wateraccord.html.

GOP 2000. Amendments to IRSA Act No. XXII of 1992. Rules and Regulations. Government of Pakistan Islamabad http://www.pakirsa.gov.pk/wateraccord.html.

GOP 2011. Agriculture Statistics of Pakistan. Economic Wing of the Ministry of Food, Agriculture and Livestock, Government of Pakistan.

IRSA 1992. Indus River System Authority Ordinance. Government of Pakistan.

IUCN 2010a. Indus Water Treaty and Managing Apportioned Rivers for the Benefit of Basin States – Policy Issues and Options. IUCN Pakistan, Karachi, p. 13.

IUCN 2010b. Pakistan Water Apportionment Accord for Resolving Interprovincial Water Conflicts – Policy Issues and Options. IUCN Pakistan, Karachi, p. 11.

IWT 1960. Indus Water Treaty. Full text and Annexure. Government of Pakistan and India.

Mustafa, D. 2010. Hydropolitics in Pakistan's Indus Basin. Special Report. United States Institute of World Peace. United States Institute of Peace. 1200 17th Street NW, Washington, DC 20036. www.usip.org.

PWAA 1991. Pakistan Water Apportionment Accord. Government of Pakistan.

Sharma, B.R., U.A. Amarasinghe and A. Sikka 2008. Indo-Gangetic River Basins: Summary Situation Analysis. IWMI, New Delhi, India. 25 July 2008.

PART VI

CHINA

16. China's political system, economic reform and the governance of water quality in the Pearl River Basin

Andre Silveira

16.1 SUMMARY

China's economic reform and opening up process initiated in 1978 has supported a more decentralized style of governing public affairs, including water resources management. With the aim of promoting economic growth, provincial and municipal authorities were given enhanced economic and administrative powers. This has encouraged greater competition among jurisdictions to attract investment. Some of the pitfalls of such fragmentation have been exposed by the difficulties in cross-boundary work to control water pollution, monitor water quality and safeguard drinking water security. This is illustrated through the analysis of the Pearl River Basin's case. The institutional structures in place display features that seem to have negative consequences for the capacity of the governance regime to adjust and address evolving social and environmental challenges.

16.2 INTRODUCTION

This chapter explores key aspects of China's political system and political economy and their influence on the management of river basin systems, particularly in respect to water quality. The chapter addresses the period since the beginning of economic reform in 1978 and is composed of three main sections. The first briefly characterizes the process of economic decentralization in China and its implications for intra-governmental relations (both centre-province and inter-provincial). It is argued that three decades of economic reform under authoritarianism have established dichotomies of change and stability that continue to underpin the study of institutions in China.

The second section analyses the governance of water quality at national scale and how this is influenced by structural conditions in the Chinese political and economic system and the politics of economic reform.

Third, the concrete case of the Pearl River Basin is discussed, focusing on the ability of the current water governance system to assess and respond to threats to the river's ecological systems and the population's drinking water safety. In this context, this chapter will make use of the concept of adaptive capacity, defined as the ability of a system to adapt to anticipated or experienced change through processes of experimentation, monitoring and learning (Folke et al. 2005).

Ultimately, this account examines how a form of 'behavioural federalism', and fragmentation in the country's administrative system, is influencing the adaptive capacity of the governance system to address emerging water quality problems at both basin and national levels.

16.3 KEY FEATURES OF CHINA'S POLITICAL AND ECONOMIC SYSTEM IN TRANSITION

Chien (2010) characterizes the central–local relation of power in China as one of 'asymmetric decentralization', in which substantial economic decentralization has taken place in the absence of meaningful political decentralization.

The success of China's economic reform has been largely attributed to the agency and power gradually handed over to regional and local levels of government (that is, provincial and municipal) since an 'open door' policy was approved by the Communist Party in 1978 (Howell 2006). Throughout the 1980s important administrative and fiscal responsibilities were devolved to regional and local governments, who gradually took over decisions regarding trade, foreign investment, land use, budgetary and extra-budgetary revenues and expenditures (Chien 2010).

A key aspect of the power of local governments relates to land use management responsibilities. Land use rights became saleable items after the 1988 Constitutional amendment, which made land use rights 'assignment' the prerogative of local authorities (Yeh and Wu 1999). The product of land use rights sales has been treated as extra-budgetary revenue and as such is not shared with central authorities. As an illustration of the importance of extra-budgetary revenue, Xiamen (overlooking the Taiwan Straits in Fujian province) reported that the income from land lease fees in 2006 was close to 15 billion yuan, which was higher than the entire budgetary revenue for that year (Chien 2010). New opportunities for enrichment and career development represented robust incentives for local government cadres and decision-makers to use these newly acquired powers to advance both economic growth as measured by gross domestic product (GDP) and personal ambitions (Oi 1995, Wu 2000). Environmental protection is often considered an obstacle to GDP growth (Ma 1997), which results in low authority being assigned to environmental protection institutions (Jahiel 1998).

In parallel with economic decentralization, political centralization has remained firmly in place, with the communist party retaining the monopoly of power. The party holds complete control of administration, legislation, judiciary and media, implementing the principle of 'the party overrides the law' [党大于法] (Guo 2000).

A key manifestation of political centralization is the cadre management system of the communist party known as *nomenklatura*, under which upper level governments have authority over appointments at subordinate levels (Tsui and Wang 2004). In 1988, the party's central organs issued guidelines establishing a new annual evaluation system targeting party secretaries and government executives at the county and township level, as an attempt to regain control of local officials' actions (Huang 1996). A 'target responsibility system' was put in place, with targets related to economic growth, such as the amount of GDP growth or foreign direct investment, representing the most decisive evaluation factor. The implementation of this system is based on the signing of annual contracts or terms of responsibility specifying concrete targets subject to some negotiation. The extent to which objectives are met determines career progression, and the existence of tangible and intangible rewards and sanctions. On average, direct financial consequences may represent the addition or subtraction of as much as two months' salary (Chien 2010).

Positive and negative effects of Chien's (2010) asymmetry have been noted at national, regional and local scales, as shown in Table 16.1.

This top-down exercise of political power has made horizontal interplay between administrative sectors a difficult challenge as functional jurisdictions are defended in almost territorial terms. The implementation of environmental policy and enforcement of law is further complicated by local government's considerable autonomy and de facto economic power, which is able to act as an effective counterweight to central political directives. Competition is embedded in China's bureaucratic matrix, given that vertical lines of authority may be challenged by horizontal lines of authority at lower levels of government where decisions on implementation are taken and the distribution of abundant financial resources takes place (Lieberthal 1995).

The performance of the environmental protection bureaus (EPBs) seems to suffer from such bureaucratic competition. The EPBs are

Table 16.1 Effects of asymmetric decentralization at different scales

	Positive effects	Negative effects
National scale	Economic stability induced by 'target responsibility system'	Inability to address increasing development disparities across provinces
Regional scale	An inter-provincial cadre transference mechanism has allowed diffusion of successful experience with market reform	Intense competition has led to market fragmentation and low importance of inter-provincial trade[1]
Local scale	Flexibility for innovation and experimentation, also induced by cadre performance indicators[2]	Lack of accountability to local populations has allowed local leaders to disregard public needs and pursue personal gain

Notes:

1. In the 1990s the internal trade as percentage of GDP was only 22 per cent compared to 28 per cent in the European Community and 27 per cent in the former Soviet Union (Zheng 2006).
2. New indicators such as energy cost per unit of GDP or agricultural land conservation are being experimentally used (Chien 2010). In the case of Zhouzhuang county in the municipality of Shanghai, traditional waterways and natural landscape were preserved after performance indicators were adjusted in favour of ecological conservation and tourist promotion. In another occasion, local governments took the initiative of experimenting with, for example, programmes of privatization of State-Owned Enterprises (SOEs) (Chien 2010).

Source: Prepared by the author based on Chien 2010.

excessively dependant on their local governments, who decide their annual budget. The high priority given to GDP growth since the late 1970s, which is engrained in the target responsibility system, along with the high autonomy of local leaders in choosing the means to achieve growth objectives, is considered one of the main reasons for continued environmental degradation in China (Jahiel 1998, Du 2011).

Reflecting the contradictions of 'socialism with Chinese characteristics', Howell (2006) describes the state's behavioural patterns as 'polymorphous', illustrating how the political system induces rivalry and unity, efficiency and inefficiency, both across space and time. This interpretation builds on previous visions of the Chinese political system as a *fragmented authoritarian regime* (Lieberthal 1995).

Several authors have gone as far as suggesting that the Chinese state *behaves* as a federal political system (Montinola et al. 1995, Zheng 2006, Weingast 2009). The dynamics of central–local relations and the local autonomy provided by economic means and regulatory capacity lend some support to this characterization.

Formally, the Chinese state is certainly not federal. According to the People's Republic of China Constitution, the country is a unitary state in which all sub-national governments are subordinate to the central government, and all its administrative organs work under the leadership of the State Council.

The country's political institutions lack several typical traits of federal systems, namely distribution of powers between central and sub-national governments inscribed in constitutional documents; division of executive, legislative and judicial power at each level of government; and division of legislative power between federal state and sub-national governments. There is no clear demarcation between central and provincial governments in regard to the scope and content of their legislative authority. While the provincial people's congresses have the right to pass local laws, the Standing Committee of the National People's Congress may annul the legislation if any contradiction is found in provincial law (Zheng 2006).

It is important to note, however, that the Chinese state also has no tradition of governing according to the rule of law (Zheng 2006, Li 2007). Li notes that the general public is normally 'suspicious of the law, the courts, and lawyers, and regard authority, money and *guanxi* [关系] connections as much more important ways of resolving disputes' (Li 2007, p. 148). In regard to central–provincial relations, the enactment of national laws or the passing of central decisions is believed to be the beginning of a bargaining process between the central government and local authorities concerning their implementation (Zheng 2006). Legal fragmentation has been noted as a feature of the Chinese political system (Zou 2006).

Authors observing patterns of explicit or implicit bargaining between the centre and the provinces, suggest that the Chinese political system represents an adapted form of federalism, which has been characterized as market-preserving federalism (Montinola et al. 1995), fiscal federalism (Weingast 2009), and behavioural or de facto federalism (Zheng, 2006).

In this context, it has been argued that formal institutions are unable to, on their own, account for the significant powers enjoyed by local governments vis-à-vis the Chinese central government, and that a behavioural approach is useful to further our understanding of central–local relations in this case (Zheng 2006).

A behavioural tradition exists in the study of federalism, which aims to understand the dynamics of central–local relations; how socio-economic change at the local level impacts on intra-governmental relations; and why the power exercised by local government officials (de facto power) is often greater than the powers constitutionally granted. Economic development at the local level has been found to change the distribution of powers between national and sub-national government in China. The more sub-national jurisdictions develop their local economies, the less financially-dependent on the central government they are. Thus, while the central government witnesses a decrease in the capacity to impose its authority, the local governments gain leverage and influence in strategic choices at national level.

The fiscal dimension of China's intergovernmental relations seems to provide strong backing to a behavioural interpretation of the country's political system (Zheng 2006, Weingast 2009). In the 1980s, central revenue declined continuously, with the central government's share in total expenditure shrinking from 51 per cent in 1979 to 28 per cent in 1993 (Zheng 2006).

A new fiscal arrangement was negotiated and taxation reform was approved in 1994, representing a step forward in the institutionalization of a federal-like relationship between central and local governmental authorities. Before this reform the central government did not have a separate tax collection system and provinces were considered a simple extension of the central government. All taxes were collected by provincial governments and were then divided between the centre and the provinces after bargaining. The 1994 taxation reform formalized a national as well as a local taxation system, with limited revenue sharing and hard budget constraints. Taxes were thus divided into three categories: central, local and a shared category, which is subject to negotiation. In the case of the latter, central government first proceeds to collect taxes and then divides revenue with provinces. Local governments have also been allowed to mobilize extra-budgetary sources through foreign investment, administrative fees and land lease fees, as noted above.

The reforms have had structural consequences for central–local relations and for the exercise of power by local governments. In turn, these new institutional dynamics have contributed to intricate challenges of collective action to address environmental degradation. Heilmann (2008) highlights how economic reform and decentralization have enabled experimentation under hierarchy and the successful emergence of an adaptive form of authoritarianism. Yet, this adaptiveness and flexibility has been to some extent 'designed' for the attainment of economic growth targets and often operates in disfavour of environmental protection objectives.

Local governments consistently try to avoid expensive wastewater treatment investments by searching for alternative sources of supply (such as water transfer and underground water extraction) and pass on pollution problems to downstream neighbours (Xie et al. 2009). The EPBs located in rural and often upstream areas tend to have less financial means and human resource capacity to face water pollution challenges. This, coupled with provincial autonomy and competition, makes the increasing pollution from small towns and non-point sources, often located in the upper reaches of densely populated river basins, particularly difficult to address under the current political and economic situation (Xie et al. 2009). Compensation mechanisms, combining market and government support through fiscal transfer 'upstream', are under consideration. Ongoing cross-provincial experiments for the protection of drinking water sources at catchment scale include a programme to protect the Miyun Reservoir watershed, which supplies the city of Beijing (Peisert and Sternfeld 2005).

Significant regulatory efforts continue to be made to mitigate serious environmental and social consequences and promote coordinated action between various affected jurisdictions and relevant administrative sectors, but their implementation and overall effectiveness remain limited or unclear.

16.4 WATER QUALITY GOVERNANCE AT NATIONAL SCALE

The institutional structures of water governance (that is, law, policy and organizational architecture) will be explored in this section, specifically those related to water quality.

China is often credited with having put in place a good set of laws and regulations for water management and environmental protection (Lee 2006). In the legal framework directing water management and governance, one naturally finds provisions with distinct levels of authority: at the highest point in the legislative hierarchy is the constitution, which defines, for example, that the state owns all water and land resources. Relevant national laws are the Water Law, Water Pollution Prevention and Control Law (WPPCL); Water and Soil Conservation Law, Flood Control Law and Fisheries Law, and their implementation guidelines. These are prepared by the ministries concerned and then approved by the State Council and the National People's Congress. Administrative regulations, such as the Regulation of River Channels and the Regulation of Flood Prevention, may be passed at the national level by ministries for implementation purposes. In addition, local laws may be approved by provincial governments and provincial people's congresses.

As far as water quality is concerned, two national laws address the subject. The Water Law, first approved in 1988 and revised in 2002, contains important provisions in respect to the protection of water resources. The Water Law and the Water and Soil Conservation Law are explicitly concerned with management of water and soil at catchment scale and their implementation is a responsibility of the Ministry of Water Resources (MWR).

In its turn, the implementation of the WPPCL, first approved by the National People's Congress in 1984, with revisions in 1996 and in 2008, is the responsibility of the Ministry of Environmental Protection (MEP). The WPPCL does not specify means for local environmental protection bureaux to finance their work and thus reduce their dependence on local government priorities.

Various authors have noted that the legal framework has much room for improvement (Xie et al. 2009, Du 2011). Two critical problems remain, in particular lack of coordination between relevant water laws and incomplete enforcement of laws and regulations (Song et al. 2010).

The 11th Five Year Plan (2006–2010) established the priority of enhancing legal enforcement records, which are consistently identified as a key weakness in the Chinese legal system. In respect of environmental and water laws, MEP and MWR often inspect implementation on the basis of performance evaluations. In 2006 the MEP created regional supervision centres and the 2008 revision of the Law on Prevention and Control of Water Pollution (LPCWP) now defines stricter penalties against infringing parties. Du (2011) notes, however, that the new rule for calculating pollution fines is ambiguous (LPCWP 2008). While it establishes that fines may be up to 30 per cent of the direct loss caused, a definition of direct loss is not provided, making implementation of the new provision difficult. Although larger fines are now possible, polluters can still negotiate and receive fines that are insignificant for their business volume. This offers little incentive for a change in behaviour.

As far as institutional arrangements for water management are concerned, the current system is typically referred to as 'the nine dragons' (Yan et al. 2006) – a mythical representation of the nine or more ministries involved in the water governance system. Although mismatches amongst policies and amongst plans are inevitable, fragmentation between water quantity and water quality management, and also within the water quality management system, is considered excessively high (Song et al. 2010). This phenomenon has both horizontal and vertical manifestations.

In respect of horizontal fragmentation, responsibilities over water quality management are shared between several administrative sectors, with emphasis on two ministerial agencies, namely water resources and environmental protection. Formally, the MWR is responsible for water resources protection while the MEP is in charge of water pollution control and prevention. This division of labour is based on the agreement that the MWR responsibilities involve the water bodies themselves, and pollutant carrying capacity calculations; while the MEP controls and prevents pollution from industrial and municipal sources. Responsibilities for water pollution control also lie with the Ministry of Agriculture, which is responsible for non-point agricultural sources while the Ministry of Transportation is responsible for prevention and control of pollutants discharged by ships. The Ministry of Housing and Urban Rural Development, former Ministry of Construction, is responsible for water supply, drainage and water saving (Wang et al. 2007).

The National Development and Reform Commission (NDRC) also plays an important role in the process of approving plans and the budgets assigned to their implementation, in accordance with economic transition priorities. These include 'river basin water resources comprehensive plans' and specialized plans such as the 'river basin water resources protection plan' prepared by the MWR and the 'water pollution control and prevention plan' prepared by the MEP. Comprehensive plans are longer term plans with no specific planning cycle while specialized plans are approved on a more regular basis according to the five-year national planning cycle.

The costs of maintaining such a fragmented water quality management are considered high, particularly in terms of the monitoring systems. Two different water quality monitoring systems are operated by MWR, in accordance with Article 32 of the Water Law, and by MEP in accordance with Article 25 of the LPCWP. Neither law clearly demarcates MWR and MEP responsibilities in respect of water quality monitoring. No data sharing has occurred and differences in water quality data published by each agency have generated controversy. For example, in the case of the Huai River, data from the period from 1995 to 2001 showed an average difference of 11 per cent in the water quality data (Song et al. 2010). Although Article 25 of the LPCWP, as revised in 2008, currently gives the MEP the authority to set water environment monitoring norms, publish information and organize a monitoring network in conjunction with the MWR, the extent to which this provision has been implemented is unclear and no regular exchange of monitoring data or joint database existed in 2011. This has proved detrimental to institutional capacity to respond to evidence of water quality decline, in the case of both accidents and the slower processes of deterioration caused by human activities.

As far as vertical fragmentation is concerned, water management is still primarily conducted on the basis of administrative boundaries rather than at the basin level. Each jurisdiction seeks to attain its own objectives, having no obligation to consider downstream effects of its actions. River basin-wide objectives related to pollution control have no specific funding, and any investments rely on local budgets. However, given that water pollution discharges tend to occur in the lower reaches of a jurisdiction, and thus mostly affect jurisdictions downstream, pollution control tends not to be a priority in local government budgeting processes.

The Water Law establishes an administrative principle that combines river basin-wide management with administrative management based on individual jurisdictions. The law recognizes the legal status of river basin commissions and assigns them some responsibility, particularly in regard to water bodies flowing across provincial boundaries, and establishes that a 'plan for a region within a river basin shall be subordinated to the comprehensive river basin plan' (Art.15). Nevertheless, vesting appropriate authority in river basin commissions is fundamental to their ability to manage and mediate the interests of individual provinces. Article 12 of the Water Law attributes some responsibilities of management and supervision to river basin commissions, but the Water Law does not specify responsibilities of river basin commissions in relation to local governments and provincial administrative agencies.

In reality, the river basin commissions' administrative rank remains lower than that of provincial governments (Figure 16.2), which enjoy the same rank as central ministries, rendering the authority of river basin commissions insufficient to mediate the different provincial interests successfully. Commissions are effectively regional agencies of the MWR, and not a platform gathering all interested parties in the management and use of the river system (Shen 2010). Finally, in respect of water quality management, river basin commissions have the authority to monitor water quality but have no authority in respect of pollution prevention and control work (Xie et al. 2009).

A new national water policy document, entitled *Accelerating the Water Conservancy Reform and Development* and often referred to as Central Document n.1, has been official state policy since January 2011, following approval by the Central Committee of the Chinese Communist Party and the State Council. The policy establishes three 'red lines' symbolizing new controls over water resources development and utilization; new targets for water use efficiency in relation to GDP and industrial production; as well as new restrictions for pollution discharge and more stringent targets for pollution loads in relation to different water function zones (Articles 19 to 21). A new water resource management responsibility and cadre assessment system is created by Article 22 with a view to hold local government officials (above county level) accountable for the state of the water environment. It also emphasizes that water quantity and quality monitoring capabilities need to be reinforced to provide support for the assessment. The legal status of the policy in relation to laws in place is not clear, but the policy seeks to reinforce implementation of water-related laws through the political system in place, that is Communist Party structures.

Work to implement the new policy is underway. In January 2012, the State Council released its *Opinions on Implementing the Strictest Water Resources Management System* introduced in Central Document n.1. In this context, the central government also announced that 1.8 trillion yuan (about £180 million) have been earmarked for the setting up of a nationwide information system for water resources by 2015 (Peng 2012). The document requires local governments to strengthen emergency management of drinking water sources; improve emergency plans for drinking water pollution accidents and establish reserve drinking water sources. This supports in part the National Plan for Environmental Emergency Response (approved in 2006) and a National Emergency Response Law (in force since 2007). Still, the focus of the new planning and regulatory instruments remains on response and after incident treatment, rather than routine pollution prevention (Xie et al. 2009).

It is uncertain whether the new Central Document n.1 will succeed in changing the priorities of local governments. The State Council press release introducing the new Document n.1 concludes with a statement emphasizing how implementation of the new policy will support 'long-term, stable and rapid economic development' (PRC State Council's Information Office, 2011). This retains water conservancy as an instrument of the more important goal of rapid economic development. It is thus unlikely that water pollution control objectives set in the new policy will become a priority in situations where high economic growth is perceived to be at risk.

16.5 WATER QUALITY GOVERNANCE IN THE PEARL RIVER BASIN

This section will summarize the case of the Pearl River Basin to illustrate institutional dynamics in episodes concerned with the monitoring of water quality at catchment scale. This, however, requires a brief introduction to the basin's natural and socio-economic features.

The Pearl River Basin (Figure 16.1) is the fourth largest in China as far as drainage area is concerned. It covers a total area of 442,527 km^2 in southern China and is formed by convergence of the Xijiang ('the West River'), the Beijiang ('the North River'), and the Dongjiang ('the East River'). The basin extends over six provinces (Yunnan, Guizhou, Guangxi, Guangdong, Jiangxi and Hunan), and includes two special administrative regions (Hong Kong and Macau) located in the delta.

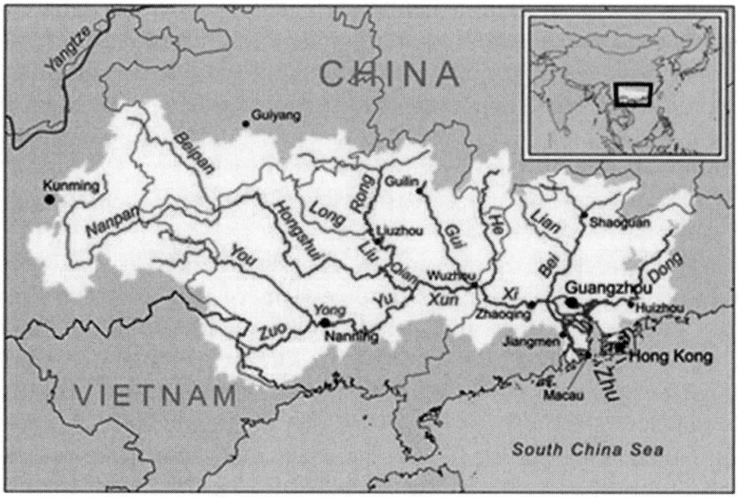

Figure 16.1 The Pearl River Basin

It is heavily influenced by the south-west monsoon, characterized by a wet season that runs from April to September, which accounts for about 80 per cent of total annual rainfall. In the province of Guangdong, at the lowest reaches of the basin, average water availability per capita has been 1,927 m^3, well below the world average of 8,210 m^3 (Sadhwani et al. 2009).

Excluding the population of Hong Kong (estimated at about 8 million) and Macau SAR (about 500,000), the Pearl River Basin was estimated by the Pearl River Water Resources Commission (PRWRC) to have 168 million inhabitants in 2000 (PRWRC 2012). Intense industrialization and urbanization started with the onset of economic reforms in 1979 when two of the first four special economic zones (SEZs) were created in the Pearl River Delta: the SEZ of Shenzhen was created in the land adjacent to Hong Kong, and the SEZ of Zhuhai in the land adjacent to Macau.

In the jurisdictions located in the lower reaches and delta, namely Hong Kong, Macau and Guangdong, the GDP per capita in 2000 was US$ 25,586, US$ 24,246 and US$ 2983 respectively. In the upper and middle reaches, namely in the provinces of Guangxi, Yunnan and Guizhou, the figures in the same year were US$ 1073, US$ 956 and US$ 617 respectively, showing a sharp decline in the upstream direction (Xu and Yeh 2011).

Access to safe drinking water is not fully guaranteed in the region. Pollution affects drinking water sources in both urban and rural areas. In Guangdong, the province in the lowest reach of the basin, Chinese authorities reported in 2007 that about 40 per cent of the province's rivers were too polluted to be used as sources of drinking water. About 16 million residents faced water shortages on account of pollution in that year (Yeung 2007). In 2001, about 10 per cent of the monitored sections registered water quality grade IV, V or worse; this percentage reached 25 per cent in 2005 and dropped again to 15.1 per cent in 2009 (MEP 2001, 2005 and 2009). Only grade I, II or III water may serve as a drinking water source. The major pollution sources are sewage, pesticides and industrial wastes in the form of ammoniacal nitrogen and petroleum (MEP 2009). There is great variety in the type and intensity of anthropogenic change in different provinces of the basin, corresponding to considerable disparities in economic development.

The number of environmental accidents has been on the rise, with 912 accidents in the five year period from 2006 to 2010, but 542 in 2011 alone (Peng 2012). It was found that 48 per cent of these happened during production processes and 14 per cent were caused by traffic accidents (Peng 2012). In the period from 1992 to 2008, Guangxi and Guangdong, occupying approximately 80 per cent of the Pearl River Basin area, were the first and fifth provinces most affected by water pollution accidents (Yuan et al. 2011).

In January 2012, one of the most serious water pollution accidents in the Pearl River during the last three decades came to public attention. A total of 21 tonnes of cadmium spilled into the Longjiang, a tributary of the Pearl River located in the province of Guangxi. The spill, caused by metal smelters, was equivalent to about two thirds of the total amount of cadmium discharged nationwide in 2010, and it has considerably endangered drinking water sources and the health of local populations (Shi 2012).

As far as institutional structures and dynamics are concerned, the main organization responsible for the management of water resources at catchment scale is the Pearl River Water Resources Commission (PRWRC), set up in 1979 as a regional office of the MWR. The commission's structure also comprises a Water Resources Protection Bureau (PRWRPB), whose management was originally intended to be shared by a director appointed by the MWR and a director appointed by the MEP. For several years now, the MEP has not appointed any director to the post, which represents a key point of official contact with the water administration at the river basin scale (Interview PRWRPB 2012).

One of the priorities of the PRWRC has been the creation of an information sharing mechanism for the water resources departments and the environmental protection departments of all the provinces in the basin, namely Yunnan, Guizhou, Guangxi, Guangdong, Hunan and Jiangxi. Such a mechanism would facilitate the work of water resources protection

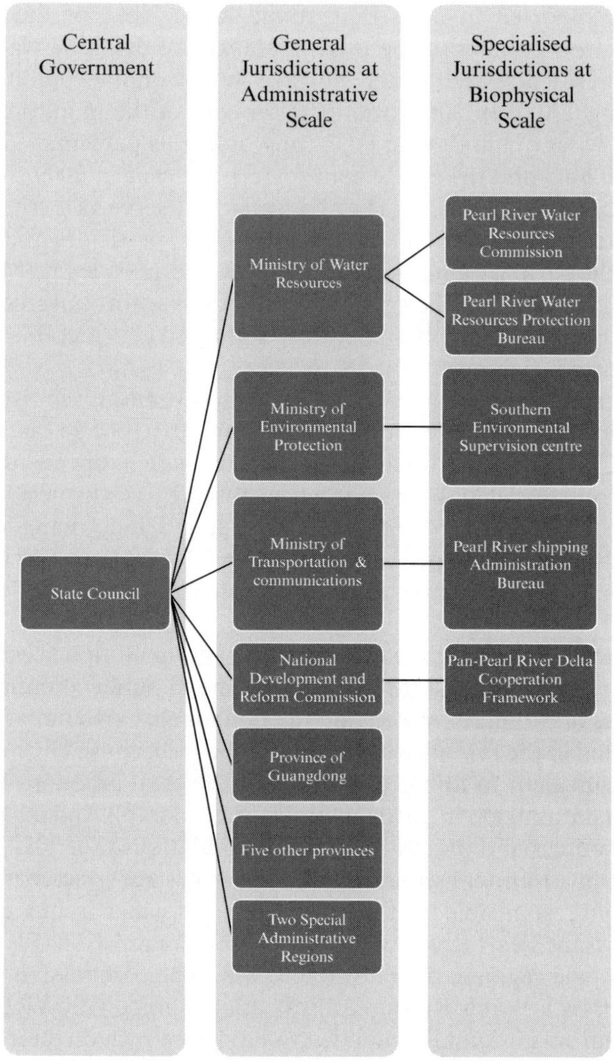

Figure 16.2 Jurisdictions involved in the governance of the Pearl River

and pollution control, with access to an integrated set of data and information on pollution discharges and on pollutant carrying capacity in different river sections. This initiative has met with resistance from environmental protection leaders, and was not given the official backing of the Ministry of Environmental Protection at the national level. However, two neighbouring provinces that suffer chronic pollution in their shared watercourses, the provinces of Guizhou (upstream) and Guangxi (downstream) have now established an information exchange system involving their water resources and environmental protection departments. Pollution sources in Guizhou were endangering drinking water sources in Guangxi but the situation has reportedly improved since the establishment of this mechanism under the coordination of the PRWRC (Meeting Report PRWRC 2010; Li 2011).

Water pollution control is financed and managed at the local scale by local governments and their environmental protection bureaux, with no involvement from river basin commissions. This is visible in the way some of the water pollution incidents have been dealt with in the Pearl River Basin. The PRWRC is required by ministerial regulations to perform investigative monitoring of the site of an accident, but its presence was not welcomed by provincial authorities and environmental protection bureaux controlling the area of an accident in 2005 (Meeting Report PRWRC 2010). In such cases, access of the PRWRC to an accident site may depend on the existence of personal connections between government leaders and officials. The PRWRC team was allowed access to another accident site in the province of Guangxi on another occasion, largely on account of personal links (Meeting Report PRWRC 2010).

Trans-jurisdictional problems, particularly trans-provincial water pollution, continue to be particularly difficult for the environmental protection apparatus to address effectively (Huan 2011). Since 2006, regional supervision centres for environmental protection have been established in order to supervise local governments in the implementation of policies, rules and standards; to check environmental law enforcement; to settle environmental disputes in trans-provincial areas and river basins; to handle environmental accidents; and to deal with complaints related to trans-provincial pollution. They coordinate the work of local environmental protection agencies, although they lack the authority to issue instructions on their daily work (Zhang 2006, Huan 2011). The South China Supervisory Centre was created in 2006 with a staff of 34, and, although its objectives concern river basin-wide water pollution, its jurisdiction excludes three important provinces within the Pearl River Basin, namely Yunnan, Guizhou and Jiangxi (Huan 2011).

16.6 CONCLUSION

This chapter illustrates how the competitive nature of China's political and economic system influences the management of water quality issues at river basin scale. Traditional vertical lines of authority, controlling how power is exercised by provincial and local ministerial departments, have been cut through by horizontal authority gained as a result of the strong financial and economic power that provincial and lower levels of government have gathered throughout the process of economic reform and decentralization.

Controlling up to 70 per cent of all government spending (Landry 2008), local governments with higher revenue and financial autonomy exercise more bargaining power vis-à-vis the central government and may be able to ignore certain directives that compromise economic growth and their position in relation to other neighbouring local governments. The situation does not favour the protection of common pool resources with high mobility such as cross-jurisdictional water resources, which require more collective action and collaborative institutional arrangements than competition amongst territorial and functional jurisdictions.

Patterns of inter-provincial and inter-municipal rivalry have emerged in the search for fast economic growth opportunities and preferential investment policies from the central government, with pollution costs being passed on to downstream communities. In addition, cross-sectoral fragmentation has resulted in limited capacity to monitor the quality of water bodies and river systems and to understand both gradual and sudden changes in its chemical and biological conditions.

There is limited ability to consider environmental processes at the river basin scale and to seek management objectives, measures and funding at this scale. Current river basin organizations are not invested with sufficient political weight to induce change in the behaviour of sub-national governments, polluters and users of the water environment, or require the politically powerful provincial governments to sit around the same table when significant water quality problems arise in trans-provincial water bodies.

The development of competent monitoring systems and decision support systems is key to the capacity of the governance system to respond and adapt to environmental challenges. However, political and economic conditions of the country, and its impact on intra-governmental relations, have hindered the development of shared monitoring systems.

The establishment of regional supervision centres for environmental protection and the approval of the new Central Document n.1 *Accelerating Water Conservancy Reform and Development* show that Chinese central authorities are attempting to address gaps in the country's water governance system. However, the emphasis on rapid economic development, which incentivises the current behaviour of local leaders, remains in place. It is unclear how the assessment of local government leaders will be implemented in an impartial manner when the data upon which assessments are based suffers from lack of transparency, with local officials often withholding data and information from other government officials and organizations. A related weakness of the new policy lies in the fact that county and township level officials are not subject to the new water resource responsibility and assessment system.

New incentives are required to actively promote systematic and transparent data and information sharing mechanisms among sub-national governments and administrative sectors. This is necessary for the development of a more adaptive and effective water governance system that is capable of safeguarding drinking water safety and protecting the health of local populations.

16.7 REFERENCES

Chien, S.-S. (2010). "Economic Freedom and Political Control in Post-Mao China: A Perspective of Upward Accountability and Asymmetric Decentralization." *Asian Journal of Political Science* 18(1): 69–89.

Du, Q. (2011). "New Developments in Water Pollution Law and Policy in China: Effective Enough to Cope with Water Pollution Conflict?" *International Journal of Rural Law and Policy* Special Edition on Water Law: Through the Lens of Conflict: 1–15.

Folke, C., T. Hahn, P. Olsson and J. Norberg (2005). "Adaptive Governance of Social-Ecological Systems." *Annual Review of Environment and Resources* 30: 441–473.

Guo, S. (2000). *Post-Mao China – from Totalitarianism to Authoritarianism?* London: Praeger.

Heilmann, S. (2008). "Policy experimentation in China's Economic Rise." *Studies in Comparative International Development* 43: 1–26.

Howell, J. (2006). "Reflections on the Chinese State." *Development and Change*: 273–297.

Huan, Q. (2011). "Regional Supervision Centres for Environmental Protection in China: Functions and Limitations." *Journal of Current Chinese Affairs* 40(3): 139–162.

Huang, Y. (1996). "Central–local relations in China during the reform era: the economic and institutional dimension." *World Development* 24(4): 655–672.

Interview with official from Pearl River Water Resources Protection Bureau (PRWRPB), January 2012.

Jahiel, A.R. (1998). "The Organisation of Environmental Protection in China." *The China Quarterly*: 757–787.

Landry, P.F. (2008), *Decentralized Authoritarianism in China: The Communist Party's Control of Local Elites in the Post-Mao Era*, Cambridge: Cambridge University Press.

Law on Prevention and Control of Water Pollution (promulgated by the Standing Committee, National People's Congress, May 11, 1984, amended May 15, 1996, and February 28, 2008, effective June 1, 2008) http://www.chinaenvironmentallaw.com/wp-content/uploads/2008/03/water-pollution-prevention-and-control-law.pdf (accessed June 23, 2013).

Lee, Y.-S. (2006). "Managing Water Resources in the Delta Border Zone: Challenges and Opportunities" in A. Gar-on Yeh, Sit, Fung-shuen, Chen, Guanghang, Zhou, Yunyuan (eds), *Developing a Competitive Pearl River Delta in South China under One Country Two Systems*. Hong Kong: Hong Kong University Press.

Li, J. (2007). "Socialist Rule of Law with Chinese Characteristics." *Issues and Studies* 43(1): 115–157.

Li, X. (2011). "Cross-jurisdictional cooperation mechanisms in the Pearl River Basin", presentation in the Third Workshop of the EU-China River Basin Governance (RiBaGo) Research Network, 29 August, Beijing.

Lieberthal, K. (1995). *Governing China. From Revolution through Reform*. New York and London: W.W. Norton.

Ma, J. (1997). *Intergovernmental Relations and Economic Management in China*. New York: St. Martin's Press.

Meeting report, Pearl River Water Resources Commission (PRWRC), May 2010.

MEP (2001, 2005 and 2009). "Report on the State of the Environment" Ministry of Environmental Protection. Beijing.

Montinola, G., Y. Qian and R. Weingast (1995). "Federalism, Chinese Style: The Political Basis for Economic Success in China." *World Politics* 48(1): 50–81.

Oi, J.C. (1995). "The Role of the Local State in China's Transitional Economy." *China Quarterly* 144: 1132–1149.

Peisert C. and E. Sternfeld (2005). "Quenching Beijing's Thirst: The Need for Integrated Management for the Endangered Miyun Reservoir." *China Environ Series* 7: 33–46.

Peng, T. (2012). "China Battles More River Spills". *Chemistry World*, Royal Society of Chemistry.

PRC State Council's Information Office (2011). "Speech in the News Conference of State Council's Information Office", Beijing.

PRWRC. (2012). www.pearlwater.gov.cn. Retrieved 12 March 2012.

Sadhwani, D., J. Chau, C. Loh, M. Kilburn and A. Lawson (2009). *Liquid Assets: Water Security and Management in the Pearl River Basin and Hong Kong*. Hong Kong: Civic Exchange.

Shen, D. (2010), "Stakeholder Participation and River Basin Water Resources Management Organizations in China: Challenges and Opportunities", presentation in the Second Workshop of the EU-China River Basin Governance (RiBaGo) Research Network, 27 August, Cambridge.

Shi, J. (2012). "Dead Fish, Toxic Drinking Water . . . Now for the Human Toll." *South China Morning Post*. Hong Kong.

Song, X., W. Ravesteijn, B. Frostell and R. Wennersten (2010). "Managing Water Resources for Sustainable Development: The Case of Integrated River Basin Management in China." *Water Science and Technology* 61(2).

Tsui, K.-Y. and Y. Wang (2004). "Between Separate Stoves and a Single Menu: Fiscal Decentralisation in China." *China Quarterly* 177: 71–90.

Wang, Y., L. Li, X. Wang, X. Yu and Y. Wang (2007). *Taking Stock of Integrated River Basin Management in China*. Beijing: Science Press.

Weingast, B.R. (2009). "Second Generation Fiscal Federalism: The Implications of Fiscal Incentives." *Journal of Urban Economics* 65: 279–293.

Wu, F. (2000). "The Global and Local Dimensions of Place-Making: Remaking Shanghai as a World City." *Urban Studies* 37(8): 1359–1377.

Xie, J., A. Liebenthal, J.J. Warford, J.A. Dixon, M. Wang, S. Gao, S. Wang, Y. Jiang and Z. Ma (2009). *Addressing China's Water Scarcity: Recommendations for Selected Water Resource Management Issues*. Washington, DC: The World Bank.

Xu, J. and A.G. Yeh (2011). "Political Economy of Regional Cooperation in the Pan-Pearl River Delta" in A.G. Yeh and Jiang Xu, *China's Pan-Pearl River Delta: Regional Cooperation and Development*. Hong Kong: Hong Kong University Press.

Yan, F., D.M. He and K. Beth (2006). "Water Resources Administration Institutions in China." *Water Policy* 8(4): 291–301.

Yeh, A.G. and F. Wu (1999). "Urban Planning System in China." *Progress in Planning* 51(3): 167–252.

Yeung, J. (2007). "Guangdong gives rural population water pledge." *China Daily*.

Yuan, P., Y. Wang and J. Wang (2011). "The Analyses of Environmental Pollution Accidents from 1992 to 2008 in China and the Management Proposals", Proceedings of the 2011 International Conference on Remote Sensing, Environment and Transportation Engineering (RSETE), 2644–2647.

Zhang, T. (2006). "Environment Supervision Centers to Open." www.china.org.cn.

Zheng, Y. (2006). "Explaining the Sources of de facto Federalism in Reform China: Intergovernmental Decentralization, Globalization, and Central– Local Relations." *Japanese Journal of Political Science* 7(02): 101–126.

Zou, K.Y. (ed.) (2006). *The Party and the Law. The Chinese Communist Party in Reform*. London and New York: Routledge.

17. Watershed management in Tai Lake Basin in China

Ke Jian

Water pollution often goes beyond administrative boundaries and can constitute a threat to people's survival, social security and sustainable development, so the Chinese government at different levels and the public have begun to regard the prevention and control of water pollution not only within an administrative jurisdiction but also in a transboundary watershed context as a high priority. The Tai Lake Basin is situated in one of the most economically developed areas in China and it has witnessed rapid industrialization and urbanization. This has caused the degradation of water quality, however, at huge environmental, economic and social cost. A drinking water pollution incident in Wuxi City in 2007 caused great astonishment. To cope with the pollution crises in the Tai Lake Basin, the Chinese government at the national and provincial level launched a variety of legal and political campaigns. In 2011 the State Council of China promulgated and enforced the Regulation on Administration of the Tai Lake River Basin.

The Tai Lake, located in the intersection of Jiangsu Province, Zhejiang Province and Shanghai Municipality, is the third largest freshwater lake in China. With rapid industrialization and urbanization in the area over the past decades, the lake basin has suffered serious water pollution and ecological degradation. The accumulated water pollution finally caused a blue-green algae bloom leading to the drinking water pollution incident in Wuxi City in 2007. This alarmed both governments and the public. The central government now pays more attention to the prevention and control of water pollution in the watershed. The enactment of the Regulation on the Administration of Tai Lake Basin approved by the State Council in 2011 was the most recent initiative. Using that regulation as its starting point this chapter analyses the administrative regimes in the Tai Lake Basin and discusses the institutional arrangements in place at the watershed level.

17.1 DRINKING WATER POLLUTION INCIDENTS IN TAI LAKE BASIN

The Tai Lake area is about 2250 km^2 and the watershed area is over 36,000 km^2. Ever since ancient times the lake has been known as 'a bright pearl' for its beautiful landscapes. The watershed area has been called 'land of fish and rice' for its high yields of agricultural and aquacultural products. The Tai Lake Basin, however, has turned into one of the most industrialized and urbanized regions in China. The level of urbanization and industrialization in the Tai Lake Basin has ranked first in China with a rough gross domestic product (GDP) contribution of 13 per cent to the nation. Though the region accounts for only 0.4 per cent of the land area, it supports 36.8 million people which accounts for 3 per cent of the whole population in China. But as polluting industries consume a huge amount of water and discharge large quantities of waste water, the Tai Lake Basin has been choked with industrial waste from factories and urban sewage and agricultural runoff. From the 1980s, encouraged by the industrialization first policy, a large number of polluting industries were established in the watershed area and have thus changed the landscape dramatically. According to statistics, there have been around 2800 chemical plants built up in the northern arc of the Tai Lake, which contribute a lot to the local economy. For example, in the mid-1990s the taxes on chemical industries' enterprises accounted for four-fifths of local government revenue.[1]

The water in the Tai Lake Basin is one of the main drinking water sources for 30 million residents. It supplies drinking water to the residents in cities as Wuxi and Suzhou City.[2] However after 1992 the water quality in the Tai Lake failed to meet Grade IV, the basic requirement for potable water under China's classification standards of water quality. Pollutants containing phosphorus and nitrogen have made the lake entropic. The Tai Lake has gradually turned green and algal blooms have broken out in the lake repeatedly.

In late May 2007 Wuxi City with a population of approximately 1 million was hit by a serious drinking water pollution incident. The large bloom of algae made the water fetid to drink, causing the drinking water supply to a great number of households to be cut off for about ten days. This affected more than 1 million residents despite the local government's attempt to mitigate the crisis.[3] During the crisis of the drinking water supply, the average price of bottled water rose to six times as high as the normal price despite the municipal government ban of price hikes.[4] The incident of drinking water supply suspension put the environmental problem under the spotlight not only in China but also in the world. Although China's official media calls the drinking water pollution incident a natural disaster, no one could deny the anthropogenic factor in the catastrophe.

17.2 LEGISLATION AND POLICY RESPONSES TO WATER POLLUTION IN THE TAI LAKE BASIN

The drinking water pollution incident in Wuxi City is a typical environmental crisis in China. To deal with water pollution, the central and local governments have taken concerted action. At the end of 1998 polluters discharging pollutants to heavily polluted watersheds such as the Huai River and the Tai Lake were required to meet a new set of pollutant emission standards promulgated by the central government. At midnight on 1 January 1999 the Tai Lake authorities launched a campaign, Midnight Action, and announced that more than a thousand heavy polluting workplaces around the lake were required to reach the new pollutant discharge standards. The announcement was supposed to herald a revival in the lake's fortunes and set an example for others,[5] but the drinking water pollution incident in the Wuxi City still occurred in 2007.

To deal with rampant environmental pollution, an environmental legislative framework has been gradually established in China over the past decades.[6] Hierarchically the legislation in relation to water issues at the central level includes the Constitution,[7] the Law on Environmental Protection,[8] the Water Law,[9] the Law on the Prevention and Control of Water Pollution,[10] the Flood Control Law,[11] and the Law on Water and Soil Conservation Law.[12] As the supreme legislation, the Constitution stipulates that the state should ensure the reasonable use of natural resources and protect rare animals and plants.[13] It also requires the state to protect and improve the living environment and the ecological environment, to prevent and control pollution and other public hazards.[14] Although watershed management is not explicitly mentioned in these laws, the water-related laws along with the administrative regulations on water issues enacted by the State Council have laid down the fundamental legal framework for watershed management based on the principles in the constitution.

The drinking water incident in Wuxi City directly generated the central government's concern on water pollution in the Tai Lake Basin. In response to the incident, the State Council, chaired by Premier Wen Jiabao, set a target to clean the Tai Lake by 2012.[15] By October 2007, it was officially reported that the government had shut down or given notice to over 1300 factories around the Tai Lake. Despite all the efforts to cope with the serious pollution in the Tai Lake Basin, however, the prevention and control of water pollution is still far from being successful. The drinking water pollution incident in Wuxi indicates that the prevention and control of water pollution in the Tai Lake Basin will be a long-term, tough task.

The administrative obstacles to cooperation and coordination not only aggravate the already poor water quality, but also threaten sustainable development and social stability in the Tai Lake Basin. In response the authorities, academics and the public have been calling for a strengthening of water protection and conservation management on a watershed basis and urging the passing of legislation on transboundary water basin management in the Tai Lake Basin.[16] The State Council finally promulgated the Regulation on Administration of the Tai Lake River Basin on 7 September 2011 to come into effect on 1 November 2011.[17] The regulation is designed to strengthen the protection of water resources and prevention, control water pollution in the Tai Lake Basin, ensure the prevention and control of flood and drought, protect the security of water use for daily life, production and improve the ecological environment of the Tai Lake Basin (Article 1). The regulation is comprehensive and covers water issues such as water safety, water conservation, water pollution prevention and control. It also aims to consolidate and institutionalize integrated and holistic measures for the administrative management at the watershed level. To establish the administrative regimes that will combine watershed management with regional management, the regulation provides that the state is responsible for establishing and improving the coordination mechanism of the Tai Lake Basin management and coordinating major water issues at the watershed level (Article 4). The regulation provides a crucial framework for overall planning and scientific development in the Tai Lake Basin (Article 3).

17.3 INSTITUTIONAL ARRANGEMENTS

Based on relevant provisions in the 2002 Amendment of Water Law (2002 WL) and the 2008 Amendment of the Law on Prevention and Control of Water Pollution (2008 LPCWP), the Regulation on Administration of Tai Lake River Basin provides more specific operating institutional arrangements for transboundary watershed management.

Watershed planning, which goes beyond regional interests and examines economic development and water protection as a whole, is an important legal instrument in coping with transboundary water problems. The 2008 LPCWP provides that relevant departments under the State Council and local governments at different levels should work together to include water environment protection into their plans and adopt ways and measures to prevent and control water pollution (Article 3). To strengthen water protection and conservation in the Tai Lake Basin, it is important to take into account the update of economic development plans and focus on the maintenance of watershed ecological security. The regulation requires the State Council to set up and improve supervision and coordination mechanisms for major overall plans, programmes, projects and activities in the Tai Lake Basin.

Given that serious pollution is already occurring, the regulation concentrates on protection as a prior principle. As required by the regulation, the allocation and scheduling of water resources in the Tai Lake Basin should take into account the living needs of the residents as well as the needs of industrial production, ecological utilization and shipping (Article 15). Local governments are required to adjust their economic structures and industrial development by optimizing industrial layout. They are required to phase out backward industries with high water consumption and pollution and promote clean production and technology (Article 28 to Article 34).

Ecological protection of watersheds is essential for water protection and conservation. Designed to promote water circulation and improve the environmental capacity of the water in the Tai Lake Basin, the regulation requires local governments to protect the ecological environment and promote watershed ecosystem restoration of the Tai Lake (Article 33), including measures such as:

(A) Constructing ecological shelters reasonably (Article 47);
(B) Protecting wetlands, promoting ecological restoration and enhancing aquatic resources (Article 48).

To promote water protection and conservation in the Tai Lake Basin, the regulation establishes the institution of water function zoning. The State Council had earlier approved the Overall Program of Comprehensive Management of the Tai Lake Basin in 2008 and launched the Water Function Zoning of the Tai Lake Basin in 2010. The regulation establishes the competent entities entitled to designate water function zones. For main water bodies in the Tai Lake Basin, the Ministry of Water Resources (MWR) should, jointly with the Environmental Protection Agency (EPA) and Zhejiang and Jiangsu provinces and Shanghai Municipality, designate water function zones in accordance with the integrated watershed planning, the protection of water resources planning and the economic and social development requirements with the approval of the State Council. For other water bodies in the Tai Lake Basin, water function zones are designated by the provincial water conservancy bodies with approval of the provincial governments. The designation must be reported to the MWR and the EPA for filing (Article 19).

In accordance with the regulation, plans for farming, shipping, tourism and other development and utilization of water resources in the Tai Lake Basin must comply with water protection requirements in the designated water function zones (Item 1, Article 20). Meanwhile, the activities of production, construction and other development and utilization in Tai Lake Basin, must be approved by the relevant local governmental administrations (Item 2, Article 20). Furthermore, the regulation requires that the water administrative department of local governments above the county level in the Tai Lake Basin and the Tai Lake Basin Management Agency should strengthen the supervision and inspection functions in the water function zones, and regularly release information to the public. If the water quality in the water function zones fails to meet the required standards, local governments are required to promptly take measures to control the pollution (Item 1, Article 21).

17.3.1 Water Monitoring and Information Sharing

As laid down in the 2008 LPCWP, national water quality and pollutant discharge standards should be established by the EPA (Item 1, Article 11). As for environmental or pollutant items that are not specified in the national standards, the provincial and other authorized governments may set their own regional standards. The environmental or pollutant discharge standards are specified in the national standards and the provincial government can establish more stringent regional standards. All regional standards should be reported to the Ministry of Environmental Protection for filing (Item 2, Article 11). To unify water quality standards, the Directive Guidance on the Prevention and Solution of the Trans-Province Water Pollution Disputes by the EPA in 2008[18] introduces a new way to set transboundary water quality standards, which stipulates that neighbouring provinces should enforce the same water quality standards in their border regions and the involved provinces should negotiate to reach consensus, otherwise the EPA will intervene and set the boundary water standards.

A unified monitoring and information-sharing mechanism is the premise to promote water protection and conservation at the watershed level. The regulation provides that the state should establish unified monitoring systems and information-sharing mechanisms in the basin. The Tai Lake Basin Management Agency should also cooperate with the relevant administrative agencies in Zhejiang Province, Jiangsu Province and Shanghai Municipality (Item 1 and Item 2, Article 54). Moreover, the monitoring of inflow and outflow water quality is an indispensable process in promoting transboundary water management and identifying the responsibilities of local governments for water protection and conservation. Although the EPA is responsible for monitoring and disclosing water quality information, relevant river basin agencies are also responsible for the monitoring and releasing of water quality information at the watershed level.[19] Joint monitoring on the boundaries of adjoining provinces in major river basins is now also supported by the EPA and involved provinces. Under the direction of the EPA, neighbouring provinces regularly take water samples in transboundary areas and carry out joint monitoring. The latest trend in the legislation is that water resource protection agencies at the watershed level are required to be responsible for monitoring water quality within the boundaries of provinces in major river basins.[20] The regulation provides that environmental protection authorities at the provincial level should be responsible for water environment quality monitoring and supervisory monitoring of pollution sources within their administrative districts. It also requires that the Tai Lake Basin Management Agency should be responsible for water environmental quality monitoring of provincial-boundary water bodies and the cross-sections of the main rivers flowing into the Tai Lake, as well as in water function zones and water diversion projects (Item 3, Article 54).

Information disclosure and sharing is key to successful watershed management. To enhance disclosure and the sharing of information on transboundary water pollution mechanisms such as joint meetings and regular information exchanges have been established. Joint meetings and information exchanges can be helpful to increase environmental information transparency and to encourage upstream polluters to fulfil their obligations to prevent and control water pollution. For example, to deal with the water pollution in the Tai Lake Basin, the provincial governments of Jiaxing and Suzhou have set up the mechanisms of joint meetings on pollution prevention and early-warning of pollution on a regular basis. In 2008, the regulatory file of the Trial Measures for the Disclosure of Environmental Information was promulgated by the Ministry of Environmental Law.[21] However the administrative power to issue information regulated in the regulation is still controlled by departments despite the attempt to strengthen cooperation between the environmental protection departments and the water conservancy departments and promote the role of the lake basin agency. The regulation provides that water quality information in the Tai Lake Basin should be disclosed by provincial environmental protection departments while hydrology and water resource information should be disclosed by the Tai Lake Basin

Administration Agency in conjunction with provincial water conservancy departments. The Regulation also requires that annual monitoring reports of the Tai Lake Basin should be jointly issued by the EPA and the MWR. If necessary the Tai Lake Basin Management Agency may be authorized to issue relevant water information (Item 4, Article 54).

Due to the drinking water pollution incident, the regulation pays close attention to the monitoring of drinking water. It provides that the local government bodies above the county level in the basin should organize real-time monitoring of drinking water sources, water supply facilities, as well as the water quality of the residential water points. During the period of blue-green algae blooms, the frequency of monitoring and monitoring points was increased in order to keep abreast of the water quality conditions. The moment anomalies are detected the local governments above the county level must immediately take preventive and controlling measures and provide timely warning information to the public (Article 13).

17.3.2 Key Water Pollutant Emissions and Local Governments

Total-amount-control of key water pollutant emissions was introduced by the 2008 LPCWP as a complement to water quality standards of pollutant emissions.[22] This not only to provide a helpful instrument to prevent and control water pollution within a certain administrative region, but also makes it possible to determine the relevant target responsibilities of the local governments in the transboundary context. The Interim Measures for the Administration of Discharge Permits of Key Water Pollutants in the Huai River Basin and the Tai Lake Basin in 2001 made the Tai Lake Basin one of the two basins that first enforced total-amount-control of key water pollutant discharges. All polluters must obtain a key water pollutants discharge permit, otherwise the discharge of any key water pollutants is prohibited.

The total-amount-control of key water pollutant emissions was introduced by the regulation (Item 1, Article 25). The regulation states that all polluters in the basin should not exceed the approved total-amount-control targets of key water pollutant emissions (Item 1, Article 28). The institution includes three steps to achieve the goal of total-amount-control of key water pollutant emissions (Item 2, Article 25):

(A) Tai Lake Basin Management Agency determines carrying capacities of key water pollutants in water bodies in the basin taking account of the natural purification capacity of water bodies in specific water functions zones then provides assessments of the permissible total-amount-control levels of the key water pollutant emissions to the environmental protection departments of the provinces and the municipality in accordance with water quality control requirements.

(B) The environmental protection departments at the provincial level must take account of the assessments and put forward reduction and control programmes in accordance with the set water quality objectives and requirements of the overall plan for the prevention and treatment of water pollution. The reduction and control programmes must be reported to the provincial and municipal governments for the examination and approval of the EPA.

(C) The provincial governments must then break down the total figures for emissions of key water pollutants and determine the particular levels for each pollutant that must be reached by each of the cities and counties in the Tai Lake Basin. The county-level governments must break down control indicators that cannot be exceeded by each pollutant discharger (Item 2, Item 3 and Item 4, Article 25). In addition, for other water pollutant emissions, the environmental protection departments at the provincial level must set control targets within their administrative jurisdictions that will be subject to examination and certification by them on behalf of the provincial governments (Article 26).

To ensure the enforcement of the total-amount-control of key water pollutant emissions, it is important to hold local governments accountable for their environmental performance.[23] The imposition of target responsibilities on local governments and their executives is meant to help to solve the problem of responsibility allocation in transboundary environmental management. In accordance with the 1989 Environmental Protection Law local governments must generally bear the responsibility for ensuring environmental quality within their jurisdictions.[24] The system of making local governments responsible for water pollution was introduced in the 2008 LPCWP. To back it up the state was required to institutionalize target responsibilities for local governments and the competent officials for water protection by examining, assessing and enforcing the results.[25]

Based on the examination and the assessment, local governments' responsibilities can be set. The regulation provides that departments such as the Development and Reform Commission, the EPA, the MWR and the Ministry of Housing and Urban and Rural Construction should examine and assess the implementation of water conservation and prevention, and control targets of water pollution of the provincial local governments annually, then report the results to the State Council. Local governments above the county level are responsible for conducting an annual assessment of the implementation of the targets of water resource conservation and prevention and control of water pollution of the lower level governments (Item 3, Article 51).

In addition, the regulation provides that development programmes and construction projects in the administrative regions that fail to achieve the targets of water resource conservation and prevention and control of water pollution should be suspended. If the targets for the total amount of key water pollutant emission reduction and control programmes, the periodic water quality control objectives of the boundary sections between administrative regions, or control sections of the rivers mainly flowing into the Tai Lake, have not been reached, the administrative procedures to review, approve and certify the proposed construction programmes and to issue permits for environmental impact assessment, water outfall and pollutant discharge pipes should be suspended (Item 1, Article 51).

17.3.3 Ecological Compensation

Ecological compensation broadly refers to economic incentive instruments encouraging or discouraging activities in relation to ecological protection. As the socialist market economy has developed in China, economic incentive instruments have been developed and utilized widely.[26] The ecological compensation provisions in the regulation are as follows:

(A) The discharger of wastewater should pay for the sewage treatment fees (Article 50);
(B) Enterprises that reduce water pollutant emissions by closing, removing, changing production or undertaking technical innovation voluntarily should be encouraged and supported by governmental policies and measures such as finance, loan and governmental procurement (Article 51);
(C) Farmers who have to convert their business and jobs for the sake of environmental protection and ecological restoration should be guaranteed support for their basic living in such ways as provide them with skill-training and incorporate them into the local government social security system (Article 52).

'Ecological compensation' refers to compensation between local governments. One of the innovative legal institutions adopted by riparian local governments in the upper reaches and in the lower reaches of the Tai Lake Basin is to sign contracts for water quality control and ecological compensation. The ecological compensation mechanism was first carried out in Jiangsu Province. The contracts generally stipulate that cities in the upper reaches should compensate those in the lower reaches if their pollutant discharge amount exceeds the contracted limits. For instance, at the end of 2008 Nanjing, capital city of Jiangsu Province, paid 18,000 yuan to Changzhou City, which in turn paid 180,000 yuan to Wuxi City in a sequence of upper to lower reaches.[27] The contracts for water quality control and ecological compensation among neighbouring provinces are broadly accepted in the Tai Lake Basin. The regulation encourages the contracts among neighbouring provinces. It provides that if upstream regions cannot achieve the goals of reduction and control programmes for the total amount of key water pollutant emissions so that they reach boundary section water quality objectives, they should compensate downstream regions. Meanwhile, if upstream regions achieve the goals successfully, downstream governments should compensate upstream regions. Compensation can be provided by fiscal transfer or other ways agreed by local governments (Article 49).

17.3.4 Incident Warning and Emergency Reactions

The necessity for incident warning and emergency reactions to be institutionalized and incorporated into transboundary water management was generated by a series of serious drinking water incidents over the past years. In order to cope with increasing environmental dangers and risks, the 2008 LPCWP provides that local governments above the county level should warn all involved governments whenever water pollution is causing or is likely to cause transboundary damage.[28] One of the main purposes of the regulation is to institutionalize the responses to water accidents and ensure the safety of drinking water supplies. The regulation gives priority to the protection of drinking water by providing a whole chapter on the issue in the legislation *Chapter Two Drinking Water Safety*. It is required that local governments above the county level in the Tai Lake Basin should determine the drinking water sources, and designate drinking water source protection zones to ensure the drinking water supply and its quality safety (Article 7). The provisions in the chapter require that local governments should be responsible for protecting drinking water sources. With regard to the serious negative impacts of the Wuxi drinking water incident, the regulation provides that local governments above the county level in the basin should plan and construct standby water sources and transboundary joint water supply projects to be ready in case of water supply emergency. Local governments should plan water supply coverage and emergency backup water sources to ensure water supply capacity for no less than seven days at the normal daily amount of water supply (Article 10). And local governments above the county level should organize the administrative departments of water conservancy, environmental protection, housing and urban and rural construction to prepare emergency reaction plans for safe water supply within their administrative jurisdictions (Item 1, Article 11).

17.4 CONCLUSION

Integrated watershed management is the main approach to water management in modern societies. On the legal basis of the state's dominant property rights over water resources,[29] the central government in China could be, compared with the federal governments in federalist systems, it is much easier for the Chinese government to exercise legitimate authority and enforce laws to promote water management at a watershed level. However the transboundary water management regimes needed to do this in practice have not been put in place. Horizontally the authorities of the two competent entities, the department of environmental protection and the department of water conservancy, are either overlapping or leave gaps in responsibilities, and vertically the different levels of administrative control are likely to conflict about watershed management. Although the State has established seven water resource protection agencies at the watershed level, they can only play a very limited role, because the Tai Lake Basin Management Agency, subordinated to the Ministry of Water Conservancy, is mainly responsible for monitoring and enforcement-reporting. China's watershed management approaches are far from integrated and the current watershed institutions have not yet been fully established.

The internalization of negative environmental externalities currently exchanged between different administrative regions is the key to strengthening transboundary water management. Therefore legislation is an indispensable solution. In the background of the environmental crisis in the Tai Lake Basin, the promulgation of the regulation indicates the Chinese government's determination to combat the water crises in the basin. But although the legislative framework has been set up and the institutional arrangements for integrated water management have been introduced by the regulation its practical effects are still uncertain. Closing the existing gap between the environmental legislation and the enforcement is a major challenge to environmental protection in China.

The most salient feature of the institutional arrangements in the regulation is that it tries to address transboundary water pollution mostly through strengthening regulation and environmental administrative powers. However, public participation is ignored in the regulation.[30] The experience of watershed management all around the world demonstrates that public accessibility, participation and supervision are indispensable for a successful watershed management. The enforcement and the implementation of the regulation will not be able to succeed without involving the public to a much greater degree than in the past.

17.5 NOTES

1. See B. Xue, S. Yao and W. Xia (2007), 'Environmental Changes in the Tai Lake during the Past Century as Recorded in Sediment Cores', *Hydrobiologia*, 581, 117–123.
2. See 'China's Third-Largest Freshwater Lake Faces Algae Threat', *China Daily*, 14 April 2008, available at http://www.chinadaily.com.cn/china/2008-04/16/content_6622243.htm (accessed 21 February 2012).
3. In response to the crisis, the government of Wuxi City asked residents to come together and surmount the crisis through the Internet. They described the crisis as a 'natural disaster' and announced that the water quality problem was in no way caused by human activities. Mr Ma Jun, an environmental activist, points out that labelling the algae bloom a natural disaster is distorting. The fact is that the serious eutrophication of the Tai Lake is mainly caused by the numerous pollutants discharged into the water. See Ma Jun (2007), 'Disaster in Tai Lake', available at http://www.chinadialogue.net/article/show/single/en/1082 (accessed 28 March 2012).
4. See 'Algae Smother Chinese Lake, Millions Panic', *MSNBC*, 31 May 2007, available at http://www.msnbc.msn.com/id/18959222/from/RS.1/ (accessed 23 February 2012).
5. See Ma Jun (2007), 'Disaster in Tai Lake', available at http://www.chinadialogue.net/article/show/single/en/1082 (accessed 28 March 2012).
6. There are nine laws in relation with environmental protection, 50 more regulations, 200 more administrative rules, and 500 more standards to regulate water pollution, available at http://www.sepa.gov.cn/ztbd/rdzl/jd/mdbd/200602/t20060216_74008.htm (accessed 13 March 2012).
7. The Constitution was adopted by the National People's Congress on 4 December 1982, amended on12 April 1988, 29 March 1993, 15 March 1999, and 14 March 2004. The text of the Constitution is available at http://english.gov.cn/2005-08/05/content_20813.htm (accessed 3 March 2012).
8. The Law on Environmental Protection was adopted by the Standing Committee of the National People's Congress on 13 September 1979, and amended on 26 December 1989.
9. The Water Law was adopted by the Standing Committee of the National People's Congress on 21 January 1988, and amended on 29 August 2002.
10. The Law on Prevention and Control of Water Pollution was adopted by the Standing Committee of the National People's Congress on 11 May 1984, and amended on 15 May 1996 and on 28 February 2008 respectively.
11. The Flood Control Law was adopted by the Standing Committee of the National People's Congress on 29 August 1997.
12. The Law on Water and Soil Conservation was adopted by the Standing Committee of the National People's Congress on 29 June 1991.
13. Item 2, Article 9 of the Constitution. However, the provision of the Constitution does not confer any rights on the individuals to invoke before the court.
14. Item 1, Article 26 of the Regulation on Administration of the Tai Lake Basin.
15. See 'Across China: Beijing', *China Daily*, 4 April 2008, available at http://www.chinadaily.com.cn/cndy/2008-04/04/content_6591233.htm (accessed 20 March 2012).
16. For example, Mr. Zhang Qun, a representative of the National People's Congress as well as director of the Shanghai municipal environmental protection bureau, has submitted a proposal to the NPC session, calling for the establishment of a central commission on the protection of Tai Lake water quality and comprehensive cooperation

among regions in the upper and lower reaches of the lake. See 'Joint Efforts against Water Pollution', *China Daily*, 10 March 2010, available at http://www.chinadaily.com.cn/china/2010-03/10/content_9563900.htm (accessed 28 March 2012).

17. No. 604 Order of the State Council of the People's Republic of China. Before the enactment, a draft regulation on the protection of water resources and ecological environment of the Tai Lake was posted on the website of the Legislative Affairs Office of the State Council to solicit opinions. The public is invited to comment on the draft regulation any time via online postings, email and letters. See 'China Solicits Public Opinion to Protect Polluted Lake', *People's Daily*, 3 June 2010, available at http://english.peopledaily.com.cn/90001/90776/90882/7010202.html (accessed 23 March 2012).
18. No. 64 Document of the Ministry of Environmental Protection, 2008, available at http://www.mep.gov.cn/info/bgw/bwj/200807/t20080711_125525.htm (accessed 1 April 2012).
19. Article 26 of the 2008 Amendment of Law on Prevention and Control of Water Pollution.
20. Moreover post monitoring providing a long-term monitoring to ensure that the polluters causing transboundary pollution can fulfil their pollution treatment objectives has been adopted by the Directive Guidance on Prevention and Solution to the Trans-Province Water Pollution Disputes issued by the Ministry of Environmental Protection in 2008.
21. No. 35 Order of the State Environmental Protection Administration (2008).
22. Article 18 and Article 19 of the 2008 Amendment of Law on Prevention and Control of Water Pollution.
23. In the past, the only negative legal consequence of illegal behaviours is administrative penalities on the responsible officials and relevant executives while the government itself would not bear any legal responsibility.
24. Article 16 of the 1989 Environmental Protection Law.
25. Article 5 of the 2008 Amendment of Law on Prevention and Control of Water Pollution.
26. For example, Jiangsu Province took the lead in carrying out the concept of managing the environment with economic incentive and market mechanism, taking the comprehensive treatment of the water environment in Wuli Lake as the important component of the urban development of Wuxi. The private funds are encouraged to enter into the sector of municipal service facilities. Market mechanism is applied in setting up share-holding company of sewage treatment; various forms of construction and operation such as BOT and TOT have been adopted; priority and incentive policies have been formulated. See 'Water Pollution Prevention and Control: Successful Cases in China', available at http://www.unep.org/gc/gc23/documents/China-Water-Pollution.pdf (accessed 2 April 2012).
27. See 'Joint Efforts against Water Pollution', *China Daily*, 10 March 2010, available at http://www.chinadaily.com.cn/china/2010-03/10/content_9563900.htm (accessed 3 March 2012).
28. Item 4, Article 19 of the 2008 Amendment of the Law on Prevention and Control of Water Pollution.
29. According to the Constitution and the Water Law in China, the state owns almost all water resources except for the water resources in ponds which are collectively constructed and owned by the villagers.
30. Similarly, in the recent cooperative agreements among the provinces in the Tai Lake Basin, no contract provides the opportunity for public accessibility and public participation.

17.6 REFERENCES

Hsu, C. Stephen (ed.) (2003), *Understanding China's Legal System*, New York University Press.

Jian, Ke (2010), 'Environmental Pollution, Justice, and Trans-Boundary Water Management in China', *CWEEP Research Paper*, available at: http://cweep.anu.edu.au/publications/rp.php.

Qin, Boqiang (ed.) (2008), *Lake Taihu, China: Dynamics and Environmental Change*, Springer.

Tai Lake Basin Management Agency, available at: http://www.tba.gov.cn/.

Varis, Ollis, Cecilia Tortajada and Asit K. Biswas (eds) (2008), *Management of Transboundary Rivers and Lakes*, Springer.

Winalski, Dawn (2009), 'Cleaner Water in China: The Implications of the Amendments to China's Law on the Prevention and Control of Water Pollution', *Journal of Environmental Law and Litigation*, **24**, 181–202.

18. China's federal river management – the example of the Han River

Lan Fang

18.1 INTRODUCTION

China is short of water and water resources are unevenly distributed. Southern China's Yangtze River Basin is rich in water resources, northern China's Yellow River Basin possesses 60 per cent of China's landmass, but holds merely 20 per cent of the nation's water resources. To alleviate the northern and northwestern water shortage, the massive South-to-North-Water-Diversion (SNWD) project is being constructed. The project consists of three routes. The East Route Project (ERP) transfers water from the downstream Yangtze River to East China's Shandong Province through the Beijing-Hangzhou Grand Canal and other parallel river ways. The West Route Project (WRP) will transfer water from the upstream of the Yangtze River on Qinghai-Tibet Plateau to the upstream of the Yellow River. Due to its construction difficulties and huge costs, the WRP is still being planned. The Middle Route Project (MRP) diverts water from the Han River, the largest tributary of the Yangtze River, to North China's Henan and Hebei Provinces and Beijing and Tianjin Municipalities. The MRP will commence operations in 2015 and ultimately transfer 9.5 billion m^3 annually, benefiting a population of 35 million.

Though the MRP will bring great benefit to North China it has created a number of disputes between national and provincial governments, and between provincial and provincial governments. An example is the Han River, which has disputes about water distribution, migrant relocation, institutional management and conflict resolution. The Han River rises in southern Shaanxi Province and flows southeastward to Hubei Province. It contributes nearly 70 per cent of the water stored in Danjiangkou Reservoir from where it will be transferred to North China. Although it provides a rich and clean water resource, southern Shaanxi is, however, the most under-developed region of the province. In 2009 the disposable incomes of urban and rural residents were 12,651 yuan and 3284 yuan respectively, just 74 per cent and 64 per cent, of China's average. Both the local government and residents have very strong incentives to develop industries to improve living standards. This will become much harder to achieve after the construction of SNWD, since the need to keep the Han River clean will limit potential expansion of the local manufacturing and mining industries. For instance, nonferrous metal mining industry used to be a key and profitable industry in this region before the implementation of SNWD project. It was obvious, however, that most of the local manufacturing and mining companies operated with low quality techniques, on a small-scale and produced heavy pollution. Because it was serving as a water source of the SNWD, southern Shaanxi had to improve local environmental standards. Thousands of small industries were closed because they failed to meet the requirements of the upgraded environmental regulations. Nearly 15,000 jobs were lost in that region. For instance, more than 200 small mining factories were closed in southern Shaanxi's Ankang City. The agricultural sector surrounding the city is also being affected significantly. Two thirds of cultivated land has been converted to forest to reduce erosion under the Government's Sloping Land Conversion Program. The restrictions are such that farmers can no longer rely on income they generate themselves on-farm. Additionally 713,000 farmers lost their land altogether and are seeking employment in non-agriculture sectors. All these factors have caused a reduction in annual GDP of 14.5 billion yuan. Though the city and its region received some compensation projects from the central government it was far from covering their sacrifices.

In addition to the monetary compensation issue, the rest of Shaanxi Province also suffers from a severe water shortage and needs water from Han River as well. A proposal for a project titled Han-to-Wei River Project was submitted by the Shaanxi provincial government and approved by the central government. This project will transfer water from upstream Han River to thirsty central Shaanxi.

In a neighboring province, Hubei, industry and agriculture located in the middle and lower catchment of the Han River and the middle catchment of Yangtze River are also being affected by the implementation of the national SNWD and Shaanxi Province's internal water diversion project. Danjiangkou Dam is located in Hubei Province. It was built in 1958 with a height of 162 meters. To better serve the SNWD the dam wall was raised to 176.6 m in 2009. That will increase the capacity to control floods along the middle and lower Han River and also increase the amount of water that can be transferred to the north. However that will lead to a significant reduction of water flowing in the Han River downstream of the Danjiangkou Dam and significantly reduce the dilution capacity of the river, which will aggravate pollution. Besides this, Hubei Province has to implement its share of what will be the world's largest migrant relocation project (displaced by the MRP). These negative impacts, conflicts about water distribution with upstream Shaanxi Province, worsening water quality, huge relocation costs and the sacrifices being imposed on migrants, caused Hubei Province to seek support from the central government. The province submitted a proposal to the national State Council requesting approval to build a canal to transfer water from upstream Yangtze River to the Han River, so that the water loss of Han River can be partly compensated. This proposal is still awaiting the approval of State Council. Both Shaanxi and Hubei provinces are competing to take large volumes of water from the Han and Yangtze Rivers to feed their people and industries. Based on Chinese Water Law, the water resource ownership belongs to the nation and is controlled by the central government. Each province negotiates with the central government and there are few successful experiences of effective negotiation between provinces. Another example of the central government's role is in response to the problems in Han River. The national Ministry of Water Resources recently announced that China's first strict water resource management pilot plan will be implemented in the Han River Basin (Li Shantong, Xu Xinyi, 2004).

18.2 SHAANXI PROVINCE'S CONTRIBUTION TO THE SNWD PROJECT

Shaanxi Province is located in the Northwest China. It is divided into three regions. Northern Shaanxi's topography is characterized by the Loess Plateau where the energy industry (particularly driven by coal and natural gas) is booming. The development goal of this region is to build a national level energy and heavy chemical industry base, but its water resource is only about 25 per cent of China's average. Central Shaanxi's topography is centered on the Guanzhong Plain, which contains the most developed economy, education and densely population of the province. The water resource per capita is less than 18 per cent of China's average. Southern Shaanxi's topography is dominated by the Qingling-Daba Mountains, home to the best vegetation and well-conserved biodiversity in the province. Southern Shaanxi possesses abundant water resources, nearly 1.8 times of China's average and seven and ten times the average of northern and central Shaanxi (Water Resources Bulletin of Shaanxi Province, 2010). Its abundant water resources and well-conserved ecosystem makes the water quality of the Han River which comes from this region the best of all the Chinese big rivers. For this reason it is one of the best water source rivers for the SNWD.

18.3 CENTRAL AND NORTHERN SHAANXI'S WATER SHORTAGE SITUATION

Southern Shaanxi's Han River's water is being transferred to north China but the central and northern parts of Shaanxi province are also suffering from severe water shortage.

As mentioned before, central Shaanxi is the most developed and densely populated area of the province. It is the core region of the nations' newly established Guanzhong-Tianshui Economic Zone. Such a national level economic zone indicates the region will benefit from more preferential policies as well as investments from the central government. Water is, however, a key bottleneck for economic expand of this region. It is estimated that the water deficit of central Shaanxi will be up to 2.6 billion m^3 by 2020. Seeking a sustainable way to secure and support central Shaanxi's economy has become a top priority for the provincial government.

18.4 PROPOSAL TO TRANSFER WATER FROM HAN RIVER TO CENTRAL SHAANXI'S WEI RIVER

The strong and urgent water demand of central Shaanxi drove the provincial government to submit a water transfer project proposal titled 'Han-to-Wei River Project (HWRP)' to the Ministry of Water Resource. It proposed to transfer water from water rich southern Shaanxi's Han River to central Shaanxi's Wei River. The engineering work mainly includes digging an 82-km-long water conveyance tunnel through Qinling Mountain so as to divert water from the Han River to the Wei River. The HWRP project will eventually transfer approximately 1.5 billion m^3 by 2030, which implies that the water deficit of central Shaanxi could be reduced to 0.99 billion m^3 after the project is completed. Since this water transfer project will be constructed within the territory of the province, its provincial government thought that inter-provincial water rights would not become a big issue (*Shaanxi Daily*, 2011). In fact, it has created a hot dispute with neighboring province Hubei. However the central government approved Shaanxi province's water transfer project and construction started in late 2011.

18.5 IMPACT OF THE SNWD ON HUBEI PROVINCE

The upstream Shaanxi Province has made a great contribution to the SNWD, but so also has the downstream Hubei Province. Hubei Province is located in the middle and lower catchment of the Han River and the middle catchment of Yangtze River. It is a water-rich province known for centuries as the 'Province of a Thousand Lakes'. The economic aggregate of Hubei Province is larger than that of Shaanxi Province.

Han River draws on a catchment of 63,000 sq. km, accounting for one-third of the area of Hubei Province. A group of cities, Shiyan, Xiangyang, Suizhou, and Xiaogan, are along the Han River and are well known as national bases for the automobile and manufacturing industries and commodity grains. The Han River plays an extremely important role in Hubei Province's industrial and agricultural activities as well as in its long history.

Danjiangkou Dam is located near Danjiangkou City in Hubei Province and also borders Xichuan County in Henan Province. Originally built in 1958, to better serve the SNWD the dam wall was raised to 176.6 m in 2009. That will not only increase the capacity to control floods along the middle and lower Han River, but also increase the amount of water that can be transferred to the north. However the extra height added to the dam resulted in the inundation of 302 sq. km. This has resulted in 328,000 people being displaced in Hubei and Henan. Of these Hubei province is responsible for relocating 180,000 migrants. Such an ambitious migrant resettlement project is a huge expenditure for China. In addition many important cultural sites will also be flooded.

The raised dam wall will also lead to a significant reduction of flow in the Han River downstream of the Danjiangkou Dam. A report titled *Han River Basin Comprehensive Development Research Report*, released by Hubei provincial government, announced that the national SNWD's MRP will take 9.5 billion m^3 water annually from the Han River after the project is completed, reducing the Han River's average annual flow by 30 per cent. Such a large reduction will also aggravate pollution. The report further predicts significant environmental impacts and says that the dilution of the middle and lower reaches of river will reduce its self-purification capacity. The water level drop will mean that water delivery capacity will reduce by 47 per cent, affecting 71 large and medium-sized irrigation areas and 241 irrigation stations along the middle and lower reaches of Han River. It will also cause considerable negative impacts for the shipping industry.

In addition to the impacts of the SNWD scheme itself Hubei will also be affected by Shaanxi Province's HWRP internal diversion project, which is expected to take 1.5 billion m^3 annually from the Han River before it enters Hubei Province. All these problems and threats caused Hubei Province to seek support from the central government for a project titled Yangtze to Han River Project II (YHRP II, in Chinese

Yin-Jiang-Bu- Han). The YHRP II proposes to alleviate the worsening water situation at downstream Han River in Hubei by building a canal to transfer water from upstream Yangtze River to the Han River.

18.6 HUBEI PROPOSAL TO TRANSFER WATER FROM YANGTZE RIVER TO ITS TRIBUTARY HAN RIVER

Dating back to 2009, Hubei Province started to construct a water transfer project titled Yangtze to Han River Project I (YHRP I, in Chinese Yin-Jiang-Ji-Han). Hydrological information shows the annual runoff of the Han River is 56.6 billion m^3, of which 9.5 billion m^3 of water will be diverted to North China by the national SNWD project. The MRP transfer will lead to water reduction in the Han River below Danjingkou Dam. YHRP I project when complete will transfer 3.1 billion m^3 of water from middle stream Yangtze River to its tributary Han River. However, some new questions are being raised. Some experts argued that a project transferring water from the main river to its tributary might result in unexpected ecological problems. Additionally downstream Han River's water quality will get worse, since the mixture of low quality Yangtze River water will degrade the Han River's water quality and create drinking water safety issue for residents in Wuhan City, the capital city of Hubei Province (Xu Mingxiang, 2005).

Though YHRP I can, to some extent, improve the water flow in the Han River it will mainly benefit the lower reaches. The middle reaches of the Han River, where industrial activity and farmland are concentrated in the Xiangyang City group will continue to suffer water shortages and environment degradation. Added to these impacts are those that will come from the upstream province Shaanxi's water project HWRP. All these factors caused Hubei Province to propose a new water transfer project (YHRP II) to the Ministry of Water Resource (MWR) and the National Development and Reform Commission (NDRC) in 2010. The new water project that is being proposed will take water from upstream Yangtze River, where the water quality is much better than downstream, to benefit the middle catchment of the Han River. The planned ultimate water transfer volume is 10 billion m^3 annually, which is nearly equivalent to the volumes that will be transferred by the national SNWD's MRP and Shaanxi's HWRP project in total. If the YHRP II is approved by the central government, it will certainly alleviate the water shortage situation in middle section of the Han River. However, as this book goes to press, the new project is still awaiting approval.

18.7 IMPORTANCE OF COMMUNICATIONS BETWEEN GOVERNMENTS

The national government plays a dominant role in river basin management and it is important to improve its effectiveness. The Chinese central government has made great efforts to secure a sustainable financial source to support the water donation area. The establishment of the South-to-North Water Diversion Project Fund is part of this effort. The Chinese State Council released a document titled *Regulations on South-to-North Water Diversion Project Fund Raising and the Use* in 2004. According to Article 2, the fund is to be collected in the water receiving areas, which are the Beijing, Tianjin municipalities and Hebei, Jiangsu, Shandong and Henan provinces. Based on their planned level of water consumption and the investments of SNWD, their contributions are determined as follows: 5.43 billion yuan from Beijing, 4.38 billion yuan from Tianjin, 7.61 billion yuan from Hebei Province, 2.6 billion yuan from Henan Province, 7.28 billion yuan from Shandong Province, and 3.7 billion yuan from Jiangsu Province. The raised funds will be used to compensate the people in the water donation area as well as the local economies (Office of the State Council, 2004).

The Chinese State Council also issued a document titled *Regulations on the strictest water resources management system (SWRM)* in January 2012. In summary, the document is mainly about setting of the 'Three Red Lines' and implementation of 'Four Systems'. The three red lines set standards designed to reduce water consumption, improve water quality and increase efficiency with industrial and farm output measured against water use. The four systems define how these targets are to be met.

The practice of the high quality water resources management needs comprehensive support from all concerned authorities and vigorous support from all social communities extending down from the Ministry of Water Resources which is leading the campaign. Because of the Han River's special important position the pilot plan has been implemented there.

18.8 NEED FOR BETTER COMMUNICATIONS

In order to resolve the various contradictions between provinces along the Han River the central government and provincial governments are seeking to improve their communication methods to solve the water-related disputes.

As the source of the Han River, Shaanxi Province 'contributes' to

Danjiangkou reservoir more than two-thirds of the storage water. Although it is ranked as one of northern China's provinces with water shortage, Shaanxi itself will not receive SNWD water. Because of its water shortage, Shaanxi Province has proposed the HWRP project. As a result of active and effective communications with the central government, Shaanxi's project was approved. This immediately led to a hot dispute between upstream Shaanxi and downstream Hubei. Hubei Province is to suffer at least a 20 per cent water reduction from the Han River once both water projects are completed.

To seek understanding from Hubei Province, the Vice Governor of Shaanxi Province and officials and experts from the Shaanxi Provincial Department of Water Resources went to Wuhan City (capital city of Hubei) to meet people at the Yangtze Water Resources Commission and the Hubei Provincial Department of Water Resources for the first round of communications in 2009. These communications and negotiations made some progress and improved understanding between the two provinces. As a result Shaanxi revised its water diversion plan and agreed to implement the project in three phases. The revised plan suggested that in the first phase, 0.5 billion m^3 water from the Han River will be taken by 2020. At the second phase 1 billion m^3 water will be taken by 2025; and at the third phase, 1.5 billion m^3 will be transferred ultimately by 2030. Those concessions by Shaanxi will allow Hubei Province a longer period of time to reallocate its economy and ecosystem along the middle and downstream sections of the Han River. As a major 'water donation' province, Hubei proposed the new YHRP II to offset the negative impacts of the national SNWD and the Shaanxi provincial water project. Hubei Province has made a strong effort to convince the State Council to approve this proposal. In response the Vice Minister of the MWR and his staff went to Hubei Province to investigate the new water diversion project site in 2011.

Both Shaanxi and Hubei provinces are competing to take large volumes of water from the Han and Yangtze Rivers to feed their people and support their industries. Both have applied for support from the central government. Based on Chinese Water Law, water resources belong to the nation and are controlled by the central government. Each province has to work under the guidance of the central government. Consequently the main strategy pursued by competing provinces is to seek the support and understanding of the central government. Though the central government plays the dominant role in making decisions about water projects, however, active communications and negotiations between provinces are still wise and helpful, as shown by the way Shaanxi province revised and slowed its water transfer project after discussions with Hubei Province.

18.9 MORE EFFECTIVE COORDINATION BETWEEN PROVINCIAL STAKEHOLDERS IS REQUIRED

Taking account of the SNWD's background and the Han River's special importance, the Danjiangkou Agency (DA) was founded in 2004. This agency is a branch office of Yangtze River Basin Water Resources Protection Bureau which is directly authorized by the Yangtze Water Resource Commission (YWRC). The DA's duty is to protect the Han River. Specifically, it needs to fulfill the duty assigned by China's Water Law to administrate the watershed by conserving the ecosystem around the dam, controlling water pollution in and around the dam area and coordinating the demands of different provincial interests. An important function of DA is to develop an effective coordination mechanism for water resource protection and pollution control. The DA conducted a three-year investigation regarding the current conservation status of local ecosystems and environmental quality as soon as it was established. During the investigation, the agency hosted several forums among the three provinces, mainly aiming to learn the demands of the water donation areas, to evaluate the achievement of pollution control policies and to assess the current difficulties. Through these investigations and forums, the three water donation provinces, Shaanxi, Hubei and Henan as well as five related cities reached an agreement to coordinate the water conservation and water pollution control at the dam area. Based on this foundation, the mutual exchange of experience and lessons in water conservation can be conducted. In addition, the arrangement makes it possible for major issues regarding future development and protection of the water source region to be discussed among all stakeholders.

In November 2009, the 1st Water Resources Protection and Water Pollution Control Joint Meeting of Danjiangkou District (hereinafter referred to a joint meeting) was held in southern Shaanxi's Shangluo City. A joint meeting constitution was released. 'The Shangluo Declaration – Clean Water to Beijing' was also made during the meeting. This joint meeting and its outcome are examples of innovation in China's water basin management.

Two years later the second joint meeting was held in Nanyang City, Henan Province. This meeting announced that the current main task is to solve some key issues regarding water conservation and water pollution control in the water source area and enhance the relevance of the joint conference system. The meeting also promoted the sharing of administrative resources to build an effective watershed management model, and help explore a reasonable approach to fulfilling the strict pilot plan of water resource management.

The DA fulfills its duty on behalf of the YWRC at the water source area. It provides a platform and brings the water donation provinces together. Through joint meetings and some other methods important agreements and understanding have been achieved. This can be an effective method to increase understanding between different provincial stakeholders, improve the environment quality and reduce the pollution based on common activities conducted by different provincial stakeholders.

18.10 IMPORTANCE OF LEGAL AND ADMINISTRATIVE REFORMS

The Water Law of the People's Republic of China stipulates that China's water management system is a combination of watershed and administrative organizations. It states that the water distribution shares will be decided by considering the needs of local residents, industries and the environments. On one hand, the Yangtze River Water Resources Commission and the Yellow River Conservancy Commission, on behalf of the Ministry of Water Resources, are responsible for administering the river basin as a whole. In addition, each province and city has its own Department of Water Resources, Environmental Protection Agencies, and so on. This indicates that China's watershed management system is a multiple management system, which involves a number of central departments and local governments. In addition, some functions and responsibilities of the central water administrative body are assigned to branch offices or departments. For instance, the provincial department of water resource is responsible to both the MWR and the provincial government. There are many examples of overlapping related to water management among different Chinese government organizations. Regarding the management of water quality and water quantity, different government bodies often work separately, with little coordination and cooperation. The biggest administration overlap of water quality management is between the departments of water resources and environmental protection. Both are in charge of water quality but sometimes none of them can work efficiently due to the overlapping duties. In urban water management, the contradiction between the department of water facility construction and water resource is serious. Regarding aquatic organisms the biggest overlap is between forestry, agriculture and environmental protection departments (Wang Yi, 2008). Regarding a river basin's management the related authority's responsibility for water quality is not defined clearly for different regions. This makes it difficult to marshal the power needed to coordinate different interest groups and conflicts along a river. The example of Han River shows that the authority and responsibility for watershed management and regional administration powers is not well clarified. The consultation mechanism is also insufficient, which affects the efficiency of water management. Central government plays a key role and has to act as a coordinator to balance the interests of different provinces but more needs to be done to improve the current water-related legal system and regulate the behavior of different interest group along rivers.

18.11 SUMMARY

This study has tried to explore ways to improve the effectiveness of communications and negotiations between national and provincial governments and between provincial governments. Four methods can be applied to improve Chinese water management. The first is better use of the power of central government, second is to promote active communication process between national and provincial governments; third is to establish more effective coordination between provincial stakeholders. Fourth and last is to reform judicial water law and improve the functioning of watershed management. The current water management system is far from efficient. The overlapping responsibilities of different government bodies are common shortcomings and a poorly defined system of responsibilities between river basin authorities and regional administration institutions lowers the efficiency of water management and create conflicts between national and provincial governments. There is considerable scope to improve China's water management arrangements. Progress is being made but more is needed.

18.12 REFERENCES

Chinese Ministry of Water Resource (2010), 2010 China Water Resources Bulletin, China Press of Water Resource.

Development Research Center of Hubei Province (2011), Fighting for the country to establish the long-term compensation mechanisms of SNWD.

Kang Chuanyi (2011), Five factors drove the water transfer project, *Shaanxi Daily* http://www.sxdaily.com.cn/data/bsgzzj/20110815_9994568_8.htm.

Li Shantong and Xu Xinyi (2004), *South-to-North water diversion and China's development*, Beijing: Economic Science Press.

Office of the State Council (2004), Fund raising and use of management practices of SNWD.

Office of the State Council (2012), State Council on the most stringent water resources management system.

Official memories of SNWD immigrants (2011), http://news.ifeng.com/mainland/detail_2011_10/18/9924743_1.shtml.

People's Yangtze River (2011), Consultative mechanism construction and practice of Yangtze River Water Resources Protection.
Planning of transferring Yangtze River water to supplement the Han River water had been reported to the Ministry of Water Resources (2010), http://news.sina.com.cn/c/sd/2010-11-02/172121400986_2.shtml.
Shaanxi proposed project of transferring Han River's water to central Shaanxi's Wei River but Hubei rejected (2011), http://news.sohu.com/20110616/n310314066.html.
Shaanxi Provincial Bureau of Statistics (2010), 2010 Statistical Yearbook of Shaanxi Province.
Shaanxi's Water Resource Department (2010), 2010 Shaanxi Water Resources Bulletin.
Shouke Wei, Hong Yang et al. (2010), Game theory based models to analyze water conflicts in the Middle Route of the South-to-North Water Transfer Project in China, *Water Research* 44 (2010) 2499–2516.
SNWD office of China's State Council (2002), SNWD Middle Route Planning.
The South-to-North Water Transfer Project Resettlement Compensation under pressure, The Twenty-first Century Economic Report (2010), http://finance.sina.com.cn/roll/20100223/04157438599.shtml.
Wang Xiaoliang (2007), Water sent to Beijing, and Southern Shaanxi sacrifice much, *China Business Daily*, http://hsb.hsw.cn/2007-03/11/content_6147318.htm.
Wang Yi (2008), Reform of the watershed management system to promote integrated watershed management, Policy and Management Research.
Xu Mingxiang (2005), Plan and design of water diversion from the Yangtze river to the Hanjiang river project, People's Yangtze River.

PART VII

AUSTRALIA

19. The Murray–Darling Basin

Daniel Connell

19.1 INTRODUCTION

When it passed the Water Act (2007) the Australian national government took on a central role promoting water reform in the Murray–Darling Basin (MDB). This development, which is a new stage in a long incremental process, has made the respective powers and responsibilities of the national and state jurisdictions in relation to water a frequent subject of public debate. A major point of interest is the use of water trading, not just to achieve economic benefits for entitlement holders, but as a strategy to achieve a complex set of connected reforms. As in many other countries water trading has been a central element in the Australian water reform program. An additional and unusual feature of the Australian example is that water trading is explicitly linked to efforts to achieve environmental sustainability. Other elements of the larger programme include comprehensive monitoring of a wide range of biophysical and economic issues and independent auditing of a range of issues. This is linked with efforts to promote public participation in water policy.

Just over a million square kilometres in size, the MDB has a diverse range of landscapes, ecosystems, land uses, and climates ranging from the sub-tropical north to the temperate south with its long, dry summers, wet winters and snowfields (although most of the basin is naturally semi-arid). It contains the watersheds of two major rivers – the Darling and the Murray – along with their many tributaries. The basin includes over 30,000 wetlands, of which 16 are listed under the Ramsar Convention of Wetlands of International Importance. The MDB is divided between the jurisdictions of New South Wales (NSW), Victoria (VIC), South Australia (SA), Queensland (QLD) and the Australian Capital Territory (ACT) and is home to just under two million people. It supplies much of the water used by another million in South Australia, and generates approximately 40 per cent of the gross value of Australia's agriculture and pastoral production (Bohensky et al. 2009). Those three million people and various industrial activities use about 4 per cent of the water diverted from the region's rivers. The other 96 per cent is used by irrigated agriculture and constitutes about two-thirds of national rural and urban usage (Connell 2011b).

There is now general agreement among the governments with responsibilities in the MDB that most rivers are over-allocated (COAG 2004). The ecological health of 20 of the 23 major river valleys in the basin has been classified as either poor or very poor (Williams 2011). Fundamental to the basin's biophysical setting and particularly significant for its impact on salinization processes within the catchment is that it sits on an almost closed groundwater basin with only one outlet at the Murray Mouth in South Australia. The basin's rainfall is low, and is one of the most variable of the world's major river systems (CSIRO 2010). Compared with other major rivers in the world, the Murray–Darling is a low energy system with little capacity to purge itself of salts and sediments. Much of the salt mobilized into streams is not flushed out of the Murray Mouth but is redistributed elsewhere in the basin to what were previously fertile low lying areas or onto floodplains of high environmental value, often over state borders (Connell 2007).

19.2 INSTITUTIONAL CONTEXT

For over a century, irrigation development in the MDB has been primarily the responsibility of basin state governments, each operating with a high degree of autonomy within its jurisdiction. This has resulted in state-focused systems with different management arrangements, types of water entitlements and contrasting conceptions of the appropriate relationship between irrigation communities and governments. Partly as a result of the institutional and policy complexity that has resulted from over a century of parallel but largely uncoordinated development in relation to many issues, and despite the existence of a near century old inter-jurisdictional water management framework, environmental degradation of the region's rivers and catchments is intensifying and resource security is declining. In response Australia's national government is playing an increasingly central role as an agent pushing and pressuring for change.

But what powers does the national government have and how should it use them? There is general agreement that primary responsibility for water management resides with the states. This reflects the historical record and the consensus of constitutional lawyers (Fisher 2011). However the national government does have some leverage through its capacity to enforce international treaties, which apply in the case of water resources primarily to environmental issues, and even more importantly through the dependence of the states on grants from the national government for a substantial proportion of their expenditures.

The MDB's policy environment is volatile. Much of the discussion about inter-jurisdictional water management in the MDB gives the misleading impression that interaction between the Commonwealth and states is highly structured, but the reality is far more elusive. At any given time, a large number of enterprises (individuals, businesses, associations, industry groups, governments, etc.) are interacting and influencing its policy process in many different ways. In practice, decisions are not made from the top down but emerge from cycles of interaction in which the participants have varying degrees of influence but no single voice is dominant. As a result of the existence of the federal political system the cross-border integration of water management involves two levels of government, each with considerable autonomy from the other.

19.3 HISTORY OF INTER-GOVERNMENT LINKS IN THE MDB

The history of intergovernmental relations involving the River Murray is a complex story involving both cooperation and competition. Ever since Victoria was created out of the colony of New South Wales in 1850 with the border on the southern bank rather than down the middle of the main channel, there have been many examples of governments using positions of strength to gain more than they might have got from a neutral arbitrator. At the same time there have been many circumstances where biophysical, social, economic, political and legal factors have forced reluctant neighbours to cooperate. A good example of the latter was acceptance by successive New South Wales governments that they have to make concessions to Victoria because any dam across the river to supply their own irrigators and towns downstream needed to be anchored to the Victorian side of the river. Another was the way in which the new legal framework created by federation in 1901 pressured the New South Wales and Victorian governments to take downstream South Australian concerns into account in order to avoid the uncertainties involved in legal proceedings in the newly created High Court (Clark 1971). There have even been decisions that appear to be motivated by a sense that it was the right thing to do in the best interests of the nation as a whole. One example was perhaps the 1967 decision of the South Australian representative on the River Murray Commission who voted against building Chowilla Dam and for Dartmouth Dam in defiance of his premier (Connell 2007).

Since the beginning of formal intergovernmental cooperation to manage the River Murray in 1914 money from the Commonwealth Government has been an essential lubricant for the interstate negotiation process. After numerous reports and investigations the Commonwealth Government finally persuaded the three state governments to join it in approving the River Murray Waters Agreement in 1914 by offering £1 million for infrastructure investment if all agreed. (That was more than the cost of the proposed Hume Dam near Albury.)

Two main phases can be identified in the earlier history of attempts at cross-border integration in the MDB. South Australia had been pushing for a basin-wide agreement since before federation in the late nineteenth century. Positioned at the end of the system it was the driest state and most threatened by development upstream. By the end of 1915 the River Murray Waters Agreement had been incorporated into identical legislation by each of the four parliaments concerned with the southern River Murray section of the catchment. This first attempt to integrate cross-border water management in the basin contained three main elements (Clark 1971). The first involved a coordinating body known as the River Murray Commission (RMC), with four members chaired by the Commonwealth representative and supported by a small full-time secretariat. This was established to oversee implementation of the works program of dams and other infrastructure and the inter-state water sharing arrangements. As recommended by the 1902 Interstate Royal Commission, a water sharing arrangement between the three states based on proportions of available flow was agreed on for times of drought. (This reflected the great variability of the basin's climate and is a good foundation for managing the even greater variability predicted to result from climate change.) With incremental changes over time, an intergovernmental arrangement based on these three elements has been in place ever since (Connell 2007).

The second phase which commenced in the mid-1980s was driven initially by increasing concern about the salinity impacts of irrigation development along the Murray and its tributaries. By the 1970s irrigation development in Victoria and New South Wales was having significant salinity impacts on South Australia, threatening both downstream irrigation and water supplies for urban centres. The negotiations about irrigation induced salinity in the 1970s and 1980s eventually resulted in a much wider and very ambitious new arrangement, known as the MDB Initiative, which for the first time explicitly acknowledged the need for a whole-of-basin approach. The RMC was replaced by the MDB Ministerial Council (MDBMC) which brought together relevant ministers from the governments of the Commonwealth and the basin states. A Community Advisory Committee was also established to advise the Ministerial Council. The MDB Commission, led by the heads of the government agencies working to the ministers in the council and supported by an enlarged secretariat, was established to support the work of the council. In addition, for the first time, the Australian Capital Territory and Queensland were also represented as observers, making coverage of the entire basin complete (Connell 2007). As with earlier iterations, these changes were incorporated in new legislation and passed as identical acts in the parliaments of each MDB jurisdiction during 1992–93 (Kellow 1995).

In principle the new arrangements allowed for integrated policy development and management in respect of any issue upon which all jurisdictions agreed to cooperate. The limits were those imposed by what was possible in terms of politics, economics and implementation capacity rather than any restriction resulting from lack of constitutional power. But most of the activities incorporated into the new arrangements were advisory or discretionary in nature and needed the enthusiastic collaboration and cooperation of all the governments and agencies involved before they could be implemented in any significant way. This applied particularly to activities outside the River Murray corridor. In addition, the long-established unanimity principle for major cross-jurisdictional decisions still applied. This gave the power of veto to any jurisdiction that wanted to exclude an item from the agenda or which was dissatisfied with any decision made. Despite these limitations, the early years of the MDB Initiative were marked by widespread enthusiasm. Its considerable achievements included the Salinity and Drainage Strategy implemented in 1989 which significantly improved water quality in the lower reaches of the Murray, and the 'cap' on further increases in extractions in 1995 which reflected agreement by all governments that the basin's water resources were over-committed (MDBMC 1995).

19.4 BACKGROUND TO CONTEMPORARY REFORMS

By the mid-2000s, however, with the passing of the generation of policy-makers who had introduced the reforms from the late 1980s, and continuing environmental problems exacerbated by one of the most severe droughts in Australia's recorded history, the water reform process had clearly stalled and dissatisfaction with the cooperative approach was mounting. Although water sharing between the states was based on proportions of available flow, the drought required unprecedented reductions in irrigators' water allocations and created pressures for new levels of integration. One of the investigations that revealed the need for further reform was a study conducted by the CSIRO that predicted major reductions in inflows in the Murray system, and in volumes of water available for irrigation and towns, because of the increasing impact of factors including climate change, farm dams, forest plantations, and reduced leakages due to delivery systems upgrades (van Dijk et al. 2006). Significantly, all these threats were outside of the previously agreed list of responsibilities of the MDBMC (which required unanimous agreement for any additions).

In response to the apparent inability of the MDBMC to manage these and other issues, the Commonwealth Government began to assume a more central role. An important stage in the development of this approach was agreement by the Council of Australian Governments (CoAG) to the National Water Initiative (NWI) in 2004. The NWI was ostensibly national in scope but was strongly shaped by the need to manage the increasingly bitter conflicts in the MDB. Central to the NWI was the requirement for integrated water resource management plans able to address all issues of concern.

Among many other changes, implementation of the NWI meant that state governments needed to step back to some extent from their long term close relationship with irrigation communities so they could take better account of other stakeholders. For over a century Australian water management has been controlled by state government agency officials tasked to encourage rural communities to accept heavily subsidized water made available by government programs building dams and distribution systems ahead of actual community demand. As part of the nation building project extending back into the nineteenth century the aim was to use water to create new rural communities based on irrigation. The NWI was developed to give recognition to new priorities. It emphasized the need for sustainability and the legitimacy of other interests. In recent years Australian governments, both conservative and social democrat have undergone a philosophical transition from nation and community building to a much stronger focus on economic growth. But at the same time they have accepted the emerging scientific consensus that the country's major hydrological systems are in long term decline in both environmental conditions and supply reliability. These shifts in policy perspective occurred in parallel with an expansion in the number and assertiveness of stakeholder groups demanding that their views and interests be taken into account in the development of water policy.

The result was a policy designed to achieve a number of goals normally seen in conflict. The organizing devise is to be the development of comprehensive water plans that will incorporate negotiated compromises between competing interests. As specified in considerable detail in the NWI the water plans should include secure water access entitlements, statutory based planning, statutory provision for environmental and public benefit outcomes, plans for the restoration of over-allocated and stressed systems to environmentally sustainable levels of extraction, the removal of barriers to trade, clear assignment of risk for future changes in available water, comprehensive and public water accounting, policies focused on achieving water efficiency and innovation, capacity to address emerging issues and many more elements. They are also to provide for adaptive management of surface and groundwater systems with their connectivity recognized where it is significant. In addition, water plans must take account of Indigenous issues by making arrangements for Indigenous representation in water planning wherever possible and provision for indigenous social, spiritual and customary objectives wherever they can be developed. They should also include allowance for the possible existence of native title rights to water in the catchment or aquifer area (Connell 2007).

A central element of the NWI was the redefinition water licences into a form that would make them more easily tradable. This was to happen, however, only after the hydrological systems in question had been restored to sustainability by reductions in the overall level of extractions and management changes to restore environmental sustainability without compensation to existing licence holders. In the following years that approach was to prove politically unacceptable. However as will be explained below an alternative based on the purchase of entitlements on the open water market has been developed that could go a long way to achieving that goal.

19.5 NATIONAL GOVERNMENT ROLE

Initially the Commonwealth Government expected rapid reform as a result of all states and territories agreeing to a detailed implementation programme when they approved the NWI. The assumption was that reform would take place because its merits were obvious. That did not happen for a variety of reasons which included both the complexity of what was being proposed and the opposition of irrigation-dependent communities concerned about proposed reductions in water allocations to irrigators. When the expected reforms did not eventuate the Commonwealth then introduced the Water Act 2007 (Australian Government 2007). This was supported by $A10 billion in funds (later increased to more than $A12 billion) provided by the Commonwealth through the Water for the Future programme to invest in new infrastructure projects by the states to help deal with climate change and to buy back water entitlements in order to conserve the environment. (Payments to the states in return for compliance with Commonwealth Government policies, has long been one of the main strategies used to achieve state government cooperation with national initiatives, albeit with a record of very mixed success.)

The new act replaced the MDB Council and Commission which had represented all governments, with a Commonwealth agency, the MDB Authority (MDBA) and charged it with responsibility for developing and implementing a Basin Plan (finally approved in late 2012). The Commonwealth intends that the Basin Plan will lead for the first time to an integrated and comprehensive approach being taken to managing the MDB's water resources. Other key organizational elements are new monitoring and auditing activities established in agencies such as the Murray–Darling Basin Authority, the National Water Commission (NWC), the Bureau of Meteorology (BOM) and the Australian Competition and Consumer Commission (ACCC). The MDBA is developing the overarching Basin Plan and assessing preparation of the state sub-plans. The NWC will assess overall progress against the environmental and economic sustainability goals of the National Water Initiative. The BOM is to be the repository for the expanded climate data collection programme (a key role given the predicted impacts of climate change and the need to make sure that the Basin Plan and the sub-plans are sufficiently responsive). The ACCC will be responsible for the ongoing integrity and evolution of water trading. Another important agency created by the act is the Commonwealth Environmental Water Holder (CEWH). It was established to manage the water entitlements being purchased for the environment through the $A3 billion allocated for that purpose under the Water for the Future programme. The water entitlements are to be managed according to an environmental plan to be developed by the MDBA as part of the Basin Plan.

Under the Basin Plan the Commonwealth will define the high level goals and policy parameters for the Murray–Darling Basin as a whole and for each state. Each of the states will then be responsible for developing an implementation sub-plan for the systems within its borders (combining both surface and groundwater), which must be approved by the relevant Commonwealth minister. In the event of failure to agree the Commonwealth is enabled by the Water Act to prepare its own sub-plan for the state in question. Likely political opposition and capacity constraints could make that difficult to do, however, so the emphasis is on negotiating to achieve agreement. In practice, the primary resource available to the Commonwealth to facilitate the process is not the legislative provisions for enforcement but rather the funds that it is able to make available to compliant states.

The process of obtaining support from the states and the national parliament for the Basin Plan has been protracted and controversial. A key point of disagreement has been the size of the target for re-allocation back to the environment, the so-called Sustainable Diversion Limit (SDL). About a quarter of the annual average extraction for irrigation has been targeted. Originally a lower figure was nominated and even that attracted strong opposition from irrigation communities in New South Wales and Victoria. However the threat of legal action in the High Court by the government of South Australia, the state at the end of the system, caused the Commonwealth to increase the SDL to its current level of 3200 gigalitres. An apparent victory was gained by opponents to the Basin Plan in the setting of the dates for its implementation, the year 2019 with a five-year phase-in period. However it can be argued that that date is not as significant as it sounds because the purchasing program for entitlements using the three billion dollars allocated under the Water for the Future programme that will be managed by the CEWH to achieve environmental improvements, has resulted in more than half of that target already being available.

19.6 REFLECTIONS ON THE IMPACT OF THE FEDERAL SYSTEM ON POLICY IN THE MDB

Since the mid-1980s public policy in the MDB has been a frequent subject of intense debate and this has provided many opportunities to see the Australian federal system in action. The independent centres of power provided by the Commonwealth and state jurisdictions create focal points in a polycentric governance arrangement around which contending interest groups arrange themselves, moving from one to the other as their members make strategic decisions about alliances and how best to promote their goals or block those of others. In addition, to a very significant extent many conflicts that might appear at first sight to be between the national government and state governments are in fact primarily between state governments that are attempting to enlist national government support for their positions. The differences within federal systems receive strong codification in the various constitutions in place at state and national levels. Constitutions limit the power of governments in relation to each other and provide the potential for delay through litigation which often equates to a veto in practice. These delays and potential vetos can be used to gain concessions. Constitutions thereby allow politicians and their administrators to introduce additional factors into negotiations that might not be allowed if issues were processed with the tighter definition of relevance that could be insisted upon in a unitary system.

On River Murray issues, for example, there is a long history of South Australia parliamentarians – from all political parties – lobbying strongly within the national parliament for state interests against New South Wales and Victoria. An early example was Patrick McMahon Glynn, the South Australian Minister for External Territories in the national government in 1914 who persuaded Prime Minister Joseph Cook to offer a very substantial contribution to fund infrastructure along the River Murray if New South Wales and Victoria signed up to the River Murray Waters Agreement (Connell 2007). Continuing the pattern, South Australian concern about the salinity impacts of upriver development was probably the strongest factor creating pressure for the institutional reforms of the 1980s. More recently it was also the strongest advocate for the enhanced Commonwealth role encompassed in the Water Act 2007 and the Basin Plan approved in late 2012. Of course the up-river states also have members of parliament within the Commonwealth Government but the River Murray is not as important to them as it is to South Australia. This creates space for trade-offs where deals about river issues are influenced by deals about other non-river subjects.

The federal political environment is reflected in many dimensions. Government-to-government agreements, for example, are different from those between governments and non-government institutions. Governments do not impose penalties on other governments in the same way they do on non-government institutions and people. All in all it can be observed that elected politicians interact with elected politicians from other governments – national, state and local – with an extra dimension of independence that comes from them being elected in contrast with others who are appointed to boards or employed by the various government agencies.

Federal systems also have more distinct legal and administrative zones. Within jurisdictions senior figures have the motivation and capacity to integrate the activities of the officials that answer to them. With officials in other jurisdictions their power is more indirect. It is likely that incompatibilities and inconsistencies of laws and regulations will be more likely to emerge between jurisdictions than within them. In addition, around each of the semi-autonomous legal and administrative zones based on the states particular cultures and patterns of behaviour are likely to develop. These tend to be self-perpetuating. The history of the various programmes designed to integrate water trading across the three southern states in the MDB shows how easy it is for such efforts to be frustrated by state agency officials and non-government stakeholders who will not necessarily share the overall benefits of system-wide productivity gains and reduced transaction costs.

In any political system where discretionary decisions can influence who gets what, when, where and how there will be lobbyists working for special treatment. Around each jurisdiction there develops distinct lobbying and pressure group activity. To maximize their leverage lobbyists often go jurisdiction shopping and play one government off against the other. In a unitary system this is harder to do because ultimately the various agencies all answer to the same senior leader directly in charge. Another dimension of the policy process which is affected by the federal system is the relationship between politicians and media. Most Australian public media is still state based with radio, television and newspapers primarily focused on state capitals. Online media is potentially wider in scope but in most cases it too is community or regionally orientated. Australian politicians, even those in the national parliament, are all based on electorates that are state wide at most so they maintain a close relationship with local media.

A federal system also has more governments and more elections. At any given time elections are a looming prospect somewhere. There can often be significant political benefit for politicians to highlight, exaggerate or deliberately foster differences with politicians in other jurisdictions. They can thus claim to be defending the interests of their constituents in a hostile world. (This can be a self-fulfilling perception. Politicians thinking or acting defensively will often act in hostile ways in relation to their neighbours who can respond similarly and thereby provide justification for the original accusations).

This feature is accentuated by the variety in voting systems that are used. Different voting methods produce different results from the same voting public. An example is the Australian Senate, which is elected proportionately, compared with the Australian House of Representatives, which is elected preferentially. Since the current Australian system of Senate elections was introduced in 1948 there have been only brief periods when majorities in the lower house have been matched with similar majorities in the upper house. Consequently Australian national governments are almost invariably dependent on negotiations with other parties in the upper house to get legislation approved. Interestingly, many voters vote differently in the national elections for the two houses, even though they are conducted simultaneously, in order to maximize the opportunities for more nuanced results. This feature of the Australian Senate was a significant factor in the protracted negotiations shaping the evolution of the Basin Plan. (Each state also has voting systems for their upper and lower houses that are quite different from their neighbours and the national parliament and their elections are conducted on different dates.)

The result is a political environment in which elections somewhere are usually imminent.

19.7 CONCLUDING COMMENTS

Water policy is always political but as a result of the institutional characteristics of federal systems those politics are often played out in a very overt way. As the chapters in this book illustrate, federal political systems are in place in many different countries and most of them have rivers that cross state or provincial borders. Compared with unitary systems it can be argued that federal systems create more complex patterns of relations between decision makers. Federal systems provide room for greater ideological diversity and lower decision making predictability than do unitary systems. The participants are more varied and have greater scope for manoeuvring and bargaining compared with their peers in non-federal systems. This accentuates the opportunistic rather than the systematic planning dimension of policy development. With high frequency it can be observed that the management of rivers in federal systems is characterized by considerable intergovernmental and interagency conflict, low decision making transparency, poor accountability, high transaction costs and ad hoc deals between competing governments.

19.8 REFERENCES

Australian Government. (2007). *Water Act 2007*. Canberra: Australian Government.

Bohensky, E., Connell, D., and Taylor, B. 2009. 'Chapter 22: Experiences with integrated river basin management, international and Murray Darling Basin: Lessons for Northern Australia'. In Stone, P. (ed.), *Northern Australia Land and Water Science Review Full Report*. Canberra: CSIRO Sustainable Ecosystems.

Clark, S. 1971. Australian water law. Unpublished PhD thesis, Melbourne University.

Commonwealth Environmental Water Holder (CEWH) 2011. 'Annual Report 2010–2011'. In Department of Sustainability, Environment, Water, Population and Communities Annual Report 2010–2011, Canberra: DSEWPaC, pp. 261–293.

Connell, D. 2007. *Water Politics in the Murray–Darling Basin*. Annandale: Federation Press.

Connell, D. 2011a. 'The role of the Commonwealth Environmental Water Holder'. In Connell, D. and Grafton, R.G. (eds), *Basin Futures: Water Reform in the Murray–Darling Basin*. Canberra: ANU E-Press.

Connell, D. 2011b. 'Water reform and the federal system in the Murray–Darling Basin'. *Water Resources Management* 25 (15): 3993–4003.

Connell, D. and Grafton, R.G. (eds) 2011. *Basin Futures: Water Reform in the Murray–Darling Basin*. Canberra: ANU E-Press.
Council of Australian Governments (COAG) 2004. *Intergovernmental Agreement on a National Water Initiative*, Council of Australian Governments, Canberra, http://www.nwc.gov.au/resources/documents/Intergovernmental-Agreement-on-a-national-water-initiative.pdf.
Crase, L. 2008. 'An introduction to Australian water policy'. In Crase, L. (ed.), *Water Policy in Australia: The Impact of Change and Uncertainty*. Washington, DC: Resources for the Future Press, pp. 1–16.
CSIRO. 2010. *Climate variability and change in south-eastern Australia: A synthesis of findings from Phase 1 of the South Eastern Australian Climate Initiative (SEACI)*. Canberra, Australia: Commonwealth Scientific and Industrial Research Organisation.
Fisher, D. 2011. 'A sustainable Murray–Darling Basin: the legal challenges'. In Connell, D. and Grafton, R.Q. (eds), *Basin Futures: Water Reform in the Murray–Darling Basin*. Canberra: ANU E-Press.
Kellow, A. 1995. 'The Murray–Darling Basin'. In Doyle, T. and Kellow, A. (eds), *Environmental Politics and Policy Making in Australia*. Melbourne: Macmillan, pp. 220–238.
Marshall, G.R. 2005. *Economics for Collaborative Environmental Management: Renegotiating the Commons*. London: Earthscan.
Marshall, G.R. and Stafford Smith, D.M. 2010. 'Natural resources governance for the drylands of the Murray–Darling Basin'. *The Rangeland Journal* 32 (3): 267–282.
Molle, F. 2009. 'River-basin planning and management: The social life of a concept'. *Geoforum* 40: 484–494.
Molle, F., Wester, P., Hirsch, P., Jensen, J.R., Murray-Rust, H., Paranjpye, V.S.P., and van der Zaag, P. 2007. 'River basin development and management'. In Molden, D. (ed.), *Water for Food, Water For Life: A Comprehensive Assessment of Water Management in Agriculture*. London: Earthscan, and the International Water Management Institute, pp. 585–625.
Murray–Darling Basin Ministerial Council (MDBMC) 1988. Draft Salinity and Drainage Strategy. Discussion Paper No. 1. Canberra: MDBMC.
Murray–Darling Basin Ministerial Council (MDBMC) 1995. *An Audit of Water Use in the Murray–Darling Basin: Water Use and Healthy Rivers – working towards a balance*. Canberra, Australia: MDBMC.
Ostrom, V. and Ostrom, E. 1999a. 'Public goods and public choices'. In McGinnis, M.D. (ed.), *Polycentricity and Local Public Economies: Readings from the Workshop in Political Theory and Policy Analysis*. Ann Arbor: University of Michigan Press, pp. 75–103.
Roberts, A.M., Seymour, E.J., and Pannell, D.J. 2011. 'The role of regional organisations in managing environmental water in the Murray–Darling Basin, Australia'. *Economic Papers* 30 (2): 147–156.
Rogers, P. and Hall, A.W. 2003. Effective water governance. TEC Background Paper No. 7. Stockholm: Global Water Partnership.
Schlager, E., and Blomquist, W. 2008. *Embracing Watershed Politics*. Boulder: University Press of Colorado.
Taylor, B.M. 2010. 'Between argument and coercion: social coordination in rural environmental governance'. *Journal of Rural Studies* 26 (4): 383–393.
van Dijk, A., Evans, R., Hairsine, P., Khan, S., Nathan, R., Payday, Z., Viney, N.,

and Zhang, L. 2006. *Risks to the shared water resources of the Murray–Darling Basin*. MDBC Publication No. 22/06. Canberra, Australia: Murray–Darling Basin Commission.

Williams, J. 2011. 'Understanding the Basin and its dynamics'. In Connell, D. and Grafton, R.Q. (eds), *Basin Futures: Water Reform in the Murray–Darling Basin*. Canberra: ANU E Press, pp. 1–38.

PART VIII

CONCLUSIONS

20. Water security in cross-border regions: what relevance for federal human security regimes?

Carmen Maganda and Harlan Koff

20.1 INTRODUCTION AND RESEARCH QUESTIONS

Water security is a timely and relevant theme in global geopolitics. However, it has rarely been empirically examined in cross-border regions in relation to federal human security regimes despite its significance to the study of this concept. Most studies of border regions (that is, ISARM in Latin America or WFD in Europe) examine water governance in terms such as pollution across borders, water and agriculture, water and economic development, and so on. However, there is a dearth of systemic cross-regional comparative analyses of water and human security in cross-border regions. This chapter's originality is its implementation of a comparative cross-border analysis of water security, building on research agendas which have emerged in previous related projects focusing on cross-border human security in world regions. The authors argue that in order to better understand the causes and possible solutions of current water (in)security, we should also analyse what is being discussed (included or omitted) in broader security agendas related to regional integration processes and federal relationships. This chapter presents the current state of cross-border water security in different regions and it discusses policy and institutional responses within the framework of federal relationships.

New contexts surrounding environmental uncertainty, including important development issues related to the access to safe water for all (water security), have redefined discussions on the role of the state, which traditionally focused upon a hierarchical organization based on centralized decision-making guided almost exclusively by scientific expertise in environmental matters. The renovation of the role of the state has been analyzed in terms of the passage from government – defined as the institutional regulation of environmental questions – to governance, defined as the management of the interaction between actors, institutions, the environment and markets. This paradigm shift has been accompanied by predominant understandings of how we perceive security. Classical security debates focused exclusively on protection from military threats to territories and countries. New security discussions have instead placed human beings at the forefront of security concerns. Thus, "states", federal or not, are defined less by the political boundaries and institutional frameworks which classically characterized them and they are addressed more in terms of the citizens which legitimize their existence.

This chapter comparatively examines the emergence of "water security" as a policy paradigm and it embeds analysis in more general debates on human security. Specifically, the chapter addresses the relevance of "water security" as an operationalizable concept for governance in cross-border regions where multi-level governance/federalism characterizes water management. We argue that in order to better understand the causes of and possible solutions to current water (in)security in cross-border regions, we should also analyse what is being discussed (included or omitted) in broader security agendas in these areas related to regional integration processes.

The research on which this chapter is based develops systemic cross-regional comparative analysis of water and human security in cross-border regions. The HUMENITY research project behind this chapter (Human and Environmental Security in Border Regions: Cross-regional Perspectives) aims to examine the evolution of human and environmental security in six cross-border regions: (1) Luxembourg's Greater Region, (2) the Eurométropole, (3) the Spain–Morocco cross-border region, (4) the US–Mexico border, (5) the Mexico–Belize border, and (6) the Venezuela–Colombia border.

These cases were chosen because they include analyses of the impacts of three different regional integration schemes (European Union (EU), North American Free Trade Agreement (NAFTA), and Andean Community of Nations (CAN)). The cases include: two internal EU borders, one external EU border, one internal NAFTA border, one external NAFTA border and a historical comparison of a region which was once an internal CAN border and is now an external CAN border. In terms of federal structures, the cases were chosen because three of them represent "symmetrical federalism", defined as the presence of comparable institutional frameworks on both sides of national divides (the Eurométropole, the Spain–Morocco border and the Venezuela–Colombia border) whereas three cases represent "asymmetrical federalism", defined through the presence of dissimilar institutional frameworks (Luxembourg's Greater Region, the US–Mexico border and the Mexico–Belize border). Similarly, the cases present a mix of "cooperative federalism" where responsibilities are blurred (US–Mexico border, Mexico–Belize border and Colombia–Venezuela border) and "competitive federalism" where there is a clarification of the roles of the different levels of governance (Luxembourg's Greater Region, the Eurométropole and the Spain–Morocco border) (see Keating, 2012). The chapter presents the current state of cross-border water security in different world regions and it will discuss policy and institutional responses within the framework of regional human security.

This chapter's originality is its implementation of a cross-regional, comparative analysis of water security, building on research agendas which have emerged in previous University of Luxembourg-sponsored projects focusing on human and environmental security in world regions. Research methods include a review of the relevant academic literature, examination of government documents, and agency websites, as well as articles in local media. Some preliminary interviewing with local water officials has been conducted in the selected cases.

The project will respond to the following research questions:

- What is the relationship between border water security and border human security regimes?
- How do different federal structures impact cross-border human and water security?

"Water security" is a complex term. It aims to capture the complexity of water management and balance it between resource protection/use, the interests of political and economic actors, and the human security needs of local populations. Supporting the analytical relationship between water security and water governance, we have to give attention to the scale of analysis and water security practices as important tools to operationalize the term. Cross-border water management is a challenge in many world regions. The literature in this field is diverse. Many authors focus on legal frameworks and the intersection of governance, rights and economic interests (such Daibes-Murad, 2005, Gopalakrishnan et al., 2005). Others focus on local power relationships and the roles local authorities play, both formally and informally in negotiating water distribution (see Maganda, 2005; Walsh, 2008). Finally, numerous studies discuss regional water management with a focus on regional institutional frameworks and supranational agreements in the field of water management (see Mumme, 2003; Ingram et al., 1995). However, as stated above, water management cannot be confused with water security. Most studies of border regions (such as UNESCO's International Hydrological Programme or so-called Internationally Shared (transboundary) Aquifer Resource Management (ISARM) Programme in Latin America, or the Water Framework Directive (WFD) in Europe); examine water management in terms such as pollution across borders, water and agriculture, water and economic development, and so on. However, there is a lack of literature focusing specifically on the governance of border water resources for the achievement of water security. This is of course linked with trends in broader cross-border human security debates.

The chapter is divided into three sections. Following this introduction, the chapter's key terms and conceptual framework are presented, then regional water scarcity and regional human security regimes are examined. Finally, concluding thoughts are presented.

20.2 CONCEPTUAL FRAMEWORK AND KEY TERMS: EXPLORING WATER SECURITY AND HUMAN SECURITY AS POLICY CONCEPTS IN FEDERAL STRUCTURES

As stated in the introduction above, security paradigms have shifted considerably since the 1980s. Scholars such as Thomas (2001), Newman (2001), and Owen (2004) have all documented the historical development of this paradigm since the Brandt Commission's Report opened a debate on the relationship between development and security in 1980 and the United Nations Development Programme's 1994 Human Development Report brought the concept of human security to the forefront of international affairs. Specifically, these milestone reports questioned our very understanding of "security" all the way to its foundations. They focus on three fundamental questions: (1) What is security?, (2) For whom is security?, and (3) What threatens security?

20.2.1 The Emergence of Human Security as a Conceptual Paradigm

The immediate answers to these questions have created an inseparable link between security, development and governance. In response to question one: human security has been defined by the United Nations Development Programme (UNDP) as "freedom from fear and freedom from want" (see UNDP 1994, *Human Development Report*). As Thomas (2001) argues in her seminal work on human security, this concept entails two facets, one quantitative and one qualitative. In quantitative terms, "freedom from want" includes the provision of all material needs for human life, including food, health, education, and so on. Conversely, qualitative aspects of human security are related to "freedom from fear" through the guarantee of human rights, the protection of physical safety and autonomous control over one's life course. For Thomas, "human security describes a condition of existence in which basic material needs are met, and in which human dignity, including meaningful participation in the life of the community, can be realized. Such human security is indivisible; it cannot be pursued by or for one group at the expense of another" (Thomas, 2001, p. 162).

Of course, these discourses affect question two: for whom is security? The major shift created by this paradigm is the emphasis on the individual, or people if considered collectively, as the referent(s) for security (see Khong, 2001). No longer do security debates exclusively address territories or states but people-centered policies are promoted.

This of course, affected policy responses to question three focusing on threats to security. First, military threats to security have been redefined to include non-state actors (organized crime, cartels, terrorism, etc.) and intra-state conflict (civil wars, revolutions, etc.) Second, non-military threats to security have been incorporated into policy strategies. These include poverty, corruption, discrimination, and of course, access to water which is the focus of the next sub-section.

20.2.2 Water Security: A Fashion Concept or an Emerging Paradigm?

"Water security" is conceptually tied to the 2000 Global Water Partnership (GWP) document entitled, "Towards water security: a framework for action". It is a principle document presented at the 2nd World Water Forum in The Hague, which basically established the basis of what is now known as the Human Right to Water[1] (safe water and sanitation access for all), without ignoring the protection of nature (see citation above). A recent study conducted by Christina Cook and Karen Bakker (2012) shows the recent widespread utilization of the term "water security" in the past decade, across multiple disciplines in both policy and academic debates. Cook and Bakker (2012) provide a summary of the academic and policy literature related to water security.

Why has "water security" become such a fashionable concept? In part, its success is related to the widespread acceptance of "human security" in geopolitics as mentioned above. Moreover, no one can deny the dramatic trends related to water stress around the globe (also called "water crisis"), that affect about 1.6 billion people, almost one quarter of the world's population. These individuals face significant economic water shortages (where countries lack the necessary infrastructure to take water from rivers and aquifers); or even the relevant water and sanitation resources described in the Millennium Development Goals which aim to reduce the approximately 780 million individuals (one in five persons in developing countries) who do not have access to drinking water and the approximately 2.5 billion people – about half the population in developing countries – with no access to an adapted sanitation device (data from World Health Organization). Because of this terrible combination, about 2 million people die every year, most of them children less than 5 years of age, from diseases associated with inadequate water supply (WHO, 2012). Regional differences are truly important as the most affected populations related to the cited statistics are located in developing countries, where extreme water conditions – poverty conditions – remain, and water security will continue being an important concern.

Indeed, water stress is a concept linked to the availability and accessibility of water. This opens two interesting axes of analysis related to development:

1) The capacity of a state to ensure that its inhabitants will continue to have access to potable water (rich countries). Link to environmental security in terms of sustainability.
2) The capacity of a state to search for available, stable and continuous access to potable water (poor countries). Stronger link towards water stress and the human need for water.

Like contemporary debates on human security, water security has significantly broadened our understanding of "security" in global affairs. For the purposes of this study, we understand "water security" as a condition anywhere in the world where different populations at any level (households communities, neighborhoods, nations, etc.), enjoy access to sufficient safe water – and its related sanitation services – to meet both their short-term and long-term needs at affordable cost for a healthy and productive life, while ensuring the protection and enhancement of natural environment. Access, of course, needs to be provided by government officials. For this reason, like human security, water security is significantly related to issues of federalism and governance.

20.2.3 Federalism and Contemporary Security Agendas: An Introduction

Obviously, the emergence of security-related challenges has impacted governance responses significantly. Border areas have been especially affected by these changes in global governance and security policies. As stated above, traditional security agendas focused on the protection of physical and territorial integrity and thus, nation-states dictated the formulation of policies and their implementation.

With the emergence of new security paradigms, such as "human security" and "water security" addressed in this chapter, governance debates have become more complex and more opaque. Because "new" threats to security are transnational, regions and global governance structures, such as the United Nations have taken leading roles in defining policy responses. In the field of human security, the United Nations has defined new objectives through the Millennium Development Goals. The European Union has codified these objectives through the Cotounou Agreements, which guide development aid from the EU and its member states. Similarly, the United Nations and the European Union have addressed human trafficking as a transnational threat to human security through the Palermo Protocols and related supranational measures. Nonetheless, while human security has emerged as a global norm with important moral value, significant challenges remain for implementation because local, national and regional authorities must collaborate to provide coherent policy strategies, and this has not yet occurred.

Like human security, water security faces serious challenges to implementation. Definitional vagueness, conceptual inclusiveness and susceptibility to interpretation based on political will have contributed to ineffective responses in these fields. Among the most relevant water security fora that address this issue, one finds the GWP, World Water Forum, UN-Water, World Economic Forum (addressing Water Security since 2008), and UNESCO-IHE (since 2009 water security is one of the research axes of their Institute for Water Education). The United Nations finally established access to water and sanitation as a human right in 2010.

Within this context, federalism has emerged in security governance. Now that security policies no longer focus solely on protection but also include rights and services, sub-national and supranational actors actively partner with nation-states in security strategies. This is especially evident in cross-border regions where scholars such as Brunet-Jailly, Sabet, and Andreas have identified the importance of these partnerships as the basis for the protection of individual and collective well-being in these areas. This is the focus of the next section.

20.3 SECURITY PARADIGMS IN CROSS-BORDER REGIONS: EMPIRICAL CONSIDERATIONS ON FEDERALISM

There should be no doubt that the emergence of human security and water security as new paradigms in global affairs are important advancements towards the equitable treatment of human beings and the eradication of poverty, defined in terms of structural obstacles to human development. Thus far, this chapter has focused on these new security concepts in terms of three fundamental queries that dominate the security literature: how do we define security, for whom is security and what threatens security? Possibly the most important innovation of this literature, however, is the de-territorialization of security. How do we guarantee security across national boundaries?

This question is fundamental because national borders represent one of the most important challenges to both traditional and new security agendas in contemporary affairs. Specifically, security in borderlands is of the utmost importance as it can have consequences at the national and even the regional level (see most recently the March 2012 *coup d'état* in Mali which resulted from protracted local conflict with Touareg rebels in the country's northern border region).

As stated above, water security and human security privilege the needs of human beings over those of nation-states. Nonetheless, serious challenges exist to the implementation of these norms in border communities. In part, this is due to the definitional problems identified in the previous section of this chapter. However, a more serious governance problem exists that is directly relevant to the question of federalism.

While the previous section of this chapter described the emergence of new security concepts in global affairs, this section asks how they can be implemented effectively. This chapter argues that security strategies must be transnational in nature but national and even local in implementation for them to be successful. Thus, emerges the need for effective federalism.

In general, the literatures on border studies or security studies frame federalism in terms of multi-level governance (Marks and Hooghe, 2001). This conceptual framework first emerged through studies of the European Union (see Perkmann and Sum, 2002) but it has since been applied to other world regions (see Blatter, 2001; Scott, 1999). Overall, studies of multi-level governance describe the relationships between actors at different levels of government. However, one weakness of this literature is its lack of theorizing what defines these interactions. This is especially true in the field of security studies. Many scholars have acknowledged and discussed the pros and cons of the emergence of new security paradigms. However, few have discussed how security governance evolves in practical terms.

This chapter contends that contemporary security debates ignore important issues related to federalism. In general, scholars examine the tension between global governance and the sovereignty of nation-states. However, regional and sub-national actors are often ignored. The relationship between these actors could considerably improve the implementation of security policies. This chapter, and the research that informs it, contends that regions could be well-placed to implement water security as a part of a broader human security agenda for different reasons. Broadening paradigms at the global level improves our understanding of how to protect vulnerable populations but unfortunately the United Nations and other global organizations have not been able to convert political and academic discussions into "operationalizable" strategies. Conversely regions are ideally placed for the implementation of new security strategies for three reasons: (1) they are prominent arenas for transnational cooperation amongst geographically concentrated states, (2) they are de facto regionalizing security agendas through development programs, such as the Plan Puebla Panama (Americas) and the Euro-Mediterranean Partnership (Europe–Africa–Middle East) which implement conditionality clauses, and (3) these polities are motors for economic integration which must be further included in security debates in order to support the fight against poverty and the promotion of human development.

Our empirical research has identified clear "regional security regimes." Table 20.1 describes regional approaches to human security, environmental security and water security in the regional organizations to which the six research cases belong. It is important to note that only the Andean Community of Nations employs policy objectives that aim to achieve security for its own sake and for the improvement of the lives of the citizens of its member states. In North America, both the North American Free

Table 20.1 The characteristics of regional security agendas in Europe and the Americas

	Characteristics of Human Security	View of Environmental Security	View of Water Security
European Union	Part of External Action Objectives	Conservation	Water Quality; Scarcity in Specific Regions
North America (NAFTA and OAS)	Threat to Stability and Growth	Conservation; Displacement of Threats	Tied to Economic Activities
South America (CAN)	Threat to Human Development and Rights	Threat to Human Development and Rights; Conservation	Threat to Human Development and Rights

Source: Compiled by authors.

Trade Agreement and the Organization of American States have referred to human security, environmental security and water security only when insecurity in these fields establishes risks to economic growth or political stability in general. Finally, in Europe, human security is viewed as an essential part of the European Union's development actions but it is not prominent in continental politics. Environmental security, framed in terms of conservation and water security, similarly focuses mainly on water quality.

The research conducted for this chapter has shown, in fact, that local security agendas have actually adopted the frameworks utilized at the regional level of governance. This is shown in Table 20.2 which illustrates the characteristics of security strategies in the study's local cross-border cases. The table shows that only in the Colombia–Venezuela borderlands

Table 20.2 Human security, environmental security and water security in selected cross-border cases

	Characteristics of Human Security	View of Environmental Security	View of Water Security
The Eurométropole	Drug Trafficking;	Conservation	Water Quality; Sporadic Drought
Luxembourg's Greater Region	No Existing Debate	Cattenom Nuclear Plant	Limited Debate on Price
Spain–Morocco	Human Smuggling and Clandestine Migration	Desertification; Land Conservation	Water Scarcity in Southern Spain; Ceuta and Melilla Dependent on Morocco for Water
US–Mexico	Drug Cartels	Protected Areas	Tied to Economic Activities (Impact on Farming/ Ranching)
Mexico–Belize	Drug Trafficking; Human Trafficking	Sustainable Tourism	No Existing Debate
Venezuela-Colombia	Armed Conflict; Poverty; Trafficking; Informal Economies	Global Warming; Water Resources; Protected Areas	Human Right to Water; Impact of Water Stress on Communities

Source: Compiled by authors.

does one find well-defined human, environmental and water security agendas. This is the case because the Andean Community has created officially recognized border integration zones where specific security projects based on these paradigms have been implemented. Conversely, local border security agendas in North America are related to economic activities, just like the continent's regional regimes and in Europe, these agendas are only partially developed because operationalized concepts have not yet emerged at the regional level.

What is most interesting about the Colombia–Venezuela border is that it provides an opportunity for historical comparison because Venezuela pulled out of the CAN in 2006. The border integration zones were officially recognized in the late 1990s and implemented just after 2000. Scholars of these areas, such as Socorro Ramirez and José Maria Cadenas (2006), Gloria Naranjo Giraldo (2009) and Rosalba Linares de Gomez (2005), have all shown that during the short period in which Colombia and Venezuela supported the CAN's border integration zones, local populations along the border enjoyed greater human and environmental security. The numbers of homicides and violent crimes decreased, local ethnic communities received greater access to education and health services and water basin management was characterized by further sharing of information. Since 2006, this border has been characterized by militarized separation and this climate has fostered armed conflict, trafficking and informal economic activities.

Venezuela's decision to withdraw from the CAN halted what seemed to be a positive experiment in cross-border security governance that was being coordinated at the regional level. For this reason, this chapter posits that regional security agendas are well-positioned to implement water security strategies in cross-border communities as part of broader human security paradigms that provide services to vulnerable populations while at the same time protecting them from cross-border threats to their physical integrity. These regional regimes, however, need local actors to be active participants in the formulation of community-based security polities and their implementation.

As stated in the introduction, the cases chosen for this research project include cross-border communities characterized by both symmetrical and asymmetrical federalism and those characterized by both cooperative and competitive federalism (see Table 20.3 and the definitions above). The empirical research conducted thus far demonstrates that the interaction of these variables affects the emergence of human and water security as central elements of local governance debates.

Much of the literature on federalism and water management (see Anderson 2012) or human security (see Thomas, 2001) contends that the

Table 20.3 Federal relationships in selected cross-border cases

	Symmetry of Federal Relationships	Nature of Federal Relationships
The Eurométropole	Symmetrical;	Competitive
Luxembourg's Greater Region	Asymmetrical	Competitive
Spain–Morocco	Symmetrical	Competitive
US–Mexico	Asymmetrical	Cooperative
Mexico–Belize	Asymmetrical	Cooperative
Venezuela–Colombia	Symmetrical	Cooperative

Source: Compiled by authors.

needs of vulnerable populations are not met because of a lack of institutionalization in federal security strategies. Following this logic, one would expect, human and water security policies to emerge in the European cases included in this study and possibly in the North American ones. Why then, has the Venezuela–Colombia border emerged as the case with the most complete attention to human and water security?

This research argues that cross-border cooperation at the local level is the result of symmetrical power relationships along with institutional freedom that permits local actors to cooperate in joint efforts to address human and water security threats. As Table 20.3 shows, only the Venezuela–Colombia case is characterized by symmetrical and cooperative federalism. This was especially true before 2006 when Venezuela pulled out of the CAN. However, even today, without a supranational umbrella under which transnational policies can be validated, local authorities continue to cooperate both formally and informally in human and water security actions that are community-based. Consequently, this research suggests that we should shift our attention away from the tension between nation-states and global governance structures in federal debates, focusing more on fostering symmetrical, cooperative relationships between local authorities with support from regional organizations.

20.4 NOTE

1. July 2010 U.N. General Assembly resolution on the right to water and sanitation. Formal recognition initiated by the Bolivian representation and supported by the work carried out by the independent expert "safe, clean drinking water and sanitation are integral to the realization of all human rights". There was a deep concern that almost worldwide there were populations with no access to clean water. However, the assembly resolution received 122 votes in favor, zero against and 41 abstentions.

20.5 REFERENCES

Anderson, G. (2012), Water management in federal regimes: comparative reflections. Paper presentation. Federal Rivers Wokshop. University of Oxford, April 15–17, 2012.

Andreas, P. (2009), *Border Games: Policing the US-Mexico Divide*, 2nd ed. Ithaca: Cornell University Press.

Blatter, J. (2001), 'De-bordering the world of states: Toward a multilevel system in Europe and a multipolity system in North America. Insights from border regions', *European Journal of International Relations* 7, 175–209.

Brunet-Jailly, E. (ed.) (2007), *Borderlands*. Ottowa: University of Ottawa Press.

Cook, C. and Bakker, K. (2012), 'Water security: Debating an emerging paradigm', *Global Environmental Change* 22 (2012) 94–102.

Daibes-Murad, F. (2005), *A New Legal Framework for Managing the World's Shared Groundwaters: A Case Study from the Middle East*. Water Law and Policy Series. London: IWA Publishing.

Global Water Partnership (GWP) (2000), 'Towards water security: A framework for action'. Retrieved online: http://www.gwptoolbox.org/images/stories/Docs/water%20security_2000_doc_78_en.pdf (consulted 4 March 2012).

Gopalakrishnan, C., Tortajada, C., and Biswas, A. (eds) (2005), *Water Institutions: Policies, Performance and Prospects*. Water Resources Development and Management Series. New York: Springer.

Ingram, H. and Laney, G. (1995), *Divided waters: Bridging the U.S.–Mexico border*. Arizona: The University of Arizona Press.

Keating, M. (2012), 'Intergovernmental relations and innovation: from co-operative to competitive welfare federalism in the UK'. *British Journal of Politics & International Relations* 14 (2), 214–230.

Khong, Y.F. (2001), 'Human security: A shotgun approach to alleviating human misery?' *Global Governance* 7 (3), 231–236.

Koff, H. (2009), 'Las migraciones internacionales: Seguridad internacional sin seguridad humana' in Adriana Gonzalez Gil (ed.) *Lugares, procesos y migrantes*. Brussels: PIE-Peter Lang, 29–50.

Linares de Gomez, R. (2005), 'Zona de Integración Fronteriza (ZIF) y su dimensión territorial en la frontera Táchira (Venezuela) y Norte de Santander' Revista *Aldea Mundo*,18 (20), 45–54.

Maganda, C. (2005), 'Collateral damage: How the San Diego–Imperial Valley Water Agreement affects the Mexican side of the border', *Journal of Environment and Development* (SAGE Publications) 14 (4) (December), 486–506.

Marks, G. and Hooghe, L. (2001). *Multi-level governance and European integration*. Boulder, CO: Rowman & Littlefield.

Mumme, S. (2003), 'Revising the 1944 Water Treaty: Reflections on the Rio Grande drought crises and other matters', *Journal of the Southwest* 45 (45), Winter.

Naranjo Giraldo, G. (2009), 'Políticas migratorias y condiciones de seguridad humana de la población desplazada, refugiada y migrantes irregulares en la frontera Colombia-Venezuela'. Paper presented at the 2009 Conference of the Consortium for Comparative research on regional Integration and Social Cohesion (RISC), Medellín, Colombia, 3–5 November.

Newman, E. (2001), 'Human security and constructivism', *International Studies Perspectives* 2 (3), 239–251.

Owen, T. (2004), 'Proposal for a threshold-based definition human security – conflict, critique and consensus: Colloquium remarks and a proposal for a threshold-based definition', *Security Dialogue* 35, 373–387.
Perkmann, M. and Sum, N-L. (eds) (2002), *Globalization, Regionalization, and Cross-Border Regions*. Houndmills: Palgrave.
Ramirez, S. and Cadenas, J.M. (eds) (2006), *Colombia–Venezuela: Retos de la convivencia*. Bogota: Universidad Nacional de Colombia.
Sabet, D. (2008), *Cleaning the Waters: Nonprofits and their Networks Along Mexico's Northern Borders*. Tucson: University of Arizona Press.
Scott, J. (1999), 'Europe and North American contexts for cross-border regionalism', *Regional Studies* 33 (7) 605–617.
Solana, J. (2004), 'A secure Europe in a better world: The European security strategy' in Friedrich-Ebert-Stiftung (ed.) *Civilian Perspective or Security Strategy? European Development Policy Confronting New Challenges in European Foreign and Security Policy*. Berlin: Friedrich-Ebert-Stiftung, 52–57.
Thomas, C. (2001), 'Global governance, development and human security: exploring the links', *Third World Quarterly* 22 (2) 159–175.
United Nations Development Programme (UNDP) (1994), *Human Development Report 1994*. New York: Oxford University Press.
Walsh, C. (2008), *Building the Borderlands: A Transnational History of Irrigated Cotton on the Mexico–Texas Border*. College Station: Texas A&M University Press.
World Health Organization (WHO) (2012), *Progress on drinking water and sanitation*. Report 2012 – update. http://www.who.int/water_sanitation_health/publications/2012/jmp_report/en/index.html.

21. Water resources management in federal systems

George Anderson

This volume has looked at the management of water resources in nine federal countries, as well as China. The chosen federations (Australia, Brazil, Canada, India, Mexico, Pakistan, South Africa, Spain and the United States) represent a good cross-section of the world's estimated 28 federations and include most of those that have large territories or complex river systems.[1] They are also highly diverse political systems facing very different water management demands.

Water system governance is challenging whether in unitary, federal or international contexts, so it is useful to bear in mind comparisons with unitary and international alternatives when considering what is special about federal arrangements. Federalism is a spatial form of governmental organization and water systems themselves are spatially defined, but political and hydrological boundaries often do not coincide. Moreover, in federal regimes, both the central and the state governments can share powers affecting water system management, sometimes along with water management agencies that have a good deal of autonomy and direct participation of stakeholders. Thus federalism poses special issues for water governance, which this volume has explored. While the constitutional arrangements around water management vary considerably, successful integrated water management seems to require both orders of government to work cooperatively, often with some clear leadership at the federal level. However, many federations fall short of success because competition trumps cooperation and the federal government may be unable or unwilling to show leadership.

21.1 NATURE AND VARIETY OF FEDERAL SYSTEMS

Federal systems have at least two orders of government – central or "national" and regional (which may be called states, provinces or other things[2]) – each of which is directly responsible to its population and has constitutionally defined responsibilities or powers to legislate that give it some genuine autonomy. Some federations are decentralized while others are centralized; some have few states and others have many; some have parliamentary regimes and others are presidential-congressional. There are dualist federations, which are those with limit the overlap in federal and state legislative powers and where each government administers its own programs, and integrated federations, in which the two orders can pass laws on many of the same subjects (with federal laws prevailing in cases of conflict) and where the states largely administer federal laws in these areas of concurrency. Even this distinction blurs in practice, with many federations having at least some elements of both types. In all federations federal governments make fiscal transfers to the states or share taxes with them. Central governments may use this "spending power" to apply conditions to such transfers and thereby influence the activities of state governments, even in areas outside federal legislative competence.

While such constitutional, fiscal and political arrangements are important for how a federal system actually operates, larger social, geographic and economic factors are obviously critical as well. Multi-ethnic religious or linguistic federations tend to have a more devolutionary political dynamic than more homogeneous federations. The dynamics of a federation can change radically over time (pendulum swings between centralization and decentralization[3]) and they can confound the well-laid plans of constitutional authors (thus Canada has become devolved and Australia centralized – the opposite of their founders' intentions). Politicians at both levels compete for votes and promote policies in ways that may reshape the federal order quite dramatically. And to the extent that they are bound by rules, these may include not only the legal rules enforced by the courts but also what Owen Saunders, in his chapter, calls the "political constitution", which reflects a political consensus on the roles of the two orders of government.

21.2 FEDERALISM AND WATER MANAGEMENT

Water resources give rise to a whole variety of issues that must be managed by governments. These relate overwhelmingly to the *actual use* of the resource and to its physical management – who gets what share for what use, water quality, flood control, hydropower, and ecosystem sustainability – rather than to realizing economic rents, however potentially valuable these may be.[4] This gives water management issues a distinctive political character in that stakeholders are usually focused directly on the resource itself, not on a second-order issue, such as revenue generation and sharing.[5]

Of course, the political saliency of water is often strongly tied to water scarcity and variability, which drive many water conflicts. Where water is surplus, the political pressure is usually lower, although issues such as hydro-power construction, flood control and pollution can be volatile.

The nature of the issues also differs: flood control and drought are about risk management, while supply is more about quantity. The political saliency of water varies greatly across regions because it is so largely tied to specific watersheds. Thus water issues usually have a strongly regional or local character, which has major implications for how the politics of water plays out in federal regimes.

The water policy community has a long-standing consensus in favour of integrated water resource management, especially within major river basins. This can be challenging to achieve across sectors because governments typically have had long established departments of agriculture, energy, industry, and health (and perhaps irrigation) to which departments of the environment are often relatively recent additions; each of these has its own agenda and stakeholder constituencies. If there is a water ministry, it often has quite a limited mandate, which is in strong contrast, for example, with some other sectoral ministries, such as those responsible for oil and gas, that have broad powers and a limited need to bring along other departments. So the push for IWRM (and for river basin authorities) has really been for a restructuring of governments in a way that would rebalance competing sectoral interests within a framework of sustainable and sound ecological practices. This has been tough enough to achieve in unitary regimes, especially when the political stakes are high, but it is an even greater challenge in federal systems, where there can be silos at two levels of government and weak intergovernmental arrangements for coordination.

21.3 THE ALLOCATION OF WATER-RELEVANT POWERS IN FEDERATIONS

There is a significant literature that sets out principles for allocating powers in devolved regimes.[6] Some cases are straightforward (thus public goods that are indivisible nationally, such as defence, should be provided by the central government), but many are not, because there is both a national and regional dimension to an area of responsibility. The concept of "subsidiarity", which has been popularized in the European Union, has severe operational limits: it calls for action by the centre only where actions by the individual members of the union are not sufficient – but this is, of course, a highly subjective judgement. Moreover, populations in different federations vary in how much they prioritize their national identity versus regional or other group identities and these identity patterns shape views regarding the national and regional communities and the appropriate powers of the two orders of government.

Thus it should not be surprising that the constitutional allocation of powers relating to water in the nine federations under review shows no great consistency and varying degrees of coherence. Their constitution documents were written at very different times (ranging from 1789 for the United States to 1996 for South Africa) and with very different philosophies and styles of drafting. For example, the US Constitution has a very brief treatment of powers (it lists only 18 federal powers, most of which are concurrent with the states; the states are assigned "residual" powers that are not federal), while the Indian Constitution lists 97 central, 66 state and 47 concurrent powers in great specificity. The older constitutions of US and Canada make no explicit mention of "water" or "rivers", so the two orders of government get their powers over water from a variety of constitutional headings, ranging from the residual power in the US to natural resource powers in Canada, to trade and commerce powers and a grab-bag of individual powers in both countries. The US federation originated in the 13 colonies "coming together" to form a federation and part of its pre-federal history was a tradition of local control of resources. The same tradition applied to the original provinces of Canada. This emphasis on a strong state role has largely continued in both federations, even if the federal governments have substantial constitutional powers to intervene in water resource management, should they so wish and which they have occasionally drawn upon. Australia, which is the other long-standing federation under review, makes explicit the strong role of the states in that section 100 of its Constitution explicitly prohibits the federal government from abridging reasonable state use of rivers for conservation or irrigation.

In three of our nine federations, the central government is fully empowered by the constitution to manage water resources. In Mexico "lands and waters" are federally owned and the states have no constitutional authority over rivers or water; however, states can own significant water rights and land use is a municipal matter so effectively there is some diffusion of power. In Pakistan, the federal government is responsible both for international relations, including the critical Indus River Treaty, but also for the Pakistan Water and Power Development Authority, which centralized water infrastructure development in 1959; formal provincial constitutional authority is largely limited to local management of the irrigation system, but politically the provinces play heavily on water politics. South Africa has an "integrated" structure, where important legislative powers are almost all centralized or concurrent, limiting the provinces to legislating within federal framework laws (though they do have exclusive competence for potable water and sanitation services).

The constitutions of Spain and Brazil distinguish between rivers or watersheds under federal versus state competence. The Spanish Constitution of 1978 is designed to promote comprehensive watershed management by ensuring that the central government is responsible for river basins that extend into more than two autonomous communities (ACs), while AC governments are responsible for basins that are entirely within their territory. (Spain had a long history of major, multi-purpose water infrastructure projects.) By contrast, the Brazilian Constitution actually breaks up responsibility for watersheds because states are responsible for tributaries that are entirely within their boundaries even though they may flow into larger rivers that cross state boundaries and are thus a federal responsibility. Local water services are a municipal responsibility.[7]

The Indian Constitution gives the federal Parliament the power to assert legislative authority over some rivers. In India, "water" (including supplies, irrigation, canals, storage, power) is a state responsibility, but the union government may legislate on "inter-state" and international rivers (which is potentially a very broad power). (In Nigeria, the federal government has the exclusive right to legislate regarding water from sources Parliament may declare to affect more than one territory, while waterpower is a concurrent responsibility and local water services are a state responsibility.)

Powers related to treaties and spending have been of particular importance for water management in several federations. In most of these, the federal government has the power to enter into and enforce treaties, including those relating to international rivers. In some cases (such as Australia, which has no international rivers) the treaty-making power applies equally to internal rivers. Canada is an exception in that the federal government cannot bind the provinces by entering into a treaty that affects their competence (though it does have the power to enforce the old Imperial treaties which are very important for water resource management with the US). In some federations, the federal government's role is strongest in relation to international rivers.[8]

Most federations give the federal government the power to spend on any object, even those that are within the exclusive legislative responsibility of the states. This "spending power" can be very powerful in that federal governments use their fiscal resources to promote their objectives, often through jointly financed schemes in which federal grants are conditional on state matching funds (in whatever proportion). Thus the spending power can extend the influence of federal governments, and even where there may be a federal power to legislate, proceeding by offering conditional grants may be more politically palatable than direct legislation or regulation.[9] In several of our federations the federal use of the spending power has been a key method of influencing water resource management.

This overview of constitutional arrangements focuses on the legal powers of the two orders of government in relation to water resources, not to their *actual use* of those powers, which is, as we shall see, quite different.

21.4 WATER POLICY FRAMEWORKS AND POLITICS IN FEDERAL SYSTEMS

Legal and "political" constitutions provide the basic rules in which a federation's politics are played. Politicians are typically vote maximizers or champions of certain groups, while voters themselves are often indifferent to constitutional arrangements if they do not conform to their political objectives. Thus both politicians and publics can be more focused on results or political advantage than on the niceties of the constitutional division of responsibilities.

In the spirit of integrated river basin management, a classic decision tree in a country would start with a national water policy framework that would set such principles as equity, efficiency and sustainability, define key objectives and projects and create or endorse the necessary institutional frameworks, including at the basin level. Such a framework would take account of the claims of different sectors and stakeholders, determine the balance between such claims, or create processes for resolving them. Few of the federations under review have approached this ideal and we can consider each case briefly:

- *South Africa* is perhaps the closest to the ideal, with a modern policy and integrated framework. It brought forward its national water policy and law in 1998, shortly after approving its new constitution in 1996. The need for a new policy framework was urgent to address the wrongs of the apartheid regime on water allocation, but also to push for widespread access to safe drinking water, while developing more sustainable ecological practices. The provinces, which had just been formed, were consulted on the policy and law, but its formulation was strongly led by the central government. Thus the policy framework is sound, but there remain serious challenges with implementation.

- *Spain* too, has a comprehensive and modern approach to water management. It has long had national water plans, but changing political and hydrological challenges required a new national framework, which the central government brought forth in 1993 and 2001. While these plans were quite comprehensive, they remained quite centralized in approach, with the central government offering major funding for strategic water infrastructure. Its most controversial proposals – for inter-basin transfers – caused tremendous political backlash in the donor regions and forced the central government to back down. The autonomous communities pushed for greater decentralization in water management, some of which was agreed politically and overturned by the courts. Thus while Spain has a modern approach to water management, it faces serious regional conflicts, some of which are unresolved. It has given new priority to an intense dialogue on water policy between its two orders of government.
- On the face of it, *Mexico*'s National Water Law, introduced in 1992 and subsequent policy initiatives, such as creating catchment area agencies, have created a relatively modern, integrated approach to water policy. The 13 river basin authorities, covering the whole country, are in place, but they are weak relative to the federal and state governments, which have significant conflicts among them. The federal government has been slow to respond so the system is not fully integrated and the lack of an agreed framework with the states is limiting the ability to address the serious over-exploitation of the water resource.
- *Brazil*'s division of constitutional authority, which distinguishes between federal and state rivers in the same watershed, had to be overcome if there was to be a coherent governance regime. The National Water Law was a major accomplishment in such a decentralized federation, which was only achieved with the consent of the states as well as the federal government. The new policy is quite comprehensive and provides an integrated framework that respects the roles of the two orders of government. Implementation has required considerable flexibility, as well as federal assistance through its spending power to even out administrative capacity between rich and poor states. The new mechanisms have even proven capable of dealing with a controversial inter-basin transfer.
- *Australia* is the driest continent and faces a water crisis. Traditionally the states led on water policy, but because of their inability to agree on a sustainable regime for the Murray–Darling Basin, by far the country's largest, the federal government moved into the central role

in creating a new governance and policy framework for the basin, which required the active cooperation of four states and the national capital territory, incentivized by lavish federal funding. Even so, pressure from the states and stakeholders has forced the federal government to make major adjustments because of problems of implementation and some major issues are still not completely resolved. The Murray–Darling policy led into a comprehensive national plan, "Water for the Future", costing A$13 billion.

- Water policy has always been a key priority in *Pakistan* because of the country's aridity and the dependence of the vast majority of its population on the Indus Basin. Authority over water development was centralized and the massive construction of the country's water infrastructure in the 1950s and 1960s was a signal accomplishment. The 1991 Water Apportionment Accord among the provinces resolved a major tension within the federation, though there are still conflicts and weaknesses around its application. While some agencies, such as the Water and Power Development Authority with its "Vision 2025", have plans that integrate various elements, there is no agreed national water plan. With the return to democracy such a plan is being discussed with the provinces, but the political system is weak and fragmented and the new plan is unlikely to resolve many of the challenges around flooding, irrigation, groundwater, drinking water and sanitation, and hydropower that need to be addressed.
- Water resource management in India has been far more decentralized than in Pakistan, with the states dominating irrigation and water services and the Union government proving highly deferential to the states, even where it could claim constitutional authority. Within both the states and the Union government, there are also major problems with departmental silos breaking up responsibilities for water. The Union government has twice made modest attempts to develop national water policies and acts, but both foundered on the resistance of the states. The country's forums for developing coordinated Union–state policies on water (and many other subjects) are weak and the party system at the federal level is not supportive. Thus despite the need to address major problems, the prospects for a comprehensive and cooperative approach is slight. Incremental changes, perhaps incentivized by Union funds, are more likely.
- In Canada, which is well endowed with water in most regions, the federal government has been very modest in exercising its potentially significant powers over water resources. Water rarely attains much visibility in federal politics. The federal government made an attempt to develop a federal water policy in the late 1980s, but failed

to see it through. It appears likely the current government will not pursue its undertaking to develop such a policy (despite the rising need for federal leadership because of climate change and some inter-provincial disputes) and in its recent budget package ceded more responsibility over water to the provinces.

- Finally, the United States, which is the oldest federation, has gone through various phases in relation to water policy over 220 years, including swings of the pendulum from little federal intervention to virtual dominance and with issues such as irrigation, hydropower, water quality, flood control and environmental protection varying in priority and philosophical approach. However, at no time has there ever been a truly integrated national water policy and this may reflect the difficulty of aligning all parts of the federal government and Congress as much as deference to the states. The current approach is practical and responsive, with a strong emphasis on collaboration with the states.

The European Union, while not the subject of a chapter, merits mention because of its Water Framework Directive. The EU, which is a unique hybrid of federalism and confederalism with some characteristics of a treaty association, has a water directive that applies to all 27 member states and establishes standards for water bodies as well as for river basin planning. Implementation rests with member states, whose levels of compliance have varied considerably.

Nigeria, also not treated in this book, is a relatively water-abundant country, which is beginning to see issues of scarcity in some regions. The greatest water priority has been the supply of potable water, which is a state and municipal responsibility, though the federal government led the development of a National Water and Sanitation Services policy in 2000, shortly after the restoration of democracy. In 2009, the federal government brought forward a draft National Water Policy, which proposed a governance framework for river basins that includes the federal and state governments as well as stakeholders. While the federal government has clear constitutional authority over river management, consultations with the states and national assembly have been slow, so there is still no new law, three years after the policy was announced.

So what can we say of national water policy frameworks? Robust versions seem to have emerged because of a crisis that had to be addressed (Australia), or a strong consensus on a major reform by new power holders (South Africa), or a deep structural problem that could be addressed without disempowering anyone (Brazil). In all these cases, they included strong fiscal incentives from the central government and extensive federal-state dialogue. By contrast, where centralized water policy frameworks have been driven by national governments (Mexico, Spain), they have been challenged by states and have needed revision, even if there were federal fiscal incentives. However, even the more successful national water policy frameworks were limited in scope, in that they typically focused on principles and processes, leaving many hard choices on such specifics as allocations, pricing, and major infrastructure for later resolution.

If need alone dictated the creation of a national water policy framework, both Pakistan and India would have them by now. But the depth of the conflicts, the weakness of their institutions and, in India, the states' jealous protectiveness of their powers over water has prevented the creation of such frameworks to date.

Canada and the United States have had the luxury of low saliency for water policy at the national level, so politicians have seen little need to develop national policy frameworks. In Canada and the United States, water issues are typically very regional, and power has historically been quite devolved. Canada's one attempt at a national water policy was half-hearted and never implemented, while the United States has never tried.

21.5 ARRANGEMENTS FOR RIVER BASIN MANAGEMENT COMPARED

The issue of how river basins are managed arises whether there is a national policy framework or not. Such frameworks can facilitate the development of river basin organizations, but they are not strictly necessary.

The water policy community has promoted watersheds as the appropriate scale for water governance. However, identifying watersheds is itself an act of choice given shifting understandings of hydrological boundaries and well as the nesting of watersheds within watersheds. Moreover, there is an asymmetry between watersheds and "problem-sheds" or "policy-sheds" (Cohen and Davidson, 2011) in that various jurisdictional boundaries will not match those of watersheds; this can be true even in unitary countries and especially so in federations.

The idea of special river basin management organizations is certainly not new: Spain's basin authorities date back to the 1920s, the very ambitious Tennessee Valley Authority (TVA) dates from the 1930s, and France has had river basin authorities since the 1960s; however, the adoption of the idea across federal systems has been remarkably slow. Of course, there is no one "model" for integrated river basin management, either in governance or mandate. Typically modern river basin organizations have representation by the relevant governments (and perhaps departments within them) as well as stakeholders. Their functions can include water quality and environmental protection, water allocation and permits, water use charges, information systems, monitoring and inspection and capacity building. Some even have wider powers over infrastructure development, land use, hydropower, and flood control. Of course, not all of these functions need be done within a river basin so a key decision in any case is the scope of functions that an organization will have.

Of the nine federal systems under review, three have national laws providing for a comprehensive system of river basin management agencies for the whole country. In Spain, these agencies, which may fall under the jurisdiction of the centre or the autonomous communities (depending on whether the river basin encompasses multiple ACs), are well established, highly professional and have deeply entrenched processes of stakeholder participation. The South African water policy of 1998 called for river basin authorities in all the country's catchment areas, as did the Mexican federal government in 2004. However, implementation in South Africa has been slow and limited to some basins, in part because of capacity issues but also because of the difficulties of introducing effective stakeholder participation. In Mexico, national implementation has lagged and some states have created their own river basin authorities within the larger, national hydrological zones and their policies have sometimes conflicted with national policies.

Brazil's water policy provides for the possibility of interstate river basin committees on all its rivers, but to date there are only seven (plus 169 state river basin committees), all of which are in the more populated parts of the country. A key aspect of Brazilian policy is flexibility and recognition of the need for asymmetrical approaches, so the various committees vary in their composition, mandates and choice of instruments (such as water pricing). That said, the water policy provides a helpful framework within which these various arrangements are developed.

Australia and Pakistan are unusual in that for each of them one river basin is more important than all the others combined. The Indus Basin has over 90 per cent of Pakistan's population, while the Murray–Darling Basin straddles four of Australia's six states and the capital territory. Thus in each of these federations, there has been a national focus on a particular river basin and efforts to create mechanisms and policies to address the key issues in that basin. As difficult as they have been, the issues of the Murray–Darling have been more tractable in that the economic interests at play are marginal to the whole country and the federal government was prepared to pay the large fiscal price to buy a kind of peace, despite some missteps. Resolving the Murray–Darling led in due course to a broader national water policy; it also provided a possible template for managing other river basins in the country.

Pakistan faces bigger challenges. Early on, it did succeed remarkably in building the extraordinary infrastructure necessary to reroute vast quantities of water, following the Indus River Treaty. It also achieved a major, if imperfect, agreement on water allocation among its provinces. Increasingly, however, it risks being overwhelmed by the number of highly contentious issues that engage governments and stakeholders. These clearly go beyond any conceivable mandate for a river basin authority. The suggestion by Ahmad et al. (Chapter 15, this volume) that they be addressed by the Council of Common Interest – the country's highest-level political forum – is, in essence, simply a recognition that they must be addressed by the heads of the federal and provincial governments because of their importance. Once key decisions have been made, their implementation might be delegated to a river basin authority.

Pakistan's neighbour, India, has several major interstate river basins but nothing that resembles a river basin authority despite the existence, since 1956, of an act providing for their creation. It does have a mechanism for adjudicating disputes, but this is weakened by the lack of an adequate framework of guiding principles, procedural flaws and limited powers of enforcement. Integrated water resource management is also undermined by strongly siloed structures within the central and state governments, though this can be mitigated somewhat by administrative mechanisms for local coordination and the power of central funding for major projects.

The United States has no systematic approach to river basin management. Its policy-making is characterized by a combination of devolved arrangements worked out by the states within a basin and ad hoc federal responses to pressures and crises. The American system did create one of the most powerful river basin authorities ever: the Tennessee Valley Authority was established as an arm's-length agency with a broad mandate for truly integrated river basin management, including economic and social development (and an environmental approach that reflected its time). It is, in a sense, the exception that proves the rule in that both federal and state politicians resented their exclusion from so many important decisions and no subsequent river basin authority in the United States (or elsewhere) has had a similar mandate. More typical are the so-called compacts among states: these tend to focus on water allocation and can be flexible or brittle in design and they can involve the federal government both as a source of capacity and an arbiter. However, they fall far short of full-blown integrated river basin management agencies and some, such as for the Colorado River, are clearly inadequate to the needs. In part, the limited mandates of such organizations reflect the fact that federal environmental laws and regulation obviate the need for basin organizations to address many issues of water quality (and the powerful functional agencies will resist ceding authority). But more fundamentally, the politics of creating such agencies are not present: the states are too protective of their sovereignty (and often of the status quo rules), while the federal government sees little advantage in choosing sides, especially over controversial disputes on allocation, so it defers to the states.

This theme of federal deference is central to the Canadian story. The country is generally well endowed with water so that water issues, including river basin management, rarely have salience in federal politics. The federal government has been extremely reluctant to use its potentially significant powers to promote integrated management regimes, even in river basins, such as the Saskatchewan and Mackenzie rivers, that face serious issues around allocation and water quality. Basin arrangements are largely left to the provinces and have been limited in scope. Federal interventions have been most notable through environmental regulation, though even here the current government is moving to reduce the scope of federal regulation in relation to water quality and habitat protection. Its deference reflects an ideology of less federal intervention and less government generally, and perhaps a political calculation in favour of staying out of contentious inter-provincial issues.

Thus the federations under review span a spectrum from those that are clearly committed to river basin organizations to those that have little or no commitment. Even where there is a commitment, successful implementation is challenging; it requires extensive federal-state dialogue (even if federal powers are extensive),[10] flexibility in the design of arrangements for different basins (including in some cases delay in creating a basin organization), and, finally, very often the catalysing grease of federal financial incentives. It is also clear that there is a limit to the load of controversial issues that can be borne by river basin organizations: these function best when they are implementing established policies, with limited discretion on very politically contentious issues. Thus the river basins that are most in need of integrated resource management, such as the Indus in Pakistan or a number in India, face strongly competing claims that can only be resolved at the highest political level – whether by a (courageous) federal government imposing decisions or by developing a consensus among the key governments and stakeholders. Basin organizations and other water agencies can, if they have the capacity, aid understanding of the issues and options but they are unlikely to be able to resolve the issues themselves.

Some federations are failing to address the challenges of river basin management, whether through river basin organizations or through higher-level political decision-making. Occasionally a crisis, such as with the Murray–Darling, can create an environment where politicians must address the issue, but very often the problems bubble along – recognized by key stakeholders and policy experts, but failing to engage the politicians, who see little advantage in taking on the challenge. This can represent a "failure" of federalism, but federalism in most countries is deeper than the constitutional structure – it is in the societal DNA.

21.6 NOTES

1. Argentina and Nigeria are the most notable omissions, in that they are both truly federal and have complex issues of river management (see Academias Nationales de Ciencias, esp pp. 105–114 and Goldface-Irokalibe). Russia and Venezuela are less truly federal, while Ethiopia has centralized management of its rivers, which are important for both irrigation and their huge hydro potential. The European Union is not a federation but its experience is highly relevant and is discussed in the Spanish chapters as well as this one.
2. For simplicity, this chapter will refer to "states" generically, but use whatever is the established term for particular federations. For an introduction to federalism see Anderson (2008).
3. Gerlak (2005) describes five eras of federalism in the US.
4. Hydro-power generation can be an exception, at least in that it can generate substantial rents. It is striking that there are no real cases in this volume of pricing water use primarily to realize economic rents, as opposed to recovering some costs of the water management system or to using pricing to help ration use; thus it is dramatically different from petroleum and other minerals, which are non-renewable and where there is a strong political focus on maximizing revenues from rents and then allocating them. See Anderson, 2012.
5. That said, the United States has had extensive discussions around the nature of benefits of major water projects (national, regional, state) and what this should mean for cost-sharing. The federal government assumes almost 100 per cent of the cost of flood control, but typically less than 50 per cent for navigation.
6. This is particularly true of the literature on fiscal federalism. Boadway and Shah (2009) are a good example. Political scientists tend to focus more on comparative experience (Watts, 2008; Majeed et al., 2005).
7. Brazil is one of the few federations that establish municipalities as an order of government with constitutionally defined responsibilities.
8. Argentina, which is not covered in this volume, also has a weak treaty power. For a comparative review of international relations in federations see Michelmann (2009).
9. On the spending power generally, see Watts (1999).
10. Ostrom (1990) argued that effective management of a common resource does not necessarily require decisions from above (Leviathan) but can be the result of cooperative and collective action. Our review suggests the limitations of both top down and consensual approaches in federations; success seems invariably to involve elements of both.

21.7 REFERENCES

Academias Nationales de Ciencias Economicas, Ciencias Exactas, Fisicas y Naturales; Ingenieria, 2010. http://www.ancefn.org.ar/publicaciones/publicaciones/LA_CUESTION_DEL_AGUA.pdf.

Anderson, G. 2008. *Federalism: An Introduction*. Toronto: Oxford University Press.

Anderson, G. (ed.) 2012. *Oil and Gas in Federal Systems*. Toronto: Oxford University Press.

Boadway, R. and A. Shah. 2009. *Fiscal Federalism: Principles and Practice of Multiorder Governance*. Cambridge UK: Cambridge University Press.

Cohen, A. and S. Davidson. 2011. "The Watershed Approach: Challenges, Antecedents, and the Transition from Technical Tool to Governance Unit", *Water Alternatives* 4(1): 1–14. www.water-alternatives.org.

Gerlak, A. 2005. "Federalism and U.S. Water Policy: Lessons for the Twenty-First Century", *Publius: The Journal of Federalism* 36(2).

Goldface-Irokalibe, I.J. 2008. "Nigerian Perspectives", www.forumfed.org/libdocs/SpainWater2008/Joe%Goldface_en.pdf.

Majeed, A., R. Watts and D. Brown (eds) 2005. *Distribution of Powers and Responsibilities in Federal Countries*. Montreal and Kingston: McGill-Queen's University Press.

Michelmann, H. (ed.) 2009. *Foreign Relations in Federal Countries*. Montreal and Kingston: McGill-Queen's University Press.

Ostrom, E. 1990. *Governing the Commons*. Cambridge UK: Cambridge University Press.

Watts, R. 1999. *The Spending Power in Federal Systems: A Comparative Study*. Kingston: Institute of Intergovernmental Relations, Queen's University.

Watts, R. 2008. *Comparing Federal Systems (3rd edn)*. Montreal and Kingston: McGill-Queen's University Press.

Index